COMPUTATION AND INTERPRETATION OF
BIOLOGICAL STATISTICS OF FISH POPULATIONS

Bulletins are designed to assess and interpret current knowledge in scientific fields pertinent to Canadian fisheries and aquatic environments. Recent numbers in this series are listed at the back of this Bulletin.

The *Journal of the Fisheries Research Board of Canada* is published in annual volumes of monthly issues. The *Journal* and *Bulletins* are for sale by Information Canada, Ottawa K1A 0S9. Remittances must be in advance, payable in Canadian funds to the order of the Receiver General of Canada.

Les bulletins ont pour objet d'évaluer et d'interpréter les connaissances courantes dans les domaines scientifiques qui ont rapport aux pêches du Canada et les environnements aquatiques. On trouvera une liste des récents bulletins à la fin de la présente publication.

Le *Journal of the Fisheries Research Board of Canada* est publié en volumes annuels de douze livraisons. Le *Journal* et les *Bulletins* sont vendus par Information Canada, Ottawa K1A 0S9. Les remises, payables d'avance en monnaie canadienne, doivent être établies à l'ordre du Receveur général du Canada.

Editor and/Rédacteur et
Director of Scientific Information
Directeur de l'information scientifique   J. C. STEVENSON, PH.D.

Deputy Editor
Sous-rédacteur   J. WATSON, PH.D.

Assistant Editors   R. H. WIGMORE, M.SC.
Rédacteurs adjoints   JOHANNA M. REINHART, M.SC.

Production/Documentation   J. CAMP
G. J. NEVILLE/SUSAN JOHNSTON

Department of the Environment   *Ministère de l'Environnement*
Fisheries and Marine Service   *Service des pêches et des sciences de la mer*
Office of the Editor   *Bureau du directeur de la publication*
116 Lisgar Street, Ottawa, Canada   *116, rue Lisgar, Ottawa, Canada*
K1A 0H3

BULLETIN 191

(La version française est en préparation)

Computation and Interpretation of Biological Statistics of Fish Populations

W. E. Ricker

Department of the Environment
Fisheries and Marine Service
Pacific Biological Station
Nanaimo, B.C. V9R 5K6

DEPARTMENT OF THE ENVIRONMENT
FISHERIES AND MARINE SERVICE
Ottawa 1975

Computation and Interpretation of Biological Statistics of Fish Populations

ISBN-10: 1-932846-23-9
ISBN-13: 978-1-932846-23-2

Library of Congress Control Number: 2010927832

THE BLACKBURN PRESS
P. O. Box 287
Caldwell, New Jersey 07006 U.S.A.
973-228-7077
www.BlackburnPress.com

CONTENTS

CONTENTS—*Continued*

CONTENTS—*Continued*

CONTENTS—*Continued*

CONTENTS—*Continued*

PREFACE

This Bulletin is the author's third that deals with the general field of biological statistics of fish populations. The earlier ones date from 1948 and 1958, respectively, and both are long out of print. The present work began as a revision of the 1958 text, but so many changes, additions, and deletions proved desirable that it has become in many respects a new work. Even so, the text does not attempt to include all the developments in this field in recent years. The general plan and arrangement of materials is similar to that of the 1958 Bulletin, but it proved impossible to indicate which passages are new and which are quotations from the earlier volumes. However, where Examples are repeated this has been indicated. Methods which seem conceptually similar are presented in the same chapter, proceeding from the simpler to the more complex as far as possible. Some attention is given also to the historical development of each topic, and this will be considered in somewhat greater detail elsewhere. The amount of space that each topic receives varies with its importance and with its availability. Procedures described in standard western journals are not, as a rule, given detailed development: usually only the formulae most useful for estimating population statistics are quoted, and a discussion of sampling error and of the conditions which make them usable. More extended treatment is given to methods taken from obscure sources and new developments or new aspects of existing methods. This plan does not give ideal balance, but it does perhaps make for maximum usefulness within a limited compass.

In selecting illustrative examples, no attempt has been made to give representation to effort in fishery research on a geographical basis: rather, examples close at hand have usually been selected. The examples from "borrowed" data involve risks of misinterpretation, and are used here to illustrate methodology rather than as a factual treatment of the situations concerned; although, at the same time, I have tried to be as realistic as possible. Some examples have been simplified for presentation here, and others have been invented, in order to keep the text within bounds. However, the practicing biologist quickly discovers that the situations he has to tackle tend to be more complex than those in any handbook, or else the conditions differ from any described to date and demand modifications of existing procedures. It can be taken as a general rule that experiments or observations which seem simple and straightforward will prove to have important complications when analyzed carefully — complications that stem from the complexity and variability of the living organism, and from the changes that take place in it, continuously, from birth to death. Two general precautions should always be taken. Firstly, divide up any body of data into different categories, for example by the size, age, sex, or history of the fish involved, and compare statistics calculated from the two or more subsets obtained in each such category. Secondly, divide the data on the basis of time (successive hours, days, or seasons), and make similar comparisons.

From the point of view of fishery management, information from computations of the kinds described in this Bulletin provide only a part of the basic information upon which policy can be based. Sometimes, to be sure, they can provide the greater part of the necessary information. In other situations they have as yet given only equivocal answers to important questions. This is particularly true where several species are possible occupants of and competitors for an important environment, and their relative abundance may vary with the intensity of the fishery or with physical changes. The fishery administrator has also the problem (often not an easy one) of selecting an objective which his regulations are designed to serve, and this involves questions of economics and public policy not touched on here. However, there is no question that the increase of biological information has already improved, and will continue to improve, the precision and effectiveness of fishery management.

Some attempt has been made to meet the needs of the beginning student of fishery biology by working out certain examples in detail, even where this consists largely of standard mathematical procedures. To be used as an introductory text book, however, this Bulletin should be "cut down" by omitting less frequently used methods and by choosing one among several alternative procedures where these exist. The choice would depend partly on local problems and interests.

This Bulletin is of course not intended as a complete text book for fishery biologists. Methods of measuring fish, determining their age, marking, tagging, collecting, and tabulating catch statistics — all these are mentioned only incidentally, although they provide the data from which the vital statistics of a stock must be estimated. Books that treat these matters include Chugunova (1959), Gulland (1966), Lagler (1956), Rounsefell and Everhart (1953), Royce (1972), and the International Biological Program handbook edited by Ricker (1971a). Nor are we concerned here with other animals and plants of the environment, with the flow of nutrient energy which maintains an aquatic population, or with the overall productivity of bodies of water. These subjects are discussed in most textbooks of ecology, and various aspects pertinent to fish production are treated in works by Dickie and Paloheimo (1974), Gulland (1971), Moiseev (1969), Nikolsky (1965), Regier (1974), Ricker (1946, 1969b), Walford (1958), Weatherley (1972), and Winberg (1956).

The 1958 Bulletin included a list of individuals who had assisted in various ways. Some of the same people have made further contributions, and I wish to acknowledge also discussions or written comments from K. R. Allen, D. H. Cushing, J. A. Gulland, G. J. Paulik, H. A. Regier, and B. J. Rothschild.

Computer calculations in this Bulletin have been done by K. R. Allen or J. A. C. Thomson. A. A. Denbigh assisted in preparing the new figures. Mrs Barbara Korsvoll prepared the typescript and has assisted with many of the computations.

CHAPTER 1. — INTRODUCTION

1.1. THE PROBLEMS

The topics which can be considered as biological statistics of a fish population include the following:

1. The abundance of the population, usually somewhat restricted as to age or size.

2. The total mortality rate at successive ages, or even within each year.

3. The fraction of the total mortality ascribable to each of several causes. It is possible at times to distinguish (a) deaths caused by fishing, (b) deaths caused by predation other than human, (c) deaths from disease, parasites, or senility; (b) and (c) together comprise "natural" mortality.

4. The rate of growth of the individual fish. In human populations the rate of growth of individuals is not generally regarded as a vital statistic. However growth rate among fishes is much more variable than in man, and it may be even more sensitive than mortality to changes in abundance and to environmental variability.

5. The rate of reproduction, particularly as it is related to stock density.

6. The overall rate of *surplus production* of a stock, which is the resultant of growth plus recruitment less natural mortality.

Historically, age and rate of growth were the first of these subjects to receive wide attention, possibly because they require less extensive field work. Most of the methods now in use for estimating growth rate had been evolved by 1910, and their potential sources of error have received close consideration.

The development of procedures for estimating population size and survival rate started early but progressed much more slowly. In the past 25 years there has been much activity along theoretical lines, and numerous new applications. An investigator now has a number of methods from which to choose one best suited to the population he is studying, and he can increasingly use one method to check another.

At first the study of reproduction or "year-class strength" was considered mainly in relation to environmental factors, but its relation to stock density has attracted much attention in recent years.

Finally, the overall production of a fish stock, in relation to density and to rate of fishing, has interested a number of authors since the middle 1920s, and there is now a considerable body of information and a corresponding methodology.

1

1.2. DEFINITIONS, USAGES, AND GLOSSARY

The list below includes only a part of the varied terminology which has been used in fish population analysis. More extended descriptions of some terms are given in later sections. If a special symbol is associated with a term, it is shown in parentheses. Terms marked with an asterisk are not used in this book, at any rate not in a context where strict definition is called for.

ABSOLUTE RECRUITMENT: The number of fish which grow into the catchable size range in a unit of time (usually a year).

AGE: The number of years of life completed, here indicated by an arabic numeral, followed by a plus sign if there is any possibility of ambiguity (age 5, age 5+)[1].

ANNUAL (or seasonal) GROWTH RATE (h): The *increase* in weight of a fish per year (or season), divided by the initial weight.

ANNUAL (or seasonal) TOTAL MORTALITY RATE (A): The number of fish which die during a year (or season), divided by the initial number. Also called: actual mortality rate, *coefficient of mortality (Heincke).

AVAILABILITY: 1. (r): The fraction of a fish population which lives in regions where it is susceptible to fishing during a given fishing season (Marr 1951). This fraction receives recruits from or becomes mingled with the non-available part of the stock at other seasons, or in other years. (Any more or less completely isolated segment of the population is best treated as a separate stock.)
2. (C/f or Y/f): Catch per unit of effort.

BIOMASS (B): The weight of a fish stock, or of some defined portion of it.

CATCHABILITY (q): The fraction of a fish stock which is caught by a defined unit of the fishing effort. When the unit is small enough that it catches only a small part of the stock — 0.01 or less — it can be used as an instantaneous rate in computing population change. (For fractions taken of various portions of the stock, see "vulnerability.") Also called: catchability coefficient, *force of fishing mortality (Fry 1949, p. 24; in his Appendix, however, Fry defines the force of fishing mortality as equivalent to our rate of fishing, F).

CATCH CURVE: A graph of the logarithm of number of fish taken at successive ages or sizes.

CATCH PER UNIT OF EFFORT (C/f or Y/f): The catch of fish, in numbers or in weight, taken by a defined unit of fishing effort. Also called: catch per effort, fishing success, availability (2).

CONDITIONAL FISHING MORTALITY RATE (m): The fraction of an initial stock which would be caught during the year (or season) *if* no other causes of mortality

---

[1] While the above is recommended, other usages exist. Roman numerals are frequently used in North America, but their cumbersomeness seems to outweigh any advantage. Some have used either roman or arabic numerals to indicate year of life, rather than years completed. For anadromous fishes both the actual age and the age at seaward migration are frequently indicated. Several conventions are employed for this purpose, and it seems necessary to specify each time which one is being used.

operated $(= 1 - e^{-F})$. Also called: annual fishing mortality rate, seasonal fishing mortality rate.

CONDITIONAL NATURAL MORTALITY RATE (n): The fraction of an initial stock that would die from causes other than fishing during a year (or season), *if* there were no fishing mortality $(= 1 - e^{-M})$. Also called: annual natural mortality rate, seasonal natural mortality rate.

CRITICAL SIZE: The average size of the fish in a year-class at the time when the instantaneous rate of natural mortality equals the instantaneous rate of growth in weight for the year-class as a whole. Also called: *optimum size.

EFFECTIVE FISHING EFFORT (F/q): Fishing effort adjusted, when necessary, so that each increase in the adjusted unit causes a proportional increase in instantaneous rate of fishing.

EFFECTIVENESS OF FISHING: A general term referring to the percentage removal of fish from a stock, but not as specifically defined as either rate of exploitation or instantaneous rate of fishing.

EQUILIBRIUM CATCH (C_E): The catch (in numbers) taken from a fish stock when it is in equilibrium with fishing of a given intensity, and (apart from the effects of environmental variation) its abundance is not changing from one year to the next.

EQUILIBRIUM YIELD (Y_E): The yield in weight taken from a fish stock when it is in equilibrium with fishing of a given intensity, and (apart from effects of environmental variation) its biomass is not changing from one year to the next. Also called: sustainable yield, equivalent sustainable yield. (See also SURPLUS PRODUCTION.)

EXPLOITATION RATIO (E): The ratio of fish caught to total mortality $(= F/Z$ when fishing and natural mortality take place concurrently). Also called: *rate of exploitation.

FISH STOCK: See STOCK.

FISHING EFFORT (f): 1. The total fishing gear in use for a specified period of time. When two or more kinds of gear are used, they must be adjusted to some standard type (see Section 1.7).
    2. Effective fishing effort.

*FISHING INTENSITY: 1. Effective fishing effort.
    2. Fishing effort per unit area (Beverton and Holt).
    3. Effectiveness of fishing.

*FISHING POWER (of a boat, or of a fishing gear): The relative vulnerability of the stock to different boats or gears. Usually determined as the catch taken by the given apparatus, divided by the catch of a standard apparatus fishing at nearly the same time and place.

FISHING SUCCESS: Catch per unit of effort.

INSTANTANEOUS RATES (in general): See Section 1.4. Also called: logarithmic, exponential, or compound-interest rates.

INSTANTANEOUS RATE OF FISHING MORTALITY (F): When fishing and natural mortality act concurrently, F is equal to the instantaneous total mortality rate, multiplied by the ratio of fishing deaths to all deaths. Also called: rate of fishing; instantaneous rate of fishing; *force of fishing mortality (see under CATCHABILITY).

INSTANTANEOUS RATE OF GROWTH (G): The natural logarithm of the ratio of final weight to initial weight of a fish in a unit of time, usually a year. When applied collectively to all fish of a given age in a stock, the possibility of selective mortality must be considered (Section 9.4).

INSTANTANEOUS RATE OF MORTALITY (Z): The natural logarithm (with sign changed) of the survival rate. The ratio of number of deaths per unit of time to population abundance during that time, if all deceased fish were to be immediately replaced so that population does not change. Also called: *coefficient of decrease (Baranov).

INSTANTANEOUS RATE OF NATURAL MORTALITY (M): When natural and fishing mortality operate concurrently it is equal to the instantaneous total mortality rate, multiplied by the ratio of natural deaths to all deaths. Also called: *force of natural mortality (Fry).

INSTANTANEOUS RATE OF RECRUITMENT (z): Number of fish that grow to catchable size per short interval of time, divided by the number of catchable fish already present at that time. Usually given on a yearly basis: that is, the figure just described is divided by the fraction of a year represented by the "short interval" in question. This concept is used principally when the size of the vulnerable stock is not changing or is changing only slowly, since among fishes recruitment is not usually associated with stock size in the direct way in which mortality and growth are.

INSTANTANEOUS RATE OF SURPLUS PRODUCTION: Equal to rate of growth plus rate of recruitment less rate of natural mortality — all in terms of weight and on an instantaneous basis. In a "balanced" or equilibrium fishery, this increment replaces what is removed by fishing, and rate of surplus production is numerically equal to rate of fishing. Also called: *instantaneous rate of natural increase (Schaefer).

MAINTAINABLE YIELD: "The largest catch that can be maintained from the population, at whatever level of stock size, over an indefinite period. It will be identical to the sustainable yield for populations below the level giving the MSY, and equal to the MSY for populations at or above this level" (Gulland).

MAXIMUM EQUILIBRIUM CATCH (see MAXIMUM SUSTAINABLE YIELD).

MAXIMUM SUSTAINABLE YIELD (MSY OR Y_s): The largest average catch or yield that can continuously be taken from a stock under existing environmental conditions. (For species with fluctuating recruitment, the maximum might be obtained by taking fewer fish in some years than in others.) Also called: maximum equilibrium catch (MEC); maximum sustained yield; sustainable catch.

*MECHANICAL INTENSITY OF FISHING: Fishing effort (1).

4

NATURAL MORTALITY: Deaths from all causes except man's fishing, including predation, senility, epidemics, pollution, etc.

NET INCREASE (OR DECREASE): New body substance elaborated in a stock, less the loss from all forms of mortality.

PARAMETER: A "constant" or numerical description of some property of a *population* (which may be real or imaginary). Cf. statistic.

PIECES: Individual items, as in the expression "two dollars a piece" (German *Stück*). Individual fish.

PRODUCTION: 1. (sense of Ivlev). The total elaboration of new body substance in a stock in a unit of time, irrespective of whether or not it survives to the end of that time. Also called: *net production (Clarke et al. 1946); *total production. 2. *Yield.

RATE OF EXPLOITATION (u): The fraction, by number, of the fish in a population at a given time, which is caught and killed by man during the year immediately following ($= FA/Z$ when fishing and natural mortality are concurrent). The term may also be applied to separate parts of the stock distinguished by size, sex, etc. (See also "rate of utilization.") Also called: *fishing coefficient (Heincke).

RATE OF FISHING (F): INSTANTANEOUS RATE OF FISHING MORTALITY.

*RATE OF NATURAL INCREASE: INSTANTANEOUS RATE OF SURPLUS PRODUCTION.

RATE OF REMOVAL: An inexactly-defined term that can mean either rate of exploitation or rate of fishing — depending on the context (see Section 1.4.3).

RATE OF UTILIZATION: Similar to rate of exploitation, except that only the fish *landed* are considered. The distinction between catch and landings is important when considerable quantities of fish are discarded at sea.

RECRUITMENT: Addition of new fish to the vulnerable population by growth from among smaller size categories (Section 11.1).

RECRUITMENT CURVE, REPRODUCTION CURVE: A graph of the progeny of a spawning at the time they reach a specified age (for example, the age at which half of the brood has become vulnerable to fishing), plotted against the abundance of the stock that produced them.

SECULAR: Pertaining to the passage of time.

STATISTIC: The estimate of a parameter which is obtained by observation, and which in general is subject to sampling error.

STOCK: The part of a fish population which is under consideration from the point of view of actual or potential utilization.

SUCCESS (of fishing): Catch per unit of effort.

SURPLUS PRODUCTION (Y'): Production of new weight by a fishable stock, plus recruits added to it, less what is removed by *natural* mortality. This is usually estimated as the catch in a given year plus the increase in stock size (or less the decrease). Also called: natural increase, sustainable yield, equilibrium catch (Schaefer).

5

SURVIVAL RATE (S): Number of fish alive after a specified time interval, divided by the initial number. Usually on a yearly basis.

SUSTAINABLE YIELD: Equilibrium yield.

USABLE STOCK: The number or weight of all fish in a stock that lie within the range of sizes customarily considered usable (or designated so by law). Also called: *standing crop.

UTILIZED STOCK, UTILIZED POPULATION (V): The part, by number, of the fish alive at a given time, which will be caught in future.

VIRTUAL POPULATION: Utilized stock.

VULNERABILITY: A term equivalent to CATCHABILITY but usually applied to separate parts of a stock, for example those of a particular size, or those living in a particular part of the range.

YEAR-CLASS: The fish spawned or hatched in a given year. In the northern hemisphere, when spawning is in autumn and hatching in spring, the calendar year of the hatch is commonly used to identify the year-class (except usually for salmon). Also called: brood, generation.

In the above, only the kinds of "rates" are defined which are most frequently used. In general, for any process there will be an *absolute* rate, a *relative* rate and an *instantaneous* rate (Sections 1.4, 1.5).

1.3. SYMBOLS

The symbols used are those of the "international" system (Gulland 1956a) as far as possible, but quite a number of additional ones are required, of which those more frequently used are shown below. The predecessors of this bulletin (Ricker 1948, 1958a) used essentially the system recommended by Widrig (1954a, b), and their symbols are indicated below in square brackets.

a 1. a coefficient used in the Ricker recruitment curve (Section 11.6.2)
      2. the multiplier in the functional weight–length relationship (Section 9.3.1)
b 1. the slope of any line
      2. the exponent in the functional weight–length relationship (Section 9.3.1)
e 2.71828 . . .
f fishing effort
h annual growth rate
k 1. Ford growth coefficient (Section 9.6.4)
      2. a rate; used in various connections
l length of a fish
m conditional rate of fishing mortality
n conditional rate of natural mortality
q catchability [c]

r 1. availability (1).
      2. rate of accession (Section 5.3)
s standard deviation
t 1. a point in time (often used as a subscript)
      2. an interval of time (also Δt)
      3. age
u 1. rate of exploitation of a fish stock, or expectation of capture by man (μ
         of Ricker 1948)
      2. the ratio of number of recoveries to number of marked fish released
         ($= R/M$)
v expectation of natural death (v of Ricker 1948)
w weight of a fish
y instantaneous rate of emigration
z 1. instantaneous rate of immigration
      2. instantaneous rate of recruitment

A annual (or seasonal) mortality rate [a]
A' annual (or seasonal) rate of disappearance of fish
B weight (biomass) of a group of fish; for example of a year–class, or of an
      entire stock
C 1. catch, in numbers — usually for a whole year
      2. number of fish examined for tags or marks
E 1. escapement (of salmon, etc., past a fishery)
      2. number of eggs
      3. exploitation ratio ($= F/Z$)
      4. (as subscript) an equilibrium level (see Appendix III)
F instantaneous rate of fishing mortality [p]
G instantaneous rate of growth [g]
K 1. Brody growth coefficient (Section 9.6.1)
      2. any rate
      3. cumulative catch (Chapter 6)
L mean length at recruitment, in Baranov's yield equation
L_∞ asymptotic length, in the Brody–Bertalanffy growth equation
M 1. instantaneous rate of natural mortality [q]
      2. number of fish marked or tagged (also M')
N number of fish in a year–class, population, or sample
P 1. abundance of a parental stock or generation
      2. level of statistical probability
Q the constant which appears in the integration of Baranov's yield computation

R 1. number of recruits to the catchable stock
    2. number of recaptures of marked or tagged fish
    3. multiple correlation coefficient

S rate of survival ($= -\log_e Z$) [s]

S' apparent survival rate ($= -\log_e Z'$)

U instantaneous rate of "other loss" (includes emigration and, for tagged fish, the shedding of tags)

V 1. utilized stock, virtual population
    2. variance

W_∞ the mean asymptotic weight which corresponds to L_∞

Y yield, catch by weight

Z instantaneous rate of (total) mortality [i]

Z' instantaneous rate of disappearance (total losses) from a stock
$$(= F + M + U = Z + U)$$

— (over a symbol) a mean value

Σ summation symbol

1.4. Numerical Representation of Mortality

1.4.1. Total mortality rate. The mortality in a population, from all causes, can be expressed numerically in two different ways.

(a) Simplest and most realistic perhaps is the *annual expectation of death* of an individual fish, or *actual mortality rate*, expressed as a fraction or percentage. This is the fraction of the fish present at the start of a year which actually die during the year.

(b) If the number of deaths in a small interval of time is at all times proportional to the number of fish present at that time, the fraction which remains at time t, of the fish in a population at the start of a year ($t = 0$), is:

$$\frac{N_t}{N_0} = e^{-Zt} \tag{1.1}$$

The parameter Z is called the *instantaneous mortality rate*. If the unit of time is 1 year, then at the end of the year (when $t = 1$):

$$\frac{N_1}{N_0} = e^{-Z} \tag{1.2}$$

But $N_1/N_0 = S = 1 - A$; hence $1 - A = e^{-Z}$, or $Z = -\log_e (1 - A)$; hence the instantaneous mortality rate is equal to the natural logarithm (with sign changed) of the complement of the annual expectation of death.

The instantaneous rate Z also represents the number of fish (including new recruits) which would die during the year if recruitment were to exactly balance mortality from day to day, expressed as a fraction or multiple of the steady density of stock.

8

The concept of an "instantaneous" rate apparently continues to trouble students. Imagine a year of a fish's life to be divided into a large number n of equal time intervals, and let the quantity Z/n represent the expectation of death of the fish during each such interval; or, in other words, Z/n is the fraction of a large population which would actually die during each time interval one-nth of a year long. In such a relationship, Z is the instantaneous rate of mortality, expressed on a yearly basis. The interval $1/n$ year is made short (n made large) so that the change in size of population during each interval will be negligible; that is, Z/n must be a small fraction. But of course the cumulative effect of the death of Z/n of the fish over a large number of $n$ths of a year is quite important. This can be illustrated by a numerical example. Let $n = 1000$ and $Z = 2.8$. Then during $1/1000$ of a year $2.8/1000 = 0.28\%$ of the average number of fish present die. Since this is a very small number of deaths, the difference between average number and initial number can be ignored; and, of a population of, say, 1,000,000 initially, about 2800 will die and 997,200 will remain alive. During the next thousandth of the year 0.28% of $997,200 = 2793$ die and hence 994,407 survive. Repeated 1000 times, this process leaves 1,000,000 $(1 - 0.0028)^{1000} = 60,000$ survivors. The mortality for the year is therefore 940,000 fish, and the annual mortality rate is $A = 0.940$, as compared with the instantaneous rate of $Z = 2.8$. This relation is not quite exact, because 1000 divisions of the year are scarcely enough to compute the relative sizes of these two rates with 3-figure accuracy. The value appropriate to an indefinitely large number of divisions of the year is given by the relationship: $(1 - A) = e^{-Z}$ where $e = 2.71828$. In this example, for $Z = 2.8$, $A = 0.9392$, so that the approximate calculation was not far off. Obviously there is no limit to the possible size of Z, but A cannot exceed unity — that is, no more fish can die than are actually present. On the other hand, when Z and A are small they approach each other in magnitude. The table of Appendix I shows that when $Z = 0.1$ there is only 5% difference between them.

It has been suggested that mortality should not really be divided up into time periods of less than a day, because of probable diurnal fluctuations in predation, etc., and hence that a calculus of finite differences should be employed. Actually, even 365 divisions of the year is close enough to an "indefinitely large number" to make the exponential relationship between Z and A accurate enough for our purposes. A more penetrating consideration is that we are not, after all, interested in dividing up the fish's year into astronomically equal time intervals; for our purpose a physiological time scale would be more appropriate, or perhaps one based on the diurnal and seasonal variation in activity of the fish's predacious enemies. It is only when total mortality is subdivided into components whose effect may vary seasonally *in different ways*, that time by the sun becomes important.

1.4.2. SUBDIVISIONS OF MORTALITY. There can be several causes of death among the fish in a population: removals by man (fishing), predation, disease, accident, etc., each with its own rate. In practice we usually consider a division into only two types: fishing, and natural mortality (which includes everything else). Each kind of mortality has its own instantaneous rate, and the sum of these is the instantaneous total mortality rate. If F represents the instantaneous rate of fishing mortality, the expression e^{-F} represents the survival rate *if there were no natural mortality*, and $1 - e^{-F}$ is the corresponding *conditional mortality rate* if no other source of mortality existed, here represented by *m*. Similarly, if M is the instantaneous rate of natural mortality, $1 - e^{-M}$ is the conditional natural mortality rate. When fishing and natural mortality act concurrently, they are competing for the same fish, so the conditional mortality rates cannot be added. However, an expectation of death can easily be computed for each cause of mortality, as described in Section 1.5.2, and these are additive. The

9

expectation of death by fishing is known as the rate of exploitation. The three kinds of mortality rates can be summarized as follows:

	Symbol
I. Instantaneous mortality rates	
Total	Z
From fishing ("rate of fishing")	F
From natural causes	M
II. Conditional mortality rates	
From fishing	m
From natural causes	n
III. Actual mortality rates (expectations of death)	
Total	A
From fishing ("rate of exploitation")	u
From natural causes	v

1.4.3. POPULAR USAGE. For popular descriptive purposes the usefulness of u and F — rate of exploitation and rate of fishing — depends partly on the kind of fishery involved. If fishing occurs at a time when there is little or no recruitment, then a rate of exploitation of, say, 65% shows the fraction of the vulnerable stock being utilized each year; and to say that the rate of fishing is 105% means little to the layman. The situation is different, however, when fishing, recruitment, and natural mortality take place throughout the same period of time: in that event, for example, with a 65% rate of exploitation and 10% natural mortality, the year's catch equals 1.21 times the stock on hand at any given time. In such a case the rate of fishing, 121%, seems the more concrete and realistic description of the effectiveness of the fishery.

1.5.   RECRUITMENT, STOCK, AND CATCH IN NUMBERS

1.5.1. TYPES OF IDEAL FISH POPULATIONS. A useful classification of fish populations is shown below. It is similar to that proposed by Ricker (1944, 1958a), but with different numbering.

Type 1. Natural mortality occurs during a time of year other than the fishing season. The population decreases during the fishing season because of catch removals only.

Type 2. Natural mortality occurs along with the fishing; each occurs at a constant instantaneous rate, or the two rates vary in parallel fashion. This is the type which has been most used in production computations.

The above types can be further divided on the basis of when recruitment occurs:

Type A. Recruitment takes place at a time of year when there is no mortality.

Type B. Recruitment is at an even absolute (linear) rate throughout the year, or is proportional to the rate of fishing throughout the fishing season.

Recruitment types A and B can be combined with fishing types 1 or 2. All these "types" are ideal rather than real, and will be approximated rather than met by actual fisheries.

1.5.2. RELATIONSHIPS BETWEEN PARAMETERS. For *all of the above types of fisheries,* the following relationships exist between the mortality and survival rates:

Instantaneous total mortality rate:	$Z = F + M$	(1.3)
Actual total mortality rate:	$A = 1 - e^{-Z} = u + v$	(1.4)
Survival rate:	$S = e^{-Z}$	(1.5)

10

For *Type 1 fisheries* it is convenient to start the biological year at the time fishing begins, and to consider that natural mortality occurs after fishing ends. We have then the following relationships, additional to (1.3)–(1.5):

Rate of exploitation:	$u = m = 1 - e^{-F}$	(1.6)
Conditional natural mortality rate:	$n = 1 - e^{-M}$	(1.7)
Expectation of natural death:	$v = n(1 - u)$	(1.8)

For *Type 2 fisheries*, in which fishing and natural mortality operate concurrently, the following relationships hold:

Conditional fishing mortality rate:	$m = 1 - e^{-F}$	(1.9)
Conditional natural mortality rate:	$n = 1 - e^{-M}$	(1.10)
Rate of exploitation:	$u = FA/Z$	(1.11)
Expectation of natural death:	$v = MA/Z$	(1.12)

Expressions (1.9)–(1.12) also imply the following:

$$\frac{Z}{A} = \frac{F}{u} = \frac{M}{v} \qquad (1.13)$$

$$m + n - mn = A \qquad (1.14)$$

Notice particularly that expressions (1.3)–(1.5) and (1.14) do not require that fishing and natural mortality occur at rates which are proportional within the year. For example, a simple calculation will show that a 50% conditional rate of natural mortality (n), combined with a 50% conditional rate of fishing mortality (m), gives a 75% total mortality rate (A), regardless of whether the two causes of death operate concurrently, or consecutively, or in any intermediate fashion. On the other hand, differences in the seasonal incidence of the two kinds of mortality can cause striking changes in the relative magnitudes of the annual expectations of death (u and v), though the latter always add up to A. Expression (1.13) pertains only to the situation where fishing and natural mortality are distributed proportionally within the year (though it is not necessary that each be of a constant magnitude on an astronomical time scale.)

To obtain a good approximation to either the Type 1 or Type 2 fishery it is legitimate to set the limits of the fishery year in as convenient a manner as possible. For example, to increase the resemblance to Type 2 it may be possible to arrange the statistical year so that the mean time of fishing is at the middle of that year, with times of little fishing distributed as symmetrically as possible at the beginning and at the end.

If fishing is so distributed, seasonally, that neither the Type 1 nor Type 2 model is realistic, the year can be divided into two or more parts and separate values of F, M, etc., computed for each.

11

1.5.3. SINGLE AGE-GROUPS. Consider a single age-group of fish in the recruited (fully vulnerable) part of a stock. Its abundance during a year decreases from N to NS, according to equation (1.2); for example, from the point A to the point B_1 in Fig. 1.1. The *average* abundance during the year is the area of the figure under AB_1, divided by the length of the base (which is unity). In our symbols, this is:

$$\bar{N} = \int_{t=0}^{t=1} Ne^{-Z}dt = N\left(\frac{e^{-Z}}{-Z} - \frac{1}{-Z}\right) = \frac{N(1 - e^{-Z})}{Z} = \frac{NA}{Z} \tag{1.15}$$

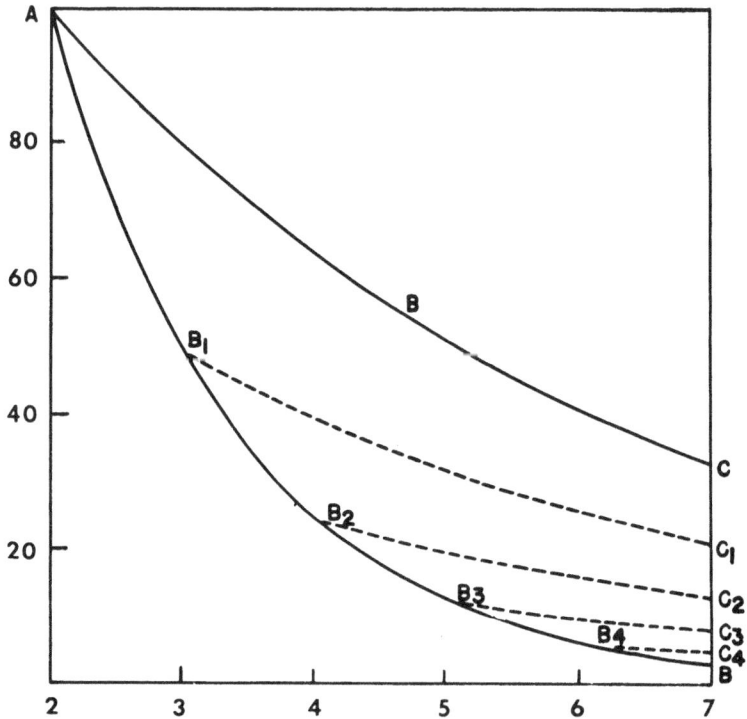

FIG. 1.1. Exponential decrease in a stock from an initial abundance of 100 at age 2, when the annual mortality rate is 0.2 (AC) and when it is 0.5 (AB). The broken lines indicate population structure during a period of transition from the smaller to the larger mortality. (Redrawn from fig. 8 of Baranov 1918, by S. D. Gerking.)

The total deaths, which equal NA by definition, are therefore Z times the average population. Since the mortality is at each instant divided between natural causes and fishing in the ratio of F to M, then natural deaths are $M/(F + M) = M/Z$ times NA, or (from 1.15) M times the average population; that is:

$$\frac{M}{F+M} \times NA = \frac{MNA}{Z} = M\bar{N} \tag{1.16}$$

12

Similarly the catch is F times the average population:

$$C = F\bar{N} = \frac{NFA}{Z} \tag{1.17}$$

This is often known as *Baranov's catch equation*. It has also been derived stochastically by Rothschild (1967).

1.5.4. SEVERAL AGE-GROUPS. A few kinds of commercial fish stocks consist of single age-groups, to which the above expressions apply directly. More commonly a stock consists of a mixture of ages, so that in order to obtain expressions for mortality, etc., of whole populations, consideration must be given to the recruits to the stock, and the manner in which recruitment occurs. We will begin by considering the equilibrium situation, first described in detail by Baranov (1918), where recruitment is the same in all years; and with the further simplification that survival rate is the same throughout life.

1.5.5. INSTANTANEOUS RECRUITMENT. Consider R recruits added to the catchable stock of a species each year. Suppose the stock is of Type A, so that the recruits become catchable during a brief interval of time, or for practical purposes all at one instant. With a constant rate of survival, S, the recruits decrease in 1 year to Re^{-Z} or RS, in 2 years to RS^2, in t years to RS^t. Under these equilibrium conditions the total population present just after recruitment in any year is found by summing the converging geometric series:

$$N = R + Re^{-Z} + Re^{-2Z} + \ldots\ldots$$

$$= R(1 + S + S^2 + \ldots\ldots)$$

$$= R\left(\frac{1 - S^\infty}{1 - S}\right) = \frac{R}{(1 - S)} = \frac{R}{A} \tag{1.18}$$

At any other time of year the population will of course be somewhat less than this. For example, at the half year it will be:

$$N = Re^{-0.5Z} + Re^{-1.5Z} + Re^{-2.5Z} + \ldots\ldots$$

$$= Re^{-0.5Z} (1 + e^{-Z} + e^{-2Z} + \ldots\ldots)$$

$$= \frac{Re^{-0.5Z}}{A} \tag{1.19}$$

Similarly, immediately *before* the annual influx of recruits the stock would be:

$$N = \frac{Re^{-Z}}{A} = \frac{RS}{A} \tag{1.20}$$

which is its least value.

The average size of the stock over the course of a year (unit time), during which it decreases from R/A to RS/A, is of course:

$$\overline{N} = \frac{R}{A} \int_{t=0}^{t=1} e^{-Zt}\, dt = \frac{R}{A} \times \frac{A}{Z} = \frac{R}{Z} \qquad (1.21)$$

1.5.6. CONTINUOUS RECRUITMENT. Consider a fishery of Type 2B, in which R recruits enter a catchable stock at a steady absolute rate throughout the year, instead of all at once. Suppose further that the stock is in equilibrium at density N, with the number of recruits just balancing the number of deaths at all times. From (1.15), the number of fish that die in the course of a year is the product of the number present times the instantaneous mortality rate:

$$\text{Total deaths} = \overline{N}Z \qquad (1.22)$$

Considering the fish on hand at the start of a year, the number of them that will die during the year is of course:

$$\text{Deaths of "old" fish} = \overline{N}A \qquad (1.23)$$

The mortality among recruits must therefore be the difference between these two, or:

$$\text{Deaths of the year's recruits} = \overline{N}(Z - A) \qquad (1.24)$$

But under equilibrium conditions the annual number of recruits must be the same as the number of deaths, i.e.:

$$R = \overline{N}Z \qquad (1.25)$$

Hence the number of recruits which die during their year of recruitment (expression 1.24) can also be written (substituting $\overline{N} = R/Z$):

$$\frac{R(Z - A)}{Z} \qquad (1.26)$$

The number of recruits which survive the year is therefore:

$$R - \frac{R(Z - A)}{Z} = \frac{RA}{Z} \qquad (1.27)$$

The development of expression (1.27) just given is that of Ricker (1944). Beverton (1954, p. 140) has developed it directly from the differential equation relating size of stock, N_t, to instantaneous mortality rate, Z, and to recruitment, R:

$$\frac{dN_t}{dt} = -ZN_t + R \qquad (1.28)$$

14

where R is the number of recruits which enter at a uniform absolute rate over a unit of time (in this case, a year). Integrating the above gives the expression for number of catchable fish at time t as:

$$N_t = \frac{R}{Z} + Ke^{-Zt} \qquad (1.29)$$

where K is an integration constant. If we consider a stock consisting of a single year's recruits, so that $N_t = 0$ when $t = 0$, the constant K is equal to $-R/Z$. Hence the general expression for the number of surviving *recruits* at time t becomes:

$$N_t = \frac{R}{Z} - \frac{R}{Z}e^{-Zt} = \frac{R}{Z}(1 - e^{-Zt}) \qquad (1.30$$

When $t = 1$ year, this number of survivors is:

$$\frac{R}{Z}(1 - e^{-Z}) = \frac{R(1 - S)}{Z} = \frac{RA}{Z} \qquad (1.31)$$

as in (1.27) above.

During their second year of life the above survivors (expression 1.27) are subject to the full mortality rate A, so that RA^2/Z die and RAS/Z survive. The total population of all ages, at the beginning of any year, is therefore found by summing the geometric series:

$$N = \frac{RA}{Z} + \frac{RAS}{Z} + \frac{RAS^2}{Z} + \dots$$

$$= \frac{RA}{Z}\left(\frac{1 - S^{\infty}}{1 - S}\right) = \frac{RA}{Z} \times \frac{1}{A} = \frac{R}{Z} \qquad (1.32)$$

But since recruitment and mortality are continuous, the population is the same at all times of year, and (1.32) represents the stock continuously on hand, $\overline{N}$.

Since (1.32) is the same as (1.21), it appears that, *regardless of the manner in which recruitment occurs, under equilibrium conditions the average stock on hand over the course of a year will be equal to* R/Z. A practical corollary is the fact that numerical examples in which recruitment is instantaneous (which are somewhat easier to construct) are for many purposes acceptable models of populations in which recruitment actually occurs along with the fishing.

1.5.7. STOCKS IN WHICH MORTALITY RATE CHANGES WITH AGE. When mortality and survival rate change with the age of the fish, whether because of a variable rate of natural mortality or variation in rate of fishing, no simple expressions for catch, etc., in the whole stock are possible: the contribution of each year-class must be summed separately. For example, with R recruits per year and continuous recruitment, the stock is:

$$\frac{RA_1}{Z_1} + \frac{RS_2A_1}{Z_1} + \frac{RS_3S_2A_1}{Z_1} + \frac{RS_4S_3S_2A_1}{Z_1} + \dots \qquad (1.33)$$

15

and the catch is:

$$\frac{F_1 R A_1}{Z_1} + \frac{F_2 R S_2 A_1}{Z_1} + \frac{F_3 R S_3 S_2 A_1}{Z_1} + \ldots \quad (1.34)$$

If, in addition, the number of recruits varies, the R terms too would have to carry separate subscripts. Numerical calculations where these parameters vary are most easily carried out in tabular form (e.g. Tables 8.2–8.4) though general formulae have been given for the situation when Z changes once (Ricker 1944, p. 32).

1.6. GROWTH AND YIELD IN WEIGHT

From the time they are hatched, the individual fish in a brood increase in size, at the same time as they are reduced in numbers. The mass of the whole brood, at a given time, is determined by the resultant of the forces of growth and of mortality. Since man is usually interested in the weight, rather than the number, of fish which he can catch, the individual rate of increase in weight must be balanced against the rate of decrease in numbers in order to obtain an expression from which to compute weight yields.

1.6.1. USE OF OBSERVED AVERAGE WEIGHTS. Possibly the simplest way to take growth into account in constructing such a population model is to combine schedules of age distribution with observed information on the average size of fish at successive ages. An example is shown in Table 10.1 of Chapter 10. This procedure presents a difficulty when any considerable deviation from the existing mortality rate is being examined. For example, as mortality rate increases, the fish caught of a given age will be smaller, on the average, because they decrease in numbers more quickly and fewer survive to the larger sizes reached later in the year. (This is distinct from any actual change in rate of growth that may occur.)

1.6.2. RATE OF GROWTH. When growth is exponential, it may be treated in the same manner as mortality. There is a *relative rate of growth*, h, and a corresponding *instantaneous rate of growth*, G. If w_t is the weight of a fish at time t, and w_0 is its weight at $t = 0$, then the equation of exponential growth is:

$$\frac{w_t}{w_0} = e^{Gt} \quad (1.35)$$

If the initial weight is taken as unity, at the end of a unit of time the weight is e^G, and it has increased by $e^G - 1$; hence:

$$h = e^G - 1$$

and

$$G = \log_e(h + 1)$$
$$= \log_e(w_t/w_0) \text{ when } t = 1 \quad (1.36)$$

For example, a fish which grew from 2 to 5 kg in unit time (say a year) would have an absolute growth of 3 kg per year. Its relative or annual growth rate is $h = 3/2$

= 1.5 or 150% per year. Its instantaneous rate of growth is $G = \log_e(5/2) = 0.916$ (on a yearly basis). Pairs of values of $h + 1$ and G are shown in columns 12 and 13 of Appendix I.

In practice, growth is not usually exponential over any very long period of the life of a fish, but any growth curve can be treated in this way if it is divided up into short segments.

1.6.3. CHANGE IN STOCK SIZE WITHIN A YEAR. The simplest way to relate growth to mortality is to calculate the mean instantaneous rate of growth (G) for each year separately, and combine it with the instantaneous mortality rate (Z) for the year, to give the instantaneous rate of change in bulk, $G - Z$. With time (t) measured in years, and putting B_0 for the initial biomass of the year-class and B_t for its biomass at any fraction t of a year later:

$$\frac{B_t}{B_0} = e^{(G-Z)t} \tag{1.37}$$

provided that the rates of growth and mortality do not change with the seasons. If the proviso holds, the average biomass of the year-class during the year can be found from:

$$\overline{B} = \int_{t=0}^{t=1} B_0 e^{(G-Z)t} \, dt$$

$$= \frac{B_0(e^{G-Z} - 1)}{G - Z} \text{ or } \frac{B_0(1 - e^{-(Z-G)})}{Z - G} \tag{1.38}$$

When $G - Z$ is negative, this expression can be evaluated from column 4 of Appendix I, putting $Z - G$ for the Z of column 1. When $G - Z$ is positive, the required values are given in column 5, and Z of column 1 is equated to $G - Z$.

If growth and mortality are not constant, but vary seasonally *in parallel fashion*, then (1.38) can be used to compute an average stock size, which can be thought of as based on the fish's physiological and ecological time scale instead of on astronomical time. Whatever time scale is used, the average biomass of the year-class, $\overline{B}$, can be multiplied by any instantaneous rate or combination of rates, to show the mass of fish involved in the activity in question, just as with mean numbers in Section 1.5:

$Z\overline{B}$ = total mortality, by weight $\tag{1.39}$

$F\overline{B}$ = weight of catch $\tag{1.40}$

$M\overline{B}$ = weight of fish that die "naturally" $\tag{1.41}$

$G\overline{B}$ = production, or total growth in weight of fish during the year, including growth in the part of the population which dies before the year is finished $\tag{1.42}$

$(G-M)\overline{B}$ = excess of growth over natural mortality $\tag{1.43}$

$(G-Z)\overline{B}$ = net increase in weight of a year-class during the year (a negative value of course indicates a decrease) $\tag{1.44}$

17

The restriction on seasonal incidence of growth and mortality may sometimes be serious, but the above expressions will be useful, at least as an approximation, in most cases. There is often some tendency for the two opposed effects to vary in a parallel fashion; for example, both growth and mortality may tend to be less in winter than in summer. During their first year of life both growth and mortality rate of a fish tend to change rapidly. Sometimes a quantitative seasonal breakdown can be obtained for both, and can be used to calculate production more accurately (Ricker and Foerster 1948).

1.6.4. CHANGE IN STOCK SIZE FROM YEAR TO YEAR. The restriction that seasonal incidence of growth and mortality be proportional is not necessary for computing the mass of the stock *from one year to the next*. That is, the weight of a year-class at age $t + 1$ is related to that at age t as follows:

$$B_{t+1} = B_t e^{G-Z} \tag{1.45}$$

regardless of how growth and mortality are distributed during the year.

In general, in the life history of a brood there will be one to several years during which $G - Z$ is positive and total bulk is increasing, followed by several years in which $G - Z$ is negative and bulk is decreasing. In an unfished population, the mean length or weight of the fish in a year-class when $G = Z$ (growth just balancing mortality) is called the *critical size* (Ricker 1945c). The same term is applied to the fish in exploited populations at the point where $G = M$, that is, where the instantaneous rate of growth is equal to the instantaneous rate of natural mortality.

1.7. FISHING EFFORT AND CATCH PER UNIT OF EFFORT

For greatest ease in estimating biological statistics, a fishery should ideally be prosecuted exclusively by one kind of gear, which should be strictly additive in effect — that is, each additional unit should increase the instantaneous rate of fishing by the same amount. Further, the investigator should have a record of all gear fished, and it should preferably fish for only one kind of fish. It usually happens that these conditions are not satisfied, and much ingenuity has been devoted to obtaining the best representative figure from incomplete or otherwise unsatisfactory data. Good reviews of some of the problems are by Widrig (1954a), Gulland (1955a), and Beverton and Parrish (1956).

1.7.1. MEASURES OF FISHING EFFORT. In general there can be more than one measure of fishing effort. A simple index is the number of vessels in use, or the number of anglers on a lake. If the vessels differ in size, their total displacement is often used, since the larger vessels usually catch more fish. If possible, number of vessels or tonnage should be multiplied by time — either days at sea, or days or hours of actual fishing, and so on. The measure of effort used will depend partly on what information is available, but the aim is always to have a figure which is proportional to the rate of fishing, F, as closely as possible, at least on a long-term average basis.

1.7.2. INCOMPLETE RECORD OF EFFORT. If records of catch are complete but records of effort are incomplete, a good plan is to compute the catch per unit effort for as much of the data as possible. This catch/effort, divided into the residual catch, will give an estimated effort figure for the latter, which can be added to the known effort to obtain a total. Sometimes effort records are complete and catch records incomplete, permitting the same procedure in reverse.

1.7.3. DIFFERENT KINDS OF FISHING GEAR. When different kinds of fishing are conducted on the same stock, the effort and catch taken by each is tabulated separately. For an overall picture, it is necessary to relate all kinds of effort to some standard unit. This is best done from a comprehensive series of fishing comparisons of the different gears under the same conditions. However, sometimes the gears are so unlike that this is impossible. If one kind of gear predominates over the others in a fishery, it may be sufficient to proceed as in the paragraph above: the effort of all other gears is scaled to terms of the dominant gear by dividing their gross catch by the catch/effort of the dominant gear. This has been done for many years for the Pacific halibut, for example (Thompson et al. 1931). When two or more very different gears are in extensive use — gillnets and traps, for example — it may be impossible to obtain a really satisfactory comparative measure of total effort from year to year, particularly if the two gears tend to select different sizes of fish, or if they are operated at different times of year.

1.7.4. VARIATION IN EFFICIENCY OF GEAR, AND GEAR SATURATION. With most kinds of gear, the fishing effort depends on the length of time it is in use, though "fixed" gears like traps often fish continuously. However, from the time they are set to the time they are lifted, some kinds of gear decrease in efficiency. For example, baits can be eaten off hooks by trash fish or invertebrates, nets can become fouled and so are more easily avoided by the fish, etc. Also, the mere fact that some fish are already caught can reduce efficiency: the fish already hooked leave fewer vacant hooks on a set-line; in most kinds of traps, fish can leave as well as enter, and a point of saturation may even be reached, so that effort depends partly on how often they are emptied; in a gillnet, the presence of some fish already caught tends to scare others away, so that saturation may be reached long before the net is full of fish (Van Oosten 1936, Kennedy 1951). The extra time needed to lift or clear a net, when fish are abundant, may appreciably decrease the time it is in the water and fishing, hence decrease the effectiveness of a "net-day" or "trap-day." Thus the catch per unit time, for many kinds of gear, tends to decrease from the time they are set to the time they are lifted, and the speed of this decrease is partly a function of the abundance of the fish.

The reverse phenomenon is also sometimes encountered: for example, in trapping for sunfishes near their spawning beds, the presence of fish in a trap appears to attract others to it, so that dozens of fish may be taken in one small trap while adjacent ones are nearly empty. Some Mississippi River fishermen are said to "bait" their traps with a mature female during the spawning season.

All such effects demand care in assessing the fishing power of a unit of gear, and standardizing it in some way.

1.7.5. VARIATION IN VULNERABILITY OF THE STOCK. Statements so far have concerned only the simple situation where the whole of a fish stock is equally vulnerable to the fishing in progress. In large-scale fisheries this is unlikely to be the true situation, for several possible reasons.

No trouble arises if a portion of a species lives completely outside the range of fishing operations and never mingles with or contributes recruits to the fished population. In that event consideration can be restricted to the vulnerable part of the stock, and the rest is ignored for purposes of current vital statistics. Other possibilities present greater problems:

1. Different portions of a fish stock, even one which is uniformly abundant throughout its range, may be fished at differing intensities in different places because of economic considerations or legal restrictions. If the various portions of the stock intermingle at any time of year, it is necessary somehow to compute average statistics of mortality, etc.

2. A situation similar to but more extreme than the above is where some parts of the range of a population contain fish too sparsely concentrated, in too deep water, or too remote from a harbor to be fished at all, yet these fish mingle with the fished stock at times of year other than the fishing season. For example, in trawl fisheries, and particularly in Danish seining, some parts of the fishing grounds are too rough to be fished without loss of gear, and these areas provide "refuges" where a part of the stock is not accessible.

Where a stock can be divided fairly sharply into a vulnerable and an invulnerable portion, each year, the fraction which is exposed to fishing is called the *availability* of the population that year (Marr 1951).

3. Catchability can also vary within a year because of seasonal physiological or behavior changes, and if a short fishing season is not exactly synchronized with this behavior each year, the result is between-year differences in catchability.

4. Fish of different sizes may be caught with varying efficiency — either as a result of selectivity of gear or because of differences in distribution or habitat. As they grow, their vulnerability to the gear in use changes.

The feature common to all the above effects is that different parts of the stock are subjected to different rates of removal by the fishery; that is, they differ in vulnerability. This complicates the estimation of vital statistics, and introduces errors which may be difficult to detect.

If these stocks are treated as though the fishery were directed against a single compact population, the effects above give to estimated vital statistics a somewhat fictitious character. One can't be sure that they are really what they seem to be. For example, some fisheries attack only the part of a stock which is fairly densely aggregated at the edge of a bank, or along a temperature boundary. Decline in catch per unit effort during a season can give an estimate of the stock *in that area* (Chapter 6), but the total population on which such fishing can draw, over the years, is considerably greater because of replenishment of the area in the off season. Again, if fish of certain sizes are more vulnerable than others, a Petersen tagging experiment (Section 3.2) is apt to overemphasize the vulnerable ones both in respect to tags put out and recaptures made; hence the estimate of rate of exploitation is too high and the population estimate is too low. However, for some purposes systematic bias of such kinds is not too great a handicap, provided it does not vary from year to year. It is secular *changes* in biological statistics that are of most interest, and changes will show up even in the biased statistics.

When there are year-to-year variations in the distribution of the fishing, or in the distribution and availability of the stock, or in the vulnerability of the stock as affected by weather or age composition, the situation is more serious. Such variability makes for changes in the estimated statistics that are not easy to distinguish from true changes in the population parameters. Comprehensive treatments of the theory of variability in these respects have been given by Widrig (1954a, b) and by Gulland (1955a).

Most of Widrig's discussion is in terms of effect No. 2, the availability, r, of the stock in different years. However, his treatment seems equally applicable to other kinds of variation in the vulnerability. Consider a static r_i', representing the ratio of the catchability of the whole stock in year i to an arbitrarily chosen standard catchability q_s; so that:

$$r_i' = q_i/q_s \qquad (1.46)$$

Then r' can be substituted for r in Widrig's computations, and the latter become applicable to a wider class of phenomena — some of which, in practice, are very difficult to distinguish from availability anyway.

1.7.6. CATCH PER UNIT EFFORT AS AN INDEX OF ABUNDANCE. When a single homogeneous population is being fished, and when effort is proportional to rate of fishing, it is well established that catch per unit effort is proportional to the mean stock present *during the time fishing takes place* (Ricker 1940) — whether or not recruitment from younger sizes takes place during that time. If the stock is not homogeneous — not all equally vulnerable to fishing — total catch divided by total effort is proportional to stock size only in special circumstances: when the *relative* quantities of fishing effort attacking different subsections of the stock do not change from year to year, or when the relative size of the stock in the different subsections does not change (Widrig 1954a).

Narrowing the discussion to *geographical* subdivisions of a population, for many kinds of fishing the vulnerability of a stock in different subareas will tend to vary approximately as stock density (fish present per unit area). If these are in direct proportion, then an overall C/f that is proportional to total stock size can be obtained by adding the C/f values for individual subareas, weighting each as the size of its subarea (Helland-Hansen 1909, p. 8, Widrig 1954b, Gulland 1955a, expression 2.4). However, if vulnerability does not vary as density, then there is no completely satisfactory substitute for a determination of absolute stock size separately in each subarea

21

in each year. This rather gloomy conclusion is indicated, in effect, by Gulland's expression (2.2). The least tractable populations are of those pelagic species which appear in varying proportions in different parts of their range in different years.

1.7.7. COMPETITION BETWEEN UNITS OF GEAR. The term "gear competition" has been used and discussed by a number of writers, but some confusion has resulted from inadequate definition. The sections above have dealt with the subject by implication, but a specific treatment may be useful. At least three kinds of effects have been included under the term:

1. A fish population is exploited by a fishery whose units of gear are scattered randomly over it, so that all fish are exposed to the possibility of capture at short intervals of time and there is no possibility of local depletion occurring. Further, the units of gear do not interfere with each other in respect to the mechanics of their operation. In such a situation, today's catch by any new unit of gear reduces tomorrow's catch by the others, and thus in a sense it may be said to "compete" with them. The competition takes the form of a faster reduction in the size of the population as a whole. As the fishing season progresses, each unit catches fewer and fewer fish (or at any rate fewer than it would have caught had there been no previous fishing that year); and the more gear present, the more rapid is this decrease in catch.

2. If fishing gear is dispersed unequally over the population, its action tends to produce local reductions in abundance greater than what the population as a whole is experiencing, leading to a different type of competition. Suppose that a population is vulnerable to fishing only in certain parts of its range (for example, only near the shore of a lake; or on only the smoother ocean bottoms). Then fishing in such areas produces a local depletion of the supply; additional nets set in the same region increase the local depletion and catch per unit effort will fall off in proportion to the *local* abundance. The magnitude of this fall will be cushioned if some fish from the rest of the stock keep wandering into the fishing area and so keep the supply there from dropping as far as it otherwise would. However, competition between units of gear is intensified because catch per unit effort reflects the size of only the immediately available restricted portion of the stock, rather than the stock as a whole.

3. Finally, if the setting of an additional unit of gear interferes directly with other gear, there exists "physical" competition between them, which is independent of population abundance, even locally. For example, too many anglers at a pool may frighten the fish; setting a new gillnet near one already in operation may scare fish away from the latter; or much fishing of a schooling fish may disperse the schools and so reduce fishing success more than proportionally to actual decrease in abundance. (There can also, of course, be physical *cooperation* between different units of gear.)

Competition of type 1 above can be considered normal and inevitable. It might be better not to call it competition at all, since the term is usually meant to suggest effects of types 2 or 3. Competition of types 2 and 3 may or may not be present in any given situation — it depends entirely on the nature of the fishery.

1.7.8. COMPETITION BETWEEN SPECIES. When more than one species is caught by the same gear, particularly the baited types (longlines and some traps), there can be competition between species for the gear. In general, the more individuals of other species present, the less efficient such gear becomes for catching the species of interest (Gulland 1964a, Ketchen 1964).

Rothschild (1967) examined the mathematical aspects of this situation for baited-hook types of gear and derived a stochastic expression for the probability of capture. When two species are involved, the instantaneous rate of capture of species 1 is estimated by:

$$\lambda_1 = \frac{-n_1 \left[\log_e(n_0/N)\right]}{N - n_0} \tag{1.47}$$

where:

n_0 = the number of empty fishless hooks

n_1 = the number of hooks carrying species 1

N = the total number of hooks

The instantaneous rate of capture of species 2 is the same as (1.47), with the subscript 1 changed to 2.

The probability of species 1 being caught on any hook in a (hypothetical) situation where species 2 is absent is:

$$P_{01} = 1 - e^{-\lambda} \tag{1.48}$$

This can be called the conditional rate of capture of species 1, and is analogous to the conditional rate of fishing mortality (Section 1.5.2).

The above expressions can also be used where three or more species are caught, by letting n_1 be the catch of any species of interest, and n_2 the catch of all other species.

Rothschild assumes that no baits are removed from the hooks by "bites" that catch no fish, which is somewhat unrealistic. His analysis can be extended to the situation where some baits are removed from hooks without capture. Suppose that in addition to its n_1 captures, species 1 has eaten n_1a_1 baits without capture; similarly species 2 has taken n_2a_2 baits without capture; and no baits are lost by any other process. The number of fishless hooks can be divided into n_e which have lost their bait and $n_0 - n_e$ which still have their bait. Hence:

$$n_1a_1 + n_2a_2 = n_e \tag{1.49}$$

In general a_1 and a_2 will be unknown, but some idea of the severity of competition can be had by assuming $a_1 = a_2 = a$; in which event, from (1.49):

$$a = \frac{n_e}{n_1 + n_2} \tag{1.50}$$

Thus a can be evaluated if a record is kept of empty hooks. We can then substitute $n_1 (1 + a)$ for n_1 and $n_0 - n_e$ for n_0 in expression (1.47) to obtain the

23

instantaneous rate of removal of baits (including captures) by species 1; call this λ_{1a}. The conditional probability of removal of baits (including captures) for species 1 then becomes:

$$1 - e^{-\lambda_{1a}} \qquad (1.51)$$

and the conditional rate of capture of species 1 is:

$$P_{01a} = \frac{1 - e^{-\lambda_{1a}}}{1 + a} \qquad (1.52)$$

Expression (1.52) of course represents a situation at the other extreme from (1.48), because it takes no account of the possibility that baits may simply fall off the hooks, or that some may be eaten by species that are never captured.

1.8. MAXIMUM SUSTAINABLE YIELD

Much of the work on vital statistics has devolved about or been stimulated by attempts to estimate the maximum equilibrium catch or maximum sustainable yield for the stock. Some of this background is necessary for appreciation of the value or significance of some of the methods which will be described.

A simple approach is shown in Fig. 1.2 (cf. Russell 1931, Schaefer 1955). The *usable stock* of a species is defined as the weight of all fish larger than a minimum useful size. This stock loses members by *natural deaths* and, if there is a fishery, also by the *catch* which man takes. The usable stock is replenished by *recruitment* from smaller size categories, and by *growth* of the already-recruited members.

If a stock is not fished, all growth and recruitment is balanced by natural mortality. If fishing begins, it tips the balance toward greater removals, and occasionally fishing may steadily reduce the usable stock until it is commercially extinct. Much more often a new balance is established, because the decreased abundance of the stock results in (1) a greater rate of recruitment, or (2) a greater rate of growth, or (3) a reduced rate of natural mortality.

Ideally, the effects of concurrent variation of all three of these rates, with respect to size of the population, should be studied in order to define equilibrium yield and compute its maximum value. In actual practice to date, it has been necessary to abstract one or two variables for consideration, keeping the others constant, or else to consider only the net result of all three. The various proposals for estimating maximum sustainable yield differ principally in respect to which of these three rates is permitted to vary with stock density, and in what way.

1. One group of methods assumes that rate of growth and rate of natural mortality are invariable. The *absolute number* of recruits is considered unvarying from year to year[2], a condition which means that *rate* of recruitment increases when the usable stock decreases, but only in a definite and narrowly-prescribed fashion. Such methods are treated in Chapter 10; their greatest usefulness has been for describing

---

[2] More exactly, the assumption is that the absolute number of recruits does not vary with stock density, but it may fluctuate from year to year in response to environmental variability.

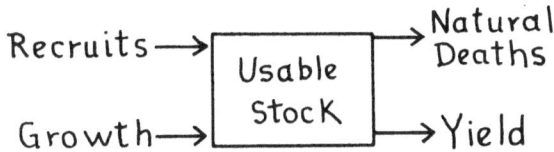

Fig. 1.2. Diagram of the dynamics of a fish stock (fish of usable sizes), when there is no fishing and when there is a fishery. (From Ricker 1958c.)

the short-term reactions of stocks to fishing, but they may have value in showing the direction in which rate of fishing should be adjusted in order to move toward maximum sustainable yield.

2. Variation in *recruitment* is approached empirically in Chapters 11 and 12. The results can be used directly to compute maximum sustainable yield in situations where, as in the method above, the rates of growth and of natural mortality do not vary with size of stock.

3. At least one author has considered *rate of growth* as the primary variable in the adjustment of a stock to fishing pressure (Nikolsky 1953), particularly for freshwater fishes having comparatively short life histories. While this does not lend itself very well to general regulation, Nikolsky suggests the determination of maximum rate of growth for each species, and regulation of abundance until something close to the maximum is achieved.

4. Finally, several authors have attempted to relate surplus production (potential sustainable yield) of a stock directly to its abundance, without any direct information on the rates of growth, recruitment, or natural mortality. Chapter 13 describes these computations.

In addition to predicting the result of increasing or decreasing rate of fishing, most of the methods outlined can also be used to predict the effect of varying the minimum size of fish which is used by the fishery.

1.9. SAMPLING ERROR

In all of the methods of estimation to be discussed in subsequent chapters, the probable size of the sampling error is an important consideration. It must be evaluated, at least approximately, before any confidence can be placed in an estimate. When a

25

computation of survival rate, for example, is calculated from recapture of only a few marked fish, or from an age-class with only a few representatives in a sample, it must be accepted with caution.

Available estimates of sampling variability or error are of two general sorts. One type depends on random distribution of the fish or random selection of all pertinent types of fish by the fishing apparatus, and is computed from the frequency distributions which are appropriate in the individual case (usually Gaussian, Poisson, binomial, or hypergeometric). Examples of variances or standard deviations calculated on this basis are expressions (3.2), (3.4), (3.6), (3.8), (5.2), (5.14), (5.15), (5.16).

For small samples the positive and negative limits demarcating zones of equal confidence are not even approximately symmetrical about the observed value. In such cases it is frequently useful to use the asymmetrical confidence limits calculated for binomial distributions by Clopper and Pearson (1934), and for Poisson distributions by Garwood (1936) or Ricker (1937). The latter are given in Appendix II here; they are especially simple to use, and can be employed as an approximation even when the binomial charts are more appropriate. Both types are available in graphical form in a paper by Adams (1951).

For larger samples a general idea of sampling variability can be had by regarding the observed ratio of (say) the marked fish to the total fish in a sample (R/C) as though it were the true ratio u which exists in the population. The expectation of marked fish to be obtained is Cu, and its variance is given by the well-known formula:

$$V = Cu(1 - u) \tag{1.53}$$

With large R, this is approximated by:

$$V = R(1 - R/C) \tag{1.54}$$

and the standard deviation is the square root of this. In the (very frequent) event that R/C is small, this means that the standard deviation of the number of marked fish retaken is a little less than its own square root. Even when R/C is not especially small, this rule is good enough for orientation, so as to have in mind the order of size of the sampling variability to be expected. Similarly, the number of fish, n, of a given age in a sample can be regarded as having associated with it approximate limits of confidence set by the normal frequency distribution with $\sqrt{n}$ as standard deviation — provided it is not too small — less than 10, say. (For small numbers the binomial or Poisson limits should be used.)

The second general type of estimate of sampling variability is calculated from some form of replication in the data themselves. Such estimates will include part or all of the variation which arises from non-random distribution of the different categories of fish in the population being sampled: effects of grouping, for example. Objective estimates of variability are involved in the methods of estimating confidence limits used in Examples 3.6 and 6.1, and could be applied to 3.7, 11.1, etc. These estimates tend to be more realistic than those based directly on random sampling theory, though of course they are not necessarily exact; they are to be preferred when available.

Limits of confidence, of either type above, should preferably be calculated for statistics whose distribution is as nearly "normal" as possible. For example, in estimating population size, N, by most of the available methods, estimates of the *reciprocal* of N tend to be distributed nearly symmetrically about their mean. Confidence limits computed from the normal curve are likely to apply fairly well to $1/N$, whereas they do not apply at all well to N (DeLury 1958). Hence computations of confidence limits should be made in the first instance for $1/N$, and then inverted to give the appropriate asymmetrical limits for N itself. Similar situations often occur where the logarithms of variates will have an approximately symmetrical or even nearly normal distribution, whereas the variates themselves do not.

No kind of estimate of sampling variability can reflect or adjust for all the systematic errors which may so easily arise from non-random fish distributions or behavior. Systematic error usually tends to be larger than sampling error, and discussions of various kinds occupy much of the text to follow. Even if not larger, systematic effects are not removed by using more observations or making bigger experiments *of the same type*, so they deserve the closest attention.

Finally, in complex situations an elementary but very useful procedure is to introduce a series of deviations of known size into the data and see what effect each has on the final result.

CHAPTER 2. — ESTIMATION OF SURVIVAL RATE AND MORTALITY RATE FROM AGE COMPOSITION

2.1. SURVIVAL ESTIMATED FROM THE ABUNDANCE OF SUCCESSIVE AGE-GROUPS

2.1.1. SIMPLE SITUATIONS. The general method of estimating survival is by comparing the number of animals alive at successive ages. Long known in human demography, this procedure became available to students of fish populations as soon as age determinations began to be made on a large scale and from representative samples. This occurred early in this century for North Sea species; the voluminous literature on the plaice *Pleuronectes platessa* contains early estimates of mortality and survival, as well as doubts concerning the representativeness of the samples available (Heincke 1913a; Wallace 1915).

If the initial number of fish of two broods, now age t and age $t + 1$, was the same, and if they have been subjected to similar mortality rates at corresponding ages, then an estimate of survival rate from age t to age $t + 1$ is obtained from the ratio:

$$S = \frac{N_{t+1}}{N_t} \tag{2.1}$$

where N represents the number found, of each age, in a representative sample. Gulland (1955a, Part III) and Jones (1956) show that (2.1) is not the best estimate of S, but is somewhat too large. However, the estimate of instantaneous mortality rate corresponding to S, which is:

$$Z = -(\log_e N_{t+1} - \log_e N_t) \tag{2.2}$$

is without appreciable bias. Since it is usually desirable to use estimates of S and Z that conform exactly to $Z = -\log_e S$, the estimates (2.1) and (2.2) are both commonly used.

In practice the samples available are often taken throughout a fishing season, so estimates from (2.1) and (2.2) pertain to the time interval approximately from the middle of one season to the middle of the next. Such estimates are represented here by $\bar{S}$ and $\bar{Z}$.

If it can be assumed that survival is constant over a period of years, a combined estimate can be made from a series of estimates of the form (2.1), by one of several methods.

2.1.2. COMBINED ESTIMATES OF SURVIVAL RATE — HEINCKE'S METHOD. In any random sample of a population the older ages will tend to be scarcer than the younger; therefore, because of sampling variability, an S estimated from them is less reliable than one from younger ages. A formula that weights successive ages as their abundance was proposed by Heincke (1913b). Let the ages representatively

29

sampled be numbered in succession *starting with zero for the youngest*, so that successive numbers of fish are N_0, N_1, N_2, etc.; ΣN is the sum of these. Heincke's estimate was of the mortality rate A:

$$A = \frac{N_0}{\Sigma N} \qquad (2.3)$$

Since $S = 1 - A$, the corresponding estimate of survival rate becomes:

$$S = \frac{\Sigma N - N_0}{\Sigma N} \qquad (2.4)$$

Note that it is not necessary to know the number of fish in each age older than that coded as 0, but only their total. *Hence this formula can be used when age determinations of older fish are unreliable.* But much obviously depends on the representativeness of the youngest age used.

EXAMPLE 2.1. SURVIVAL RATE OF ANTARCTIC FIN WHALES, BY AGE COMPOSITION. (From Ricker 1958a; data from Hylen et al. 1955.)

Age frequencies of male fin whales in Norwegian catches sampled in the 1947–48 to 1952–53 seasons is given by the above authors as follows:

Age	0	1	2	3	4	5	6+
Frequency (%)	0.3	2.3	12.7	17.2	24.1	14.1	29.5

Ages 4 and 5 are regarded as likely to be accurately determined, and they may possibly be representatively sampled, so that survival between these ages can be estimated from (2.1) as:

$$S = \frac{14.1}{24.1} = 0.585$$

Alternatively, assuming a constant survival rate, ages 5 and older can be compared with age 4 using (2.4):

$$S = \frac{14.1 + 29.5}{24.1 + 14.1 + 29.5} = 0.643$$

This gives a larger figure than the simple comparison, and might suggest that older whales really survive better than the age 4–5 group. However, strictly from these data, without considering any accessory information that may be available, there is no way to be sure that age 4 was as vulnerable to whaling as age 5, since the next younger age (3) is obviously much less vulnerable. It might be safer therefore to consider only whales of age 5 and older; again using (2.4):

$$S = \frac{29.5}{14.1 + 29.5} = 0.676$$

The effect of any increase in whaling effort over the time these stocks were being recruited would be to make this survival estimate greater than the average one prevailing at the time the samples were taken (Section 2.6; see also Hylen et al.).

2.1.3. COMBINED ESTIMATES OF SURVIVAL RATE — ROBSON AND CHAPMAN'S METHOD. According to Robson and Chapman (1961) the best estimate of S from age census data is:

$$S = \frac{T}{\Sigma N + T - 1} \tag{2.5}$$

with sampling variance estimated as:

$$S\left(S - \frac{T-1}{\Sigma N + T - 2} \right) \tag{2.6}$$

$$T = N_1 + 2N_2 + 3N_3 + \ldots$$

$$\Sigma N = N_0 + N_1 + N_2 + \ldots$$

An example of the application of these formulae is given in Example 4.3, below.

All the above formulae involving more than two ages assume that survival rate is constant at all ages, that all year-classes are recruited at the same abundance, and that all ages are equally vulnerable to the sampling apparatus. When these conditions don't apply, estimates of S are biased and confidence limits from their estimated variance are, in general, too narrow. Robson and Chapman give a χ^2 formula for testing whether these assumptions may not have been fulfilled. Actually in most stocks differences in year-class strength will be the major source of variability in samples of moderate to large size, in which case the best estimate of S will be obtained from a catch curve with equal weighting, described in Section 2.2.

2.1.4. ESTIMATES OF SURVIVAL RATE FROM A PORTION OF AN AGE SERIES. For one reason or another we may wish to estimate survival rate from only a portion of an age series. If only 2 years are involved expression (2.1) can be used. For 3 years the expression:

$$S = \frac{N_{t+1} + N_{t+2}}{N_t + N_{t+1} + N_{t+2}} \tag{2.7}$$

is possible, and similarly for 4 or more; however (2.7), like (2.1), has a small positive bias. Robson and Chapman (1961, p. 184) give suitable unbiased formulae analogous to (2.5) and (2.6). In practice, however, the method of Section 2.2 will usually be best, for reasons given above.

2.1.5. SURVIVAL RATE FROM MEAN AGE. Expression (2.5) can be derived approximately from a consideration of the mean age of the fish in a catch, still assuming constant recruitment and survival rate. If the calendar age (in years completed) of the first completely vulnerable age-group in a sample be called coded age 0, and it contains N_0 fish, the numbers of fish to be expected at subsequent ages are shown in Table 2.1. The number of fish actually observed, that corresponds to the sum of column 2 of the Table, is $N_0 + N_1 + N_2 + \ldots = \Sigma N$ of expression (2.5). Similarly the observed sum corresponding to the sum of column 3 is $N_1 + 2N_2 + 3N_3 + \ldots$

= T of expression (2.5). An estimate of the mean coded age in the population is the quotient of these, and thus is an estimate of S/A:

$$\text{Mean coded age} = \frac{T}{\Sigma N} = \frac{N_0 S}{A^2} \Big/ \frac{N_0}{A} = \frac{S}{A} \tag{2.8}$$

To obtain an expression for S from (2.8), invert both sides and add 1:

$$1 + \frac{\Sigma N}{T} = \frac{T + \Sigma N}{T} = 1 + \frac{A}{S} = \frac{1}{S} \tag{2.9}$$

Inverting the second and fourth terms of (2.9) gives (2.5), except for the minor adjustment of –1 in the denominator of the latter. Thus an expression for annual mortality rate is:

$$A = 1 - S = \frac{1}{1 + (\text{mean coded age})} \tag{2.10}$$

TABLE 2.1. Computation of mean age of the fish in a population.

1 Coded age	2 Frequency	3 Product
0	N_0	0
1	$N_0 S$	$N_0 S$
2	$N_0 S^2$	$2N_0 S^2$
3	$N_0 S^3$	$3N_0 S^3$
etc.	etc.	etc.
Totals	$N_0(1 + S + S^2 + ...)$	$N_0 S(1 + 2S + 3S^2 + ...)$
	$= \dfrac{N_0}{1 - S} = \dfrac{N_0}{A}$	$= \dfrac{N_0 S}{(1 - S)^2} = \dfrac{N_0 S}{A^2}$

2.1.6. MEAN TIME SPENT IN THE FISHERY. The mean coded age of the fish in a sample is not necessarily the same thing as the mean time they have spent in the fishery. In a Type 1 population, if recruitment is the same in successive years and occurs instantaneously at the start of coded age 0, the mean time spent in the fishery by the fish in a sample taken immediately after recruitment will evidently be equal to the mean coded age, S/A. Throughout the year this mean time will increase, until immediately before the next annual recruitment it will be $1 + S/A = 1/A$.

In a Type 2 population, suppose that the constant number R recruits of each year-class enter the vulnerable stock uniformly over 1 year's time, and let time 0 be the start of that year. In that event the total number of fish present at any moment will be, from (1.32), equal to R/Z. The sum of the products of (time since recruitment) and (number present), or tR_t, is:

$$\int_0^\infty tR_t dt = R \int_0^\infty te^{-Zt} dt = \frac{R}{Z^2} \tag{2.11}$$

The mean time that a fish spends in the fishery is (2.11) divided by R/Z, or:

$$\text{Mean time} = \frac{R}{Z^2} \Big/ \frac{R}{Z} = \frac{1}{Z} \qquad (2.12)$$

This result should not be confused with mean coded age, nor should it be applied to stocks in which recruitment is markedly seasonal (see above). However, if a Type 1 population were to be sampled throughout a year in proportion to its abundance, the mean time that the fish sampled had spent in the fully-recruited phase would be equal to 1/Z.

2.2. Simple Catch Curves

Edser (1908) was apparently the first to point out that when catches of North Sea plaice (*Pleuronectes platessa*) were grouped into size-classes of equal breadth, the logarithms of the frequency of occurrence of fish in each class form a curve which has a steeply ascending left limb, a dome-shaped upper portion, and long descending right limb which in his example was straight or nearly so through its entire length. This was soon recognized as a convenient method of representing catches graphically. Heincke (1913b) plotted a number of curves of this type and, combining them with information on rate of growth, computed mortality rates for a series of size intervals of the plaice, equating these approximately to age. Baranov (1918) later gave the name *catch curve* to the graph of log frequency against size, and elaborated the theory of estimating mortality and survival from it in the situation where fish increase in size by a constant absolute amount from year to year.

The same kind of plotting is useful for the simpler situation where age rather than length is considered.[1] Most recent authors plot log frequency against age directly, and the name catch curve has been applied to this kind of graph as well (Ricker 1948). The catch curve has a considerable advantage over the simple ratios of Section 2.1, and over arithmetic plots of abundance at successive ages, when any kind of variation in survival rate has to be examined.

The upper line of Fig. 2.1 is an example of a straight catch curve, pertaining to the bluegills (*Lepomis macrochirus*) in a small Indiana lake (Ricker 1945a). Rate of survival, S, from such a curve can be computed in two slightly different ways. The flatter the right limb, the greater is the survival rate. The difference in logarithm between age t and age $t-1$ is of course negative; it can be written with a positive mantissa and then antilogged, giving S directly. Alternatively we could follow Baranov in keeping the difference of (base 10) logarithm at its negative numerical magnitude, changing the sign, and multiplying by 2.3026, which gives the instantaneous rate of mortality, Z. A table of exponential functions will give the annual rate of survival, from the equation $S = e^{-z}$. Since we will almost always want to know Z as well as S, one

---

[1] The straightness of Edser's and Baranov's 1906 catch curve for North Sea plaice, plotted with length on the abscissa, was evidently a temporary phenomenon resulting from a recent increase in fishing effort. Plotted with age on the abscissa it would become the concave curve characteristic of such a situation (cf. Section 2.6), since rate of increase in length drops off sharply among the older fish.

method of computation is as convenient as the other. The annual mortality rate, A, is equal to 1 – S. If survival rate *during* instead of *between* successive years is desired, it can be obtained by taking tangents on the curve at each age.

The ascending left limb and the dome of a catch curve represent age-classes which are incompletely captured by the gear used to take the sample: that is, they are taken less frequently, in relation to their abundance, than are older fish. This

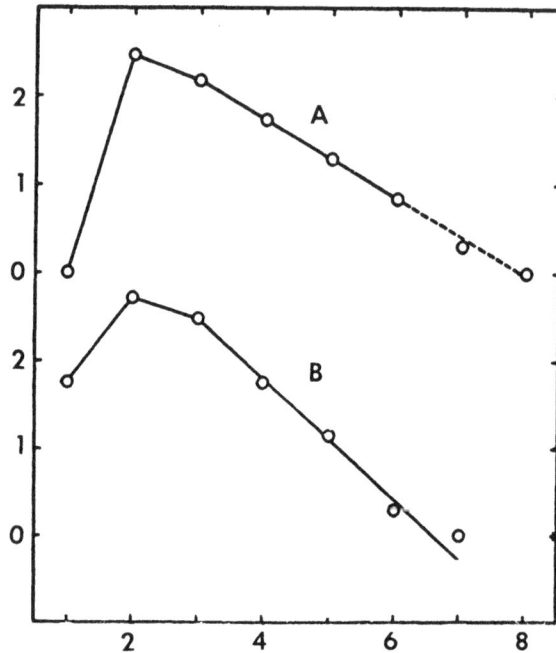

Fig. 2.1. (A) Logarithms of numbers of bluegills of successive ages, in a sample from Muskellunge Lake, Ind., 1942; (B) Logarithms of the percentage representation of successive age-classes of pilchards in the catch from California waters, season 1941–42. (Redrawn from Silliman 1943.)

may come about either because younger fish are more thickly distributed in another part of the body of water than that principally fished, or because they are less ready to take the baits or enter the nets. Other things being equal, total mortality rate will be increasing during this period of recruitment. However, it is impossible to find out anything definite about the actual mortality rate during the years covered by the left limb and dome of the curve, simply because sampling of the population is not random[2].

---

[2] It is being assumed, of course, that the sample is taken from the commercial catch. If better means of sampling are available, they will push the representative part of the sample back to earlier years, and in this way it may be possible to detect and measure otherwise-inaccessible changes in total mortality and in natural mortality. Jensen (1939) interprets some experimental trawl catches in this manner.

34

We turn then to a more promising part of the curve, the descending right limb. Straightness of this right limb, or any part of it, is usually interpreted in the manner described by Baranov, which involves the following conditions:

1. The survival rate is uniform with age, over the range of age-groups in question.

2. Since survival rate is the complement of mortality rate, and the latter is compounded of fishing and natural mortality, this will usually mean that each of these, individually, is uniform.

3. There has been no change in mortality rate with time.

4. The sample is taken randomly from the age-groups involved. (If the sample is representative of the commercial catch, this condition is implied in 2 above.)

5. The age-groups in question were equal in numbers at the time each was being recruited to the fishery.

If these conditions are satisfied, the right limb is, in actuarial language, a curve of survivorship which is both age-specific and time-specific.

Deviations from the above conditions often result in nonlinear right limbs of the catch curve. Such nonlinear curves are quite common, and in the Sections to follow we attempt to set up standards for the interpretation of some of the more likely types. Equally important is the allied question: under what conditions can a linear or nearly linear catch curve result from postulates other than the above?

EXAMPLE 2.2. TWO STRAIGHT CATCH CURVES: FOR BLUEGILLS AND CALIFORNIA SARDINES.

Catch curves having a straight right limb have already been treated adequately by Baranov and others, and need little comment. An interesting selection is presented by Jensen (1939). The bluegill example of Fig. 2.1A was selected for its close adherence to theoretical requirements; much more often fluctuating recruitment makes it necessary to use averages over a period of years to obtain a reasonably representative survival rate. Silliman (1943, p. 4) has an example of a straight catch curve, reproduced here in Fig. 2.1B. It pertains to the season 1941–42 of the fishery for California sardines (*Sardina caerulea*), and gives an estimated survival rate of about 0.20.

While straight catch curves will probably usually be interpretable in the manner proposed by Baranov and outlined in Section 2.2, two principal possible exceptions should always be kept in mind; (1) a decrease in vulnerability to fishing with age, and the consequent tendency toward increase in survival rate, will not be reflected in the catch ratio, or will be very imperfectly reflected; and (2) long-term trends in recruitment deflect the slope of a catch curve without introducing much or any curvature. Obviously, information on these topics is not to be looked for in the catch curve, and must be obtained from other sources. To illustrate, Silliman (1943) tentatively concluded that an increase in recruitment of about 130% occurred between 1925–33 and 1937–42 in the pilchard stock. If any of this increase carried over into the years when the fish of Fig. 2.1B were being recruited, the straight curve computed for those years would be too steep, i.e. would suggest a survival rate less than the true one. Some idea of the possible magnitude of this effect can be had from Silliman's data, assuming recruitment increased at a constant exponential rate for ten years. If k represents this rate, we have $e^{10k} = 2.3$, $k = 0.083$, and the annual increase is 0.087. Thus the survival rate computed from the catch curve would be less than the true rate by only about 9% of the former, even assuming the increase in recruitment to have persisted through the entire formative period of Fig. 2.1B.

Another danger in interpreting a straight catch curve lies in the possibility of a fortuitous balancing of opposed tendencies. For example, a straight curve like Fig. 2.1A could conceivably result from the combination of a normally convex curve (natural mortality rate increasing with age) with the effect of a recent increase in rate of fishing. In view of the general increase in rate of fishing in the North Sea and North Atlantic during the period 1920–35, one wonders whether the approximate linearity of some of Jensen's (1939) curves for cod, haddock, and plaice in those waters has not been achieved in this manner. Such possibilities emphasize the desirability of continuous sampling of a stock, and also the value of having information on the level of fishing effort, etc., in successive years.

2.3. NON-UNIFORM RECRUITMENT. USE OF CATCH PER UNIT EFFORT FOR ESTIMATING SURVIVAL

2.3.1. RANDOM VARIATION IN RECRUITMENT. Moderate fluctuations in recruitment from year-class to year-class, which are of an irregular character, make a catch curve bumpy, but do not destroy its general form, and thus do not greatly affect its value. Such irregularities are like those which result from random errors of sampling, but with this difference: they do not tend to disappear as sample size is increased. As a matter of fact, recruitment sufficiently uniform to make a really smooth catch curve appears to be rather rare. A good way to reduce irregularities from unstable recruitment is to combine samples of successive years. If fishing has been fairly steady, and the population consequently is presumed to be in a state of equilibrium except for the variations in recruitment, then quite a number of years can be combined in this way. Even when secular changes in mortality rate have occurred it may still be useful to combine samples of two successive years, as in this way a considerable increase in the regularity of the curve may often be obtained without too much sacrifice of information concerning the past history of the stock in question.

2.3.2. SUSTAINED CHANGE IN LEVEL OF RECRUITMENT. If recruitment changes suddenly from one steady level to a new one, and remains stabilized there, the effect on the catch curve can easily be distinguished and interpreted. As Baranov has shown, such a change shifts the position of a part of the right limb without changing its slope.

2.3.3. EXTREME VARIATION IN RECRUITMENT. Sometimes recruitment is exceedingly variable, adjacent year-classes differing by a factor of 5, 10, 25 or more; as

36

shown, for example, by Hjort (1914) for cod and herring and by Merriman (1941) for striped bass. This makes it practically impossible to use the usual type of catch curve for estimating survival rate: comparisons must be made within individual year-classes, if at all.

2.3.4. TRENDS IN RECRUITMENT. More insidious than the above is the situation where recruitment has a distinct trend over a period of years. In actuarial language, the survivorship curve obtained by sampling in a single season will then be time-specific, and will not indicate actual mortality rates over the period concerned. Such trends in recruitment are likely to be reflected in trends in catch, after a suitable interval, but not all trends in catch result from variation in recruitment. The only direct way to check on the possibility of trends in recruitment is to continue sampling over a considerable period of years, the assumption being that a trend cannot continue indefinitely in one direction. However, it will be useful to examine the exact nature of the shift in the catch curve which is produced by changing recruitment.

Examples of catch curves affected by a progressive change in recruitment are shown in Fig. 2.2, Curves B and C. For comparison, Curve A is a curve of the Baranov

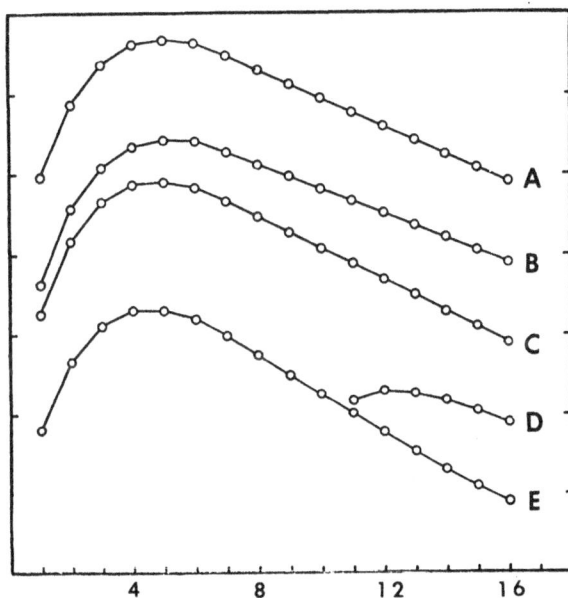

FIG. 2.2. Effect of variation in recruitment on a catch curve when there is a constant survival rate of 0.67 from age 7 onward. (A) Steady recuitment; (B) Curve based on the same data as A, but recruitment has decreased with time by 5% per year over the period of years shown; (C) Similar to B, but recruitment has increased by 5% per year; (D) Recruitment has decreased at an accelerating rate; (E) Recruitment has increased at a rate which initially was accelerating, but later flattened off. Abscissa — age; ordinate — logarithmic units.

37

type, based on uniform recruitment; its straight right limb has a slope corresponding to a survival rate of 0.670. Curve B is based on the same data, except that recruitment decreased by 5% per year over the period of years shown, i.e. it was 1.00, 0.95, 0.902, 0.857, etc., of its original value, in successive years. (The earlier years are to the right on the graph.) The right limb of Curve B is still straight, but it has a slope which corresponds to a catch ratio (apparent survival rate) of 0.705, which differs from 0.670 by 5% of the former. Similarly, when recruitment increases by 5% per year, as shown by Curve C, the line is straight with a slope corresponding to a catch ratio of 0.638, which differs from 0.670 by 5% (of 0.638). These and other examples show that deviation of the true survival rate from the apparent survival rate, when expressed as a percentage of the latter, is numerically equal to the annual percentage change in recruitment, but of opposite sign; i.e. when recruitment increases, apparent survival rate decreases.

From the above it follows that to obtain a *curved* right limb of the catch curve by varying recruitment, the rate of change in recruitment must vary from year to year. Two examples are shown in Fig. 2.2. Curve D shows the result of increasing the absolute decrease in the rate of recruitment by 0.05 each year; i.e. recruitment is 1.00, 0.95, 0.85, 0.70, etc., in successive years. A curved line is produced, but after only 6 years it terminates, because recruitment has been reduced past zero! Curve E shows the result of increasing recruitment in the same way. Here the annual rate of increase in recruitment (ratio of each year's increase to the preceding year's level) increases at first, and produces a short curved section, but soon the increase in the actual level of recruitment catches up to the increase in rate of increase, and the nearly straight section between age 7 and age 13 results. During the tenth year shown (i.e. at age 7), recruitment is 3.2 times its original level; however, to produce a line which would have the original curvature throughout its entire length for that period, recruitment at age 7 would have become many times greater.

Such computations as these illustrate the fact that in order to obtain recognizably curved right limbs by varying recruitment, the changes in recruitment would soon become so great as to produce acute symptoms in other statistics of the fishery, e.g. in total catch, average size of fish caught, relative abundance of young fish in successive years, etc. Hence we can confidently expect that the effect of any reasonable trend in recruitment will be to change the slope of the catch curve, without appreciably changing its linearity. If any significant curvature does occur, its explanation should be sought elsewhere.

In interpreting a catch curve, it would be useful to have some independent estimate of recruitment from year to year, as it might then be possible to introduce a correction for any trend which has occurred. Such information may be available from other catch statistics, particularly the catch of the youngest age-groups, per unit fishing effort. Information on the number of spawners (potential egg deposition) in successive years might also seem to offer possibilities, but actually the relation between eggs deposited and the resulting recruitment will usually be unknown, even apart from fortuitous variations; it is about as likely to be inverse as direct (Chapter 11).

2.3.5. COMPARISON OF ABUNDANCE OF INDIVIDUAL YEAR-CLASSES AT SUCCESSIVE AGES. To reduce the error caused by variable recruitment, it is natural to try to follow separate year-classes throughout their life, comparing the number present at age t with the number at age $t - 1$, and so on. However, if this is attempted with ordinary age composition data, trial computations will readily show that the presence of an exceptionally numerous year-class depresses the estimated survival rates *at all ages* in the year of its first appearance; afterward it makes them all too great for as many more years as it remains in the fishery. The geometric mean of survival rates estimated over a period of years tend toward the true value for each age (assuming the latter does not vary with time), but in practice there is usually little if any gain in accuracy over what would be provided by taking the mean of the slopes of the appropriate segments of the corresponding series of catch curves.

2.3.6. COMPARISON OF INDIVIDUAL YEAR-CLASSES ON THE BASIS OF CATCH PER UNIT OF EFFORT. A means of avoiding some of the difficulties caused by variable recruitment, of whatever type, is to compare *catch per unit effort* of individual year-classes, in successive years of their existence. The principal reason this method is not used more often is the frequently great labor necessary to obtain a reasonably representative measure of fishing effort, particularly when more than one type of gear harvests a stock, or when the same gear harvests two or more species having overlapping but not identical distributions, and concentrates its attention now on one, now on another. Furthermore, the advantages of using catch per unit effort are to some extent offset by possibilities of systematic bias that are not present in the ordinary catch curve. For example, there may be distortion resulting from changes in catchability of the fish from year to year, either from differences in distribution or behavior of the fish themselves, or from variations in the seasonal deployment of the fishing apparatus, or from its variable effectiveness because of weather conditions.

One great advantage of survival rates estimated from catch per unit of effort is the fact that they give information about the current situation: they apply to the interval between the middle (approximately) of the two fishing seasons sampled. Ordinary catch-curve methods, by contrast, give estimates which tend to lag several years behind the time the data are collected and which represent average conditions during the years of recruitment (Section 2.6).

The method of comparisons of catch per unit effort has been used principally with certain trawl fisheries, whose effort is well standardized, and where the species is available over a wide area (Graham 1938b; Jensen 1939; Gulland 1955a).

EXAMPLE 2.3. SURVIVAL OF PLAICE OF THE SOUTHERN NORTH SEA, ESTIMATED FROM CATCH PER UNIT OF EFFORT OF INDIVIDUAL YEAR-CLASSES. (From Ricker 1958a, after Gulland 1955a, p. 43.)

Gulland's data for catch of plaice (*Pleuronectes platessa*) per 100 hours of fishing by standard trawlers, at successive ages in 3 years, are given in Table 2.2. The ratio of C/f in successive seasons is an estimate of survival rate for that year for the

TABLE 2.2. Catch per 100 hours of trawling for plaice in the southern North Sea in three seasons, and survival rates estimated from this.

Age	C/f 1950–51	$\bar{S}$	C/f 1951–52	$\bar{S}$	C/f 1952–53
2	39		91		142
		....		....	
3	929		559		999
		....		....	
4	2320		2576		1424
		....		....	
5	1722		2055		2828
		0.570		0.637	
6	389		982		1309
		0.671		0.529	
7	198		261		519
		0.768		0.471	
8	93		152		123
		0.763		0.697	
9	95		71		106
		0.600		0.859	
10	81		57		61
		0.741		0.702	
11	57		60		40
		(0.576)		(0.673)	
12+	94		87		99
Geometric mean		0.665		0.642	

year-class in question. For example, the year-class of 1945, age 5 during the 1950–51 season, decreased in abundance from 1722 per 100 hours in 1950–51 to 982 in 1951–52; its estimated survival over that period was therefore $\bar{S} = 982/1722 = 0.570$. For ages above 11, where the data are lumped, an approximate $\bar{S}$ is obtained from, for example, $87/(57 + 94) = 0.576$.

Gulland notes that there are no consistent trends in the $\bar{S}$-values with age, and little difference between the 2 years shown: unweighted geometric means are 0.665 and 0.642[3].

2.4. RECRUITMENT TO THE FISHERY OVER SEVERAL AGES

2.4.1. GENERAL RELATIONSHIPS. Recruitment is here defined as the process of becoming vulnerable to the fishing in progress, whether by movement into the region

---

[3] If the logarithms of the three catch samples of Table 2.2 are plotted as ordinary catch curves, they prove to be of the "concave" type (Section 2.6), each with a break in slope whose timing corresponds fairly well with the resumption of large-scale fishing following the Second World War. The slopes of the steeper left-hand (more recent) portions of the right limbs suggest a survival rate of about 0.41, which applies to the period 1946–50, approximately. Gulland (1968, p. 310) accounts for the difference between this figure and the 0.64–0.66 of Table 2.2 on the basis that the year-classes 1946–48 were much stronger than those of several previous years. It is also true that the two types of estimate apply to different series of years, but fishing effort during 1946–50 averaged somewhat *less* than that during 1950–53 (Gulland 1968, Fig. 2).

fished or by change in size or behavior. Different types of recruitment are outlined in Section 11.1, but for the present purpose such distinctions are unnecessary.

At the risk of spending time on what may be an obvious proposition, we can consider first the effect on a catch curve of having recruitment spread over several ages. Table 2.3 shows such a population, in which total mortality rate increases from 0.3 to 0.6 during a recruitment period which is completed three years after the fish first enter the fishery. If the population at the end of 1906 be *randomly* sampled (the sample taken by the fishery will not be representative), the ratios of the older age-groups will represent the definitive survival rate 0.4, and the greater survival rates characteristic of the years of recruitment appear only among the age-groups which are as yet incompletely recruited.

TABLE 2.3. Decrease of different year-classes of a population in successive years of their life, when the total mortality rate is 0.3 at age 3, 0.4 at age 4, 0.5 at age 5, and 0.6 at all later ages.

	Year-class (year in which fry were hatched)						
Year	1898	1899	1900	1901	1902	1903	1904
	10,000						
1901							
	7,000	10,000					
1902							
	4,200	7,000	10,000				
1903							
	2,100	4,200	7,000	10,000			
1904							
	840	2,100	4,200	7,000	10,000		
1905							
	336	840	2,100	4,200	7,000	10,000	
1906							
	134	336	840	2,100	4,200	7,000	10,000
Ratio	0.4	0.4	0.4	0.5	0.6	0.7	

This proposition becomes a little less obvious when the definitive survival rate itself changes over a period of years, as shown later in Fig. 2.7. In that event the ratio of two of the older age-groups in a catch may represent a survival rate which they themselves have never actually experienced, but which is the definitive rate that used to prevail among mature fish (now long dead) at the time when the given age-groups were being recruited.

2.4.2. AGE OF EFFECTIVELY COMPLETE RECRUITMENT. Without a little study it will often be difficult to decide at what age recruitment is effectively complete, particularly with convex catch curves. It is advisable to try to duplicate any observed curve using trial values of the instantaneous rates of fishing and natural mortality in order to get some idea of the actual situation. Since the length distribution of the fish in any age-group, of most fishes, tends to be fairly close to normal, it can readily

41

be assumed that the curve of recruitment will usually have a fairly symmetrical shape: for example, the magnitude of F might be 0.01, 0.1, 0.5, 0.9, and 0.99 of its definitive value, in successive years of recruitment of a given year-class. (Asymmetry resulting from the median magnitude of F being something more or less than 0.5 will not affect our argument.) Now a facile assumption would be that the number of years from the first age to the modal age of the catch curve would represent the ascending limb of a symmetrical curve of recruitment, and that therefore an equal number of years to the right of the mode would be affected by recruitment and should be discarded in estimating survival rate.

Such an assumption would be misleading, for two reasons. First, the number of fish in the first age taken (except sometimes when it is age 0 or age 1) tends to be quite small, often of the same order of size as the number in the oldest age taken (cf. Fig. 2.1, 2.6, 2.8, 2.12, 2.13). That is, the identity of the first age to be taken is partly determined by the size of the whole sample. When the latter is of moderate size (several hundred fish), the fish from an age-group for which rate of fishing (F) is of the order of 0.01 of its definitive magnitude will probably be the first to appear; if the sample is increased eight- or ten-fold, an age-group may be represented for which F is in the neighborhood of 0.001 of its definitive value. Now at the other end of the symmetrical curve of recruitment, an age-group that is either 99.9% vulnerable or merely 99% vulnerable is for practical purposes completely vulnerable when it comes to estimating survival rate. Even 95% would be fairly satisfactory in most cases. Consequently, the distance in years from the first age to the median age of recruitment is practically always a year or two too great to be used as an estimate of the distance to which recruitment will have a distorting effect beyond the median.

A second source of error is the fact that the modal age in the catch does not necessarily coincide with the median age of recruitment. Examples show that it may be at an age either younger or older than the median, its exact position depending principally on the magnitude of the total mortality rate. When annual mortality rate is moderate or small (0.5 or less), at the beginning of recruitment at least, there are usually two adjacent ages having much the same number of fish, with the mode falling sometimes in the median age of recruitment, sometimes in the next older age. In the latter event the distance from the first age present to the modal age would be more than ever misleading, if it were considered as an estimate of the distance to which the effects of recruitment extend beyond the mode.

Considering both of the effects just described, it appears that the modal age in the catch will commonly lie quite close to the first year in which recruitment can be considered effectively complete. In the examples used here there is at most one unusable age-group intervening between the first usable age and the modal age (or the second of two nearly-equal ages), as shown by Fig. 2.8 and 2.12. When recruitment is abrupt, the first year beyond the modal age seems usable, as illustrated in Fig. 2.1, and in Fig. 2.9 the point for age 6 comes close to being usable.

2.4.3. VULNERABILITY VARYING CONTINUOUSLY WITH AGE. The question arises whether a stable or "definitive" rate of fishing, beyond a certain age, is commonly achieved at all in fish populations. Perhaps F usually continues to increase through-out life, or it might conceivably rise to a maximum and then decrease if the older fish become too large to be captured or held by the hooks or nets in use. Obviously no universal answer is possible to such a question, and to obtain information con-cerning it usually requires more than a catch curve. The subject is closely related to that of net selectivity, which is considered in Section 2.11 below.

2.4.4. AGE OR SIZE OF ARRIVAL ON THE FISHING GROUNDS. A distinction can sometimes be made between the vulnerability of the whole of the stock at a given size, and the vulnerability of that portion of it which is on the fishing grounds. In fact, the term recruitment has been used (by Beverton, etc.) in the sense of physical movement onto the fishing grounds, instead of its more common meaning of overall increase in vulnerability to capture by the gear in use.

Occasionally it is possible to classify the reduced vulnerability of smaller fish into a portion that results from their relative scarcity in places where most fishing is carried on, and a portion due to their "habit" of avoiding capture by nets, hooks, etc. For example, Rollefsen (1953) compared the sizes of Lofoten cod caught by longlining and by purse seines (Fig. 2.3). Considering the latter to be representative of the sizes of cod present (something which probably needs confirmation), it would appear that vulnerability to hooks actually *decreases* with increase in length from the smallest fish up to quite large sizes (60–110 cm or so). At the same time, the vulnerability of the stock as a whole (as distinct from that part of it which assembles on the Lofoten spawning grounds) to longline fishing increases at least up to a size of 90 cm.

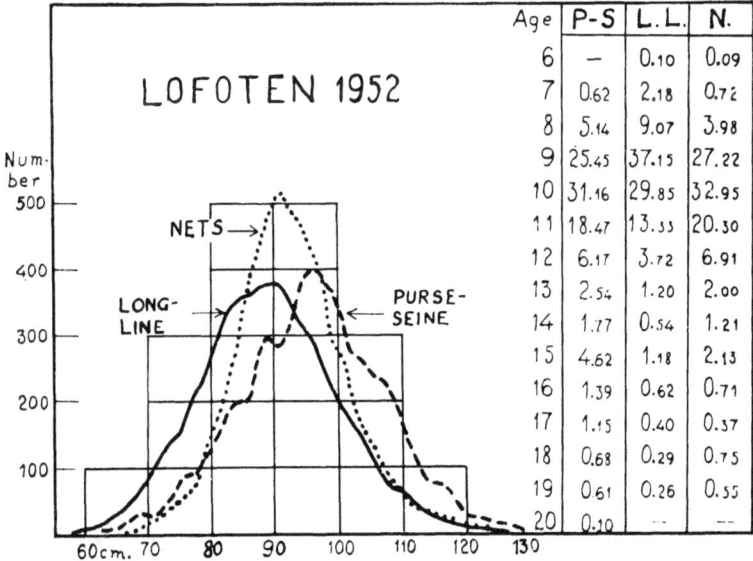

Age	P-S	L.L.	N.
6	–	0.10	0.09
7	0.62	2.18	0.72
8	5.14	9.07	3.98
9	25.45	37.15	27.22
10	31.16	29.85	32.95
11	18.47	13.33	20.30
12	6.17	3.72	6.91
13	2.54	1.20	2.00
14	1.77	0.54	1.21
15	4.62	1.18	2.13
16	1.39	0.62	0.71
17	1.15	0.40	0.37
18	0.68	0.29	0.75
19	0.61	0.26	0.55
20	0.10	--	--

FIG. 2.3. Length and age distribution of Lofoten cod taken by three kinds of gear. (From Rollefsen 1953, fig. 1.)

43

Another means of separating movement onto the fishing grounds from increasing vulnerability of fish already there is given in Section 5.8.

2.5. Change in Mortality Rate with Age

In addition to increase in fishing mortality rate by progressive recruitment, there can be other types of change in mortality. Table 2.4 shows two balanced populations, constructed on the basis that the survival rate, S, changes by the absolute figure 0.1 in each year of life of the fish, and that all mortality is the result of fishing.

TABLE 2.4. Effects of a change in survival rate with age on catch and on catch ratio, when all mortality is the result of fishing.

Age	Survival rate	Survivors	Catch	Catch ratio	Survival rate	Survivors	Catch	Catch ratio
		100,000				100,000		
1	0.9		10,000		0.1		90,000	
		90,000		1.80		10,000		0.09
2	0.8		18,000		0.2		8,000	
		72,000		1.20		2,000		0.18
3	0.7		21,600		0.3		1,400	
		50,400		0.93		600		0.26
4	0.6		20,160		0.4		360	
		30,240		0.75		240		0.30
5	0.5		15,120		0.5		120	
		15,120		0.60		120		0.40
6	0.4		9,072		0.6		48	
		6,048		0.47		72		0.45
7	0.3		4,234		0.7		22	
		1,814		0.36		50		0.47
8	0.2		1,451		0.8		10	
		363		0.22		40		0.40
9	0.1		327		0.9		4	
		36				36		

The left half of the Table is a recruitment situation. Catch ratios are consistently higher than true survival rate, the discrepancy being 35% to 50% over most of the range covered.

In the right half of Table 2.4, where mortality decreases, catch ratio is always less than the adjacent survival rates. Noteworthy is the fact that over the range of survival rates from 0.5 to 0.9 there is not much change in catch ratio. If encountered in practice, such a segment of a catch curve would probably be interpreted as substantially meeting the uniform conditions mentioned earlier, the irregularities being ascribed to small fluctuations in recruitment.

An example modelled after situations more likely to be encountered in actual investigations is shown in Fig. 2.4. The population described by these curves has an instantaneous natural mortality rate of 0.2 during ages 1 through 10. This is combined

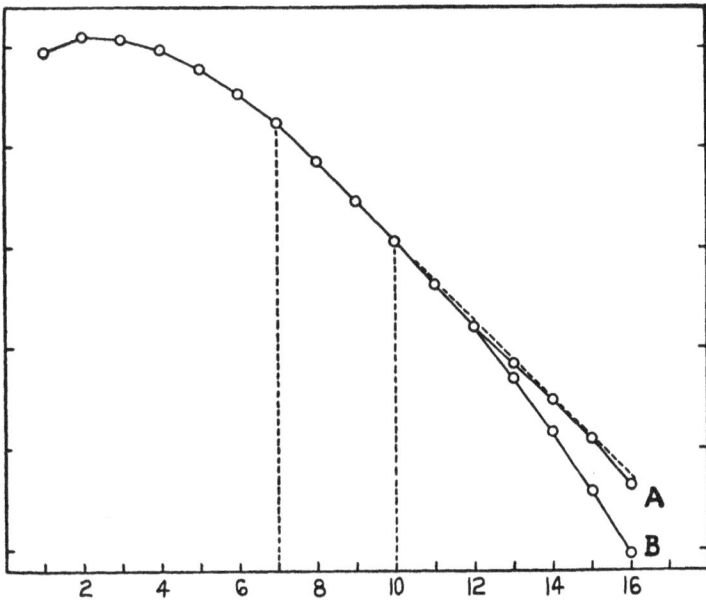

FIG. 2.4. Catch curve for a population which has a constant fishing and natural mortality rate from age 7 to age 10, followed by a decrease in rate of fishing (Curve A) or by an increase in natural mortality (Curve B). Abscissa — age; ordinate — logarithmic units.

with a rate of fishing that increases from 0.1 at age 1 to 0.7 at age 7, then remains steady for 3 more years. This latter is shown by the straight portion of the catch curve from age 7 to age 10, and, if continued, would be represented by the dotted projected line.

Three variations, after age 10, are examined. First, the rate of fishing is made to decrease by 0.1 unit during each year of age, for 6 years, the result being shown by Curve A. There are some fluctuations, but the net result differs very little from the dotted line, and would scarcely be distinguishable in an actual investigation. This means that this section of the curve gives a fair estimate of survival rate during the previous state of balance (ages 7 to 10), but does not reflect the actual survival rate, which is rising. This is illustrated more graphically in Fig. 2.5A, in which the catch ratio, R, is compared with the actual survival rate, S.

Secondly, the rate of natural mortality is made to increase from 0.2 to 0.9, as shown by Curve B of Fig. 2.4 and 2.5. The decrease in survival is faithfully reflected by the catch ratio, the latter being only inappreciably greater (Fig. 2.5B).

Finally, rate of fishing is made to decrease while natural mortality increases, so that total mortality remains steady. The catch curve for this situation has not been drawn in Fig. 2.4, since it almost coincides with Curve B. This means that the curve obtained does not represent the actual survival rate, which (since survival rate is constant) is the sloping dotted line of Fig. 2.4. Curve C of Fig. 2.5 shows the discrepancy between catch ratio and survival rate.

45

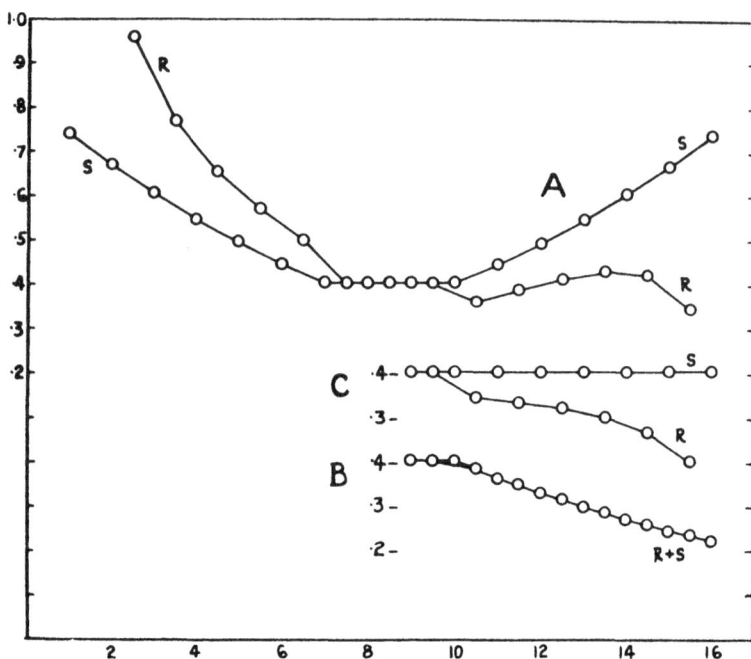

FIG. 2.5. Comparison of survival rates (S) and catch ratios (R) for the populations of Fig. 2.4. (A) Decrease in rate of fishing; (B) Increase in natural mortality; (C) Decrease in rate of fishing compensated by an equivalent increase in natural mortality. Abscissa — age: ordinate — survival rate and catch ratio.

Additional examples of the effects of continuous change in rate of fishing with age have been computed by Beverton and Holt (1956). Panels (a)–(e) of their Fig. 2 illustrate cases where F decreases, while in panels (f)–(h) it increases. In panel (i), F increases to a maximum and then decreases; this proves to be a particularly misleading situation, since the right limb of the catch curve is nearly straight, but indicates an apparent survival rate not much more than half the actual.

From the above and similar examples, the following conclusions can be drawn:

1. An increase (or decrease) in *natural* mortality rate, among the older fish of a population, is correctly represented by the catch curve, when rate of fishing is the same for all the ages involved.

2. A decrease in rate of *fishing*, among the older fish in a population, is not correctly reflected in the catch curve, and in many situations the resulting curve approximates closely to the survival rate obtaining at ages prior to the decrease in rate of fishing.

3. When rate of fishing increases with age throughout life, the catch curve is useless for estimating survival rate: in effect, the curve consists only of the portions which we have called the ascending limb and dome, and the catch ratio between successive years is always greater than the true survival rate, often very much greater.

46

4. When natural mortality increases with age, and rate of fishing decreases, the catch curve tends to represent the survival rate characterized by the *observed* natural mortality plus the *original* rate of fishing.

5. Hence, altering the wording of 1. a little, an increase in natural mortality rate, among the older fish of a population, is at least reasonably well represented by the catch curve, whether rate of fishing is steady or whether it decreases.

In so far as these conclusions involve the rate of fishing, they apply only when the latter has been stabilized for enough years that all the fish involved have been subjected to the appropriation rates for each age, throughout their life. If this is not so, there is no restriction on the type of curve which may be obtained when rate of fishing varies with age. For example, if a new fishery begins to attack a previously unexploited population, the number of fish taken at each age will be the product of the abundance at that age and the rate of fishing at that age. Thus the ratio of the number of fish taken at age t to the number at age $t - 1$ will be the product of the natural survival rate times F_t/F_{t-1}, the ratio of the rates of fishing at the two ages.

The considerations above are of particular importance in dealing with catch curves which have the right limb convex upward. Theoretically, such could result from a steady increase in rate of fishing with age; but this situation seems likely to be uncommon, except possibly in sport fisheries where there is great interest in large specimens (Section 2.4). On the basis of what has been found up to this point, a curve that is convex to the very end will ordinarily indicate an increase in natural mortality rate with age, among the older ages at least, since a decrease in rate of fishing with age does not cause much or any deflection of the catch curve in either direction. On the same basis, a concave curve could only mean that natural mortality in the population decreases with age. However, alternative explanations of curvature are available when there has been a change in mortality rate with time (Section 2.6).

EXAMPLE 2.4. SURVIVAL RATE IN AN UNEXPLOITED HERRING POPULATION: A CONVEX CATCH CURVE. (From Ricker 1948.)

Dr A. L. Tester has brought to my attention some convex catch curves of exceptional interest. During the fishing season of 1938–39 a population of herring (*Clupea pallasi*) on the east coast of the Queen Charlotte Islands, British Columbia, was exploited commercially for the first time. Five samples totalling 580 fish were taken and their ages determined. The points of Curve A of Fig. 2.6 are the logarithms of the percentage representation of each age-group. The unsmoothed curve appears generally convex, but is quite bumpy, because of the moderate fluctuations in recruitment which are encountered among herring from this general region.

To smooth out the curve and get a representative picture of the age distribution of natural mortality, there are several possible procedures. A simple freehand curve fitted to the data for 1938–39 is shown in Fig. 2.6A. As a check on the investigator's judgment the curve can be smoothed by a running average of 3, as shown in Fig. 2.6B. This procedure of course tends to flatten the dome of the curve, so that the modal point should not be considered at all in drawing a new freehand curve, and even the

47

point to either side of it will be a little depressed. Also the curve is extended one year at either end by the process. The left-hand end does not concern us, but at the right-hand end it may "improve" the picture because the point for age 12, represented by no fish in the sample, would be $-\infty$ on Curve 2.6A. Actually, of course, there are very likely a few fish of this age or even older in the population, so that the delay in the asymptotic fall of the curve suggested in Fig. 2.6B is according to expectation.

To get a better idea of the primitive distribution of natural mortality it is also possible to use data for later years, to help smooth out the curve. They have this disadvantage, that each additional year used brings the influence of the fishery farther into the catch, and accordingly fewer ages can be considered representative of the original natural mortality rate. Curve 2.6C shows the combined data for 1938–39 and 1939–40, giving each year equal weight, while Curve 2.6D is based on the combined data for the first 4 years of the fishery.

The percentage annual survival rates found by taking tangents at successive ages, on the four curves of Fig. 2.6, are shown below for ages whose relative numbers are not affected by the new fishery (or very little so):

Age	6	7	8	9	10	11	12
Curve							
A	72	63	58	52	42	28	--
B	69	66	60	52	47	31	--
C	--	--	59	48	41	29	21
D	--	--	--	48	43	32	19

The figures for age 6 are slightly less than what was determined from the actual slope, because of the proximity of age 5, for which recruitment is presumed to be somewhat incomplete. The determination of the age distribution of mortality in unexploited populations such as this is of special interest, because often it may be the only clue to the natural mortality rate under conditions of exploitation.

Under conditions of a developed fishery, the original convexity of the catch curves for British Columbia herring stocks tends to be diminished, but is still quite recognizable (Tester, 1955). In the southern part of the North Sea, Jensen (1939) also shows strongly convex curves for herring in two areas. Jensen suggests increased natural mortality or emigration among older fish, and net selectivity making younger fish more vulnerable, as possible causes of the convexity of the North Sea curves. In regard to the last, the analysis of this Section shows that net selectivity of this sort would not in fact produce any appreciable curvature, so this possibility can be ruled out. The reason is that while such nets sample the older stock less completely than the younger, they also permit more fish to survive to the older ages, and the combination of these two opposed tendencies results in a fairly straight catch curve (cf. Fig. 2.4A).

Catch curves for a number of other species under unexploited conditions have now been obtained, and all indicate an increase in natural mortality among the older fish. From northern lakes there is information for sauger (*Stizostedion canadense*),

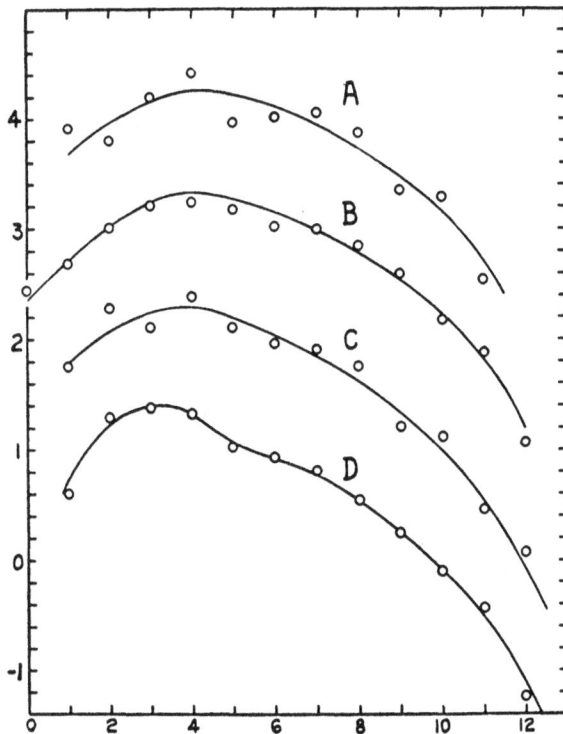

FIG. 2.6. Catch curves for a population of herring from the Queen Charlotte Islands, B.C. (A) Age composition during the first year of exploitation, 1938–39; (B) The same, smoothed by a running average of 3; (C) The combined samples of 1938–39 and 1939–40; (D) The combined samples for the first 4 years of exploitation. All curves are in terms of the logarithms of the percentage frequency at each age, set one log-unit apart on the figure, with the ordinate scale applying to Curve D. (From unpublished data of A. L. Tester.)

rock bass (*Ambloplites rupestris*), whitefish and lake trout (Ricker 1949a, Kennedy (1953, 1954b). A similar increase in natural mortality was observed in fished lakes among perch (*Perca flavescens*), black crappies (*Pomoxis sparoides*), yellow bullheads (*Ameiurus natalis*), and several other species in Indiana (Ricker 1945a); the survival rates in these instances being estimated from recoveries of marks. Although the more heavily fished bluegills in the same waters had a nearly straight catch curve (Fig. 2.1A) it is probable that originally they survived less well at the older ages then present: we must assume that no individual of any species is capable of living forever. Also, a sample of older plaice than those available to Gulland (Example 2.3) would probably behave similarly.

2.6. CHANGE IN MORTALITY RATE WITH TIME

All of the conclusions obtained in the last Section presuppose that, however they may vary with age, the rate of fishing and rate of natural mortality for any

49

given age are constant from year to year. But the effort used in a fishery can vary from year to year for a variety of reasons. Some fisheries are of recent origin, and the gear in use has been expanded since their beginning. Others have passed through a profitable phase, and now their decreased return per unit effort tends to drive off boats which formerly fished them. Economic conditions play a large part in determining what constitutes profitability and thus affect total fishing effort. Hence a consideration of secular change in rate of fishing cannot be avoided. Similar changes in natural mortality rate may possibly occur at times; their effects can readily be examined, but they are not considered here.

2.6.1. INSTANTANEOUS RECRUITMENT. Table 2.5 shows a population in which the survival rate for fish of all catchable ages is 0.7, 0.6, and 0.5 in 3 successive calendar years, then remains steady at 0.4 for 4 years. In this situation (unlike Table 2.3) the commercial catch will sample the population representatively, since recruitment to the fishery occurs abruptly. Such a random sample of the population, taken at the start of any given year, would have successive age-groups represented in proportion to the figures in the horizontal rows of the table, beginning with the youngest at the right. Each of the catch ratios shown in the last row represents the ratio of *all* of the pairs of figures in the two adjacent columns above it. Obviously then, no matter at what time year-classes t and $t-1$ are sampled, the ratio of their abundance is a measure of the survival rate which existed during the first year that year-class $t-1$ became vulnerable to fishing. *Thus the survival rates which we estimate from age-frequencies in a catch are ancient history.* They pertain to past years, to the time

TABLE 2.5. Decrease of successive year-classes in a population acted on by a survival rate which decreases for 3 years and then remains steady, but is always the same for fish of all recruited ages during any given year.

Year	S	Year-class							
		1898	1899	1900	1901	1902	1903	1904	1905
1901	0.7	10,000							
1902	0.6	7,000	10,000						
1903	0.5	4,200	6,000	10,000					
1904	0.4	2,100	3,000	5,000	10,000				
1905	0.4	840	1,200	2,000	4,000	10,000			
1906	0.4	336	480	800	1,600	4,000	10,000		
1907	0.4	134	192	320	640	1,600	4,000	10,000	
		54	77	128	256	640	1,600	4,000	10,000
Catch ratio		0.7	0.6	0.5	0.4	0.4	0.4	0.4	

when the year-classes involved were being recruited to the catchable size range, and are independent of what survival rates have prevailed since that time. In terms of the catch curve, this means that the slope of any given part of the curve will represent the survival rate which prevailed at the time the fish in question were being recruited to the fishery.

2.6.2. GRADUAL RECRUITMENT. In the example just given recruitment takes place suddenly, one age being completely vulnerable, the next younger one completely invulnerable. In practice, recruitment usually takes place less abruptly, and is often gradual. A model of that sort has been constructed in the following manner: a stock of fish which gains a uniform number of recruits each year is considered to have an unchanging instantaneous natural mortality rate of 0.2. To this is added a rate of fishing which increases for the first 6 years after the fish enter the fishery, as follows:

Year of life (starting with the first vulnerable age)................. 1 2 3 4 5 6 7+
Percentage of the definitive rate of fishing...........................0.5 5 20 45 70 90 100

These values are approximately those estimated from an actual fishery.

The definitive rate of fishing varies in successive calendar years as shown in Table 2.6. Adding 0.2 to the rate of fishing gives the definitive instantaneous mortality rate for each year, and from Appendix I the annual mortality rate and survival rate were found in the usual manner. The same statistics were estimated for each year of recruitment, at each level of (definitive) total mortality. Armed with these survival rates, a comprehensive table was prepared, analogous to Table 2.5, showing the number of surviving fish in each successive brood for a series of years sufficient to give the complete history of the period of change. Annual deaths in each age category were found by subtraction for 4 different years — the 1st, 7th, 12th and 24th — and by dividing these between fishing and natural mortality in the ratio of F to M, the number of

TABLE 2.6. Rates of fishing in successive calendar years for the model populations of Section 2.6.2.

Year	Instantaneous mortality rates			Actual mortality rate	Survival rate
	Fishing[a]	Natural	Total		
	F	M	Z	A	S
Up to 1	0.2	0.2	0.4	0.330	0.670
2	0.3	0.2	0.5	0.394	0.606
3	0.4	0.2	0.6	0.451	0.549
4	0.5	0.2	0.7	0.503	0.497
5	0.6	0.2	0.8	0.551	0.449
6	0.7	0.2	0.9	0.593	0.407
7 and later	0.8	0.2	1.0	0.632	0.368

[a]For fully-recruited age-groups.

each age-class in the catch was computed. The logarithms of these values are shown in Fig. 2.7, curves A to D.

Curve A, showing the catch after an indefinite number of years of steady survival rate 0.670, is a simple catch curve with 6 years involved in the left limb and dome (corresponding to the 6 years of recruitment) and a long straight right limb.

Curve B, based on the catch in year 7, when the survival rate of 36.8% was first achieved, shows by its partially concave right limb that survival rate has been decreasing. However, the curve is not representative anywhere of the *current* survival rate. Its steepest part, between age 7 and age 8, corresponds to a survival rate of 51%: that is, approximately the survival rate of 3 years previously (year 4 in the schedule above). For a series of years near its outer end the curve is still straight, and here represents the original survival rate of 0.67.

Curve C is based on the catch in year 12, after the 36.8% survival rate has been stabilized for 6 years. Here, for the first time, there appears a portion of the curve (age 7 to age 8) which is steep enough to represent the current rate of survival. The slope of the curve at older ages gradually decreases, and between ages 17 and 18 it still has the original slope. Between ages 7 and 11, and also 15 to 18, there is not much change in slope; consequently, even if there were considerable fluctuation in recruitment, a fairly good estimate of both the old and the new survival rate could be made from a curve such as this, simply by measuring its greatest and its least slope, on the right limb. The region between ages 11 and 15 shows the maximum curvature. (A catch curve which would have no such variation in rate of change in curvature would result if mortality rate were to change gradually over the whole series of years involved.)

Curve D is the new balanced population, which only appears after 18 years of the new mortality rate of 0.632. It is similar to A, but of course has a much steeper slope of the right limb.

The types of curve obtained during a period of transition from a larger to a smaller rate of fishing, and hence of total mortality, are shown by Curves E and F of Fig. 2.7. The change is quantitatively the same as shown by B and C, but in reverse. Starting from the balanced situation of Curve D, after 6 years' progressive decrease in mortality rate Curve E is obtained. Such a curve, if found in an actual investigation, would scarcely be interpreted as indicating a recent decrease in mortality, since the whole region up to age 11 could well be in the range of recruitment. Hence the survival rate estimated would be that indicated by the straight outer limb, and would of course be wide of the current value, but representative of the former value.

Curve F, representing conditions 11 years after the mortality rate began to decrease, and 5 years after it was stabilized at 0.670, is a convex curve entirely analogous to concave Curve C. There is the same region of maximum curvature between ages 11 and 15, with rather flat portions to either side of it. In practice, the outer end of such a curve might be interpreted as representing a state of near balance, but the region from age 7 to age 11 would again present difficulty, because of the

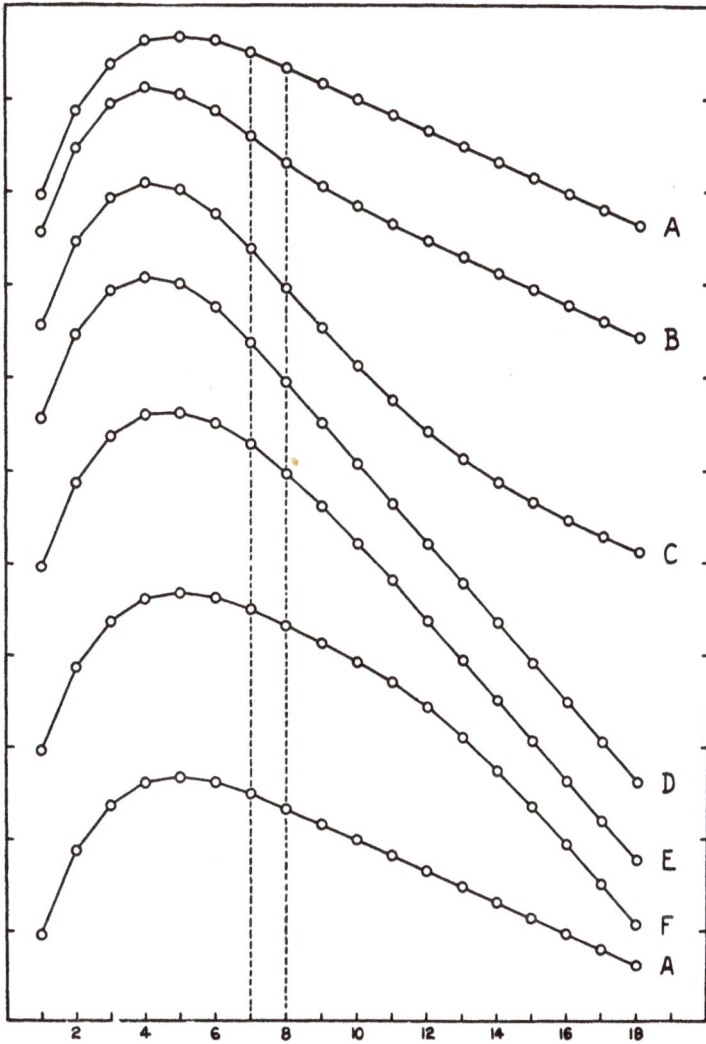

FIG. 2.7. Catch curves illustrating changes in rate of fishing with time. In every instance recruitment is complete following the first 6 ages shown, and the instantaneous rate of natural mortality is the same, 0.2, for all ages and years. (A) Constant rate of fishing of 0.2; (B) Rate of fishing has increased from 0.2 to 0.8 during the preceding 6 years; (C) Five years after B, with rate of fishing stabilized at 0.8; (D) Balanced curve for rate of fishing 0.8; (E) Rate of fishing has decreased from 0.8 to 0.2 during the preceding 6 years; (F) Five years after E, with rate of fishing stabilized at 0.2. Abscissa — age; ordinate — logarithmic units.

possibility of incomplete recruitment. Even if this were ruled out, it would be harder to estimate current survival rate here than on the same part of Curve C, because there is no point of inflection.

53

In general then, secular changes in rate of fishing result in curved right limbs of the catch curve, these being concave if fishing has increased, and convex if fishing has decreased. The latter type will usually be much harder to interpret in terms of the survival rate in past years, for two principal reasons: (1) there is danger of confusion with the type of convex curve which results from a natural mortality rate which increases with age, and (2) it is difficult or impossible to delimit the part of the curve affected by incomplete recruitment. The concave type of curve, on the other hand, is not likely to occur except as a result of increased fishing, and the point of maximum slope on the right limb will always give the most recent available estimate of survival rate.

It is difficult to express the relationships of this section in quantitative terms, but for the examples worked out to date the following statements seem to be true:

1. If the peak of recruitment is at age m, the survival rate estimated at age n on the catch curve pertains to a period approximately $n - m$ years prior to the date the sample was taken, except as noted below.

2. When the bulk of recruitment occupies a period of say $2x$ years (x years from the first *important* age to the modal age in the catch), the most recent representative survival rate observable on the catch curve will pertain to a period x years prior to the date the sample was taken.

3. If mortality becomes stabilized following a period of change it will, strictly speaking, require $2x$ years for the new stable survival rate to begin to appear in the catch curve, though for practical purposes a somewhat shorter period will usually suffice.

Obviously it will be desirable to have as much information as possible about fishing effort in past years when interpreting a catch curve. The simple fact that effort has decreased, or increased, or remained fairly steady will be of considerable value. If good quantitative estimates of effort are available, then it may be possible to interpret different segments of the curve in relation to fluctuations in the rate of fishing, or perhaps even to compute the actual rate of fishing and of natural mortality by Silliman's method (Section 7.3).

2.6.3. ILLUSTRATION. An example of the effect of an increase in rate of fishing is seen in Fig. 2.6D. Although this curve represents the average age distribution for the first 4 years of this herring fishery, there is a distinct concave section immediately to the right of the dome which reflects the much greater mortality rate that prevailed after the fishery began. (In a curve for year 4 by itself this concavity would be considerably accentuated.) By contrast, the curves for the unfished stock (Fig. 2.6A and B) were smoothly convex.

EXAMPLE 2.5. SURVIVAL OF THE LOFOTEN COD STOCK: CONCAVE CATCH CURVES. (From Ricker 1958a.)

Rollefsen (1953) presented length frequencies of cod (*Gadus morhua*) caught by three kinds of gear in the 1952 Lofoten fishery: purse seines, longlines, and gillnets.

He also tabulated the distribution of ages in the three kinds of samples (Fig. 2.3). The three gears differ considerably in the range of sizes they select, and the stock itself is a selection of the *mature* fish from the great shoals which roam the Barents Sea. Consequently the chances of obtaining a representative survival rate from these data might appear particularly unfavorable.

Logarithmic plots of the three age distributions are shown in Fig. 2.8. There is moderate, but not excessive, variation in recruitment from year to year; the year-class of 1937, age 15 in 1952, was a particularly good one. The right limbs of the three distributions are all markedly concave upward. From the analysis of Sections 2.5 and 2.6, this could either be a result of a decrease in rate of natural mortality (*not* fishing mortality) with age, or a result of a recent increase in rate of exploitation of the stock as a whole. The second alternative is much more likely.

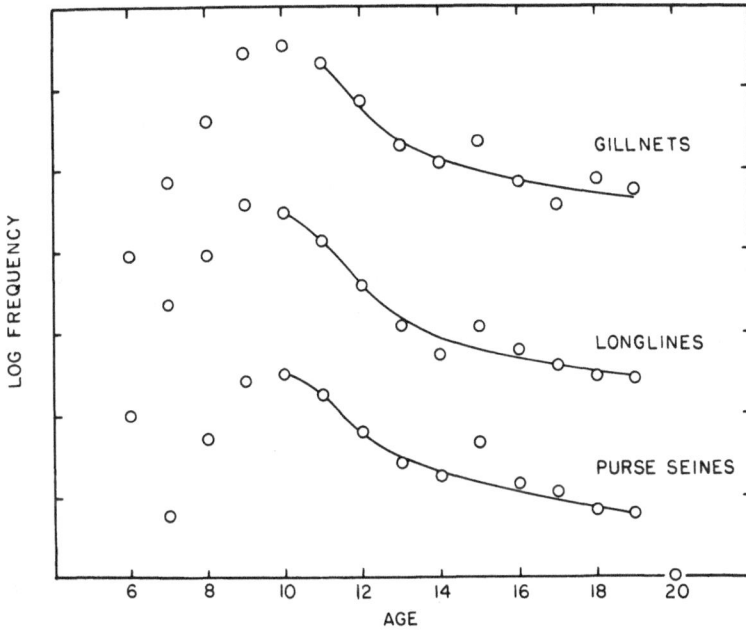

FIG. 2.8. Catch curves for Lofoten cod taken by three kinds of gear. The ordinate divisions are 1 log unit. (Data from Rollefsen 1953.)

Examples of annual survival rates computed from the slopes of the freehand lines are as follows:

Age interval	Purse seines	Longlines	Gillnets
11–12	$S = 0.33$	$S = 0.29$	$S = 0.30$
12–13	$S = 0.50$	$S = 0.40$	$S = 0.37$
13–14	$S = 0.63$	$S = 0.56$	$S = 0.60$
14–16 (avg.)	$S = 0.75$	$S = 0.76$	$S = 0.75$

55

The seines suggest a somewhat greater survival rate than the other gears, up to age 14, but the other curves would be useful to a first approximation. From age 14 onward there is little difference between the three, though of course the seine curve should be more reliable because it is based on a larger sample of the old fish. We may conclude that even knowledge of the existence of considerable net selectivity should not discourage attempts to obtain some kind of information about survival rate from age distribution.

Rollefsen points out that purse seining has been only recently introduced at Lofoten, and that it takes larger fish than the two historic methods. Insofar as the purse seine has increased the overall rate of fishing it would contribute to a (temporary) concavity of the catch curves; however the greater vulnerability of *large* fish to the seines would tend to have the opposite effect.

EXAMPLE 2.6. SURVIVAL OF LAKE WINNIPEGOSIS WHITEFISH: A SINUOUS CATCH CURVE. (From Ricker 1948.)

An interesting curve, for the whitefish (*Coregonus clupeaformis*) of Lake Winnipegosis, is shown in Fig. 2.9. The data are taken from Bajkov (1933, p. 311), who used them to compute the whitefish population of the lake by Derzhavin's method (Section 8.1); hence he presumably considered them representative. The right limb has two steep portions, separated by a period of 4 years in which it is considerably flatter. More than one kind of irregularity might produce such a curve. In terms of possible variations in fishing, the concave part of the curve would suggest an increase, and the convex part a decrease, in fishing effort over the corresponding times in past years. A second possibility is that there may have been a pronounced cyclical trend in recruitment: an increase for several years, followed by 4 years of decrease, then 2 or more

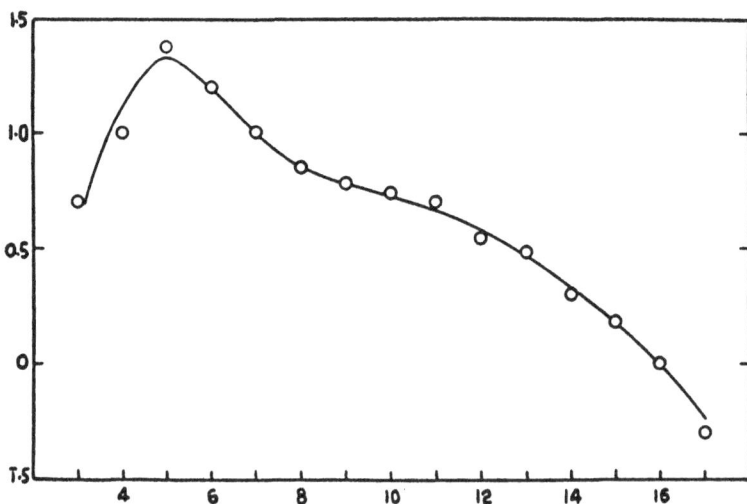

FIG. 2.9. Catch curve for the whitefish of Lake Winnipegosis, 1928. Abscissa — age; ordinate — logarithm of the percentage of the catch which occurs at each age. (From data of Bajkov 1933.)

years of increase. Finally, the two steeper parts of the curve might indicate a younger and an older range in which natural mortality is relatively heavy, separated by a period of less severe natural mortality from age 8 to age 12.

TABLE 2.7. Number of gillnets licensed on Lake Winnipegosis, Manitoba, compared with rate of survival and instantaneous rate of mortality of whitefish, as deduced from the catch curve.

Year	Nets	S	Z	Year	Nets	S	Z
1915	No data	0.55	0.60	1921	3304	0.84	0.18
1916	2745	0.63	0.46	1922	4112	0.87	0.14
1917	9535	0.68	0.39	1923	5560	0.87	0.14
1918	8580	0.72	0.33	1924	5765	0.76	0.27
1919	No data	0.75	0.29	1925	6722	0.66	0.42
1920	7730	0.80	0.22	1926	7422	0.63–	0.45+

Dr K. H. Doan courteously compiled data (shown in Table 2.7) on the number of gillnets used on Lake Winnipegosis and also found out that Dr Bajkov's samples were taken during the winter fishing season early in 1928. From the catch curve it appears that recruitment is spread over ages 3 through 5, or perhaps even 6; age 4 will be taken as the mode. Hence the slope at age t on the curve reflects the survival rate $t - 4$ years previous to 1928. Taking tangents on the curve at successive ages gives the series of survival rates (S) shown in Table 2.7, and after 1916 a suggestive inverse relationship between them and the gear used is evident. The direct relation between number of nets and instantaneous rate of total mortality (Z) is about as good; theoretically it should be somewhat better. The relationship could be "improved" by drawing the catch curve in the light of the net data; as actually drawn, sudden changes are obscured by rounding of the curve. On any system, the points for ages 16 and 17 (1915 and 1916) are wide of the expected value, which suggests a sharp increase in natural mortality rate among the oldest fish, such as is found among whitefish elsewhere. Aside from the last-mentioned effect, it would seem that fluctuations in fishing effort alone *may* be sufficient to account for the sinuous shape of this catch curve.

It would be pressing the data too far to attempt any more exact analysis. Number of nets licensed has obvious limitations as a measure of fishing effort. We should, for example, expect them to be more efficiently utilized as time goes on, since motors were introduced among the fishing fleet during the period shown, and doubtless other improvements in efficiency of utilization occurred. We should also expect more intensive utilization of nets when prices were good (1917–20, 1925–29) than when markets were slack. Some such considerations are necessary to explain why the instantaneous rate of mortality more than doubled between 1921–23 and 1925–26, whereas the number of nets was scarcely doubled. Considering that there is some natural mortality, an increase in fishing effort should be followed by a somewhat *less* than proportional increase in instantaneous mortality rate. Another factor which should be considered is the possibility of a decrease in recruitment, since in the later history of this lake the whitefish disappeared as a commercial fish.

Notice that the curve of Fig. 2.9 is one of the type which does not show the current (1928) survival rate, since fishing effort was increasing right up to the time the sample was taken. The steepest slope of the curve, corresponding to S = 0.63, represents the survival rate about 2 years earlier.

TABLE 2.8. Fishing effort (hours of trolling), catch, and age of the catch of Opeongo lake trout.

Year	Fishing effort (hours)	Catch	Catch by ages														
			3	4	5	6	7	8	9	10	11	12	13	14	15	16	17
1936	2,030	2,600	30	95	128	233	474	665	478	260	118	57	19	25	10	15	11
1937	2,240	2,700	0	4	34	198	650	1025	555	176	38	4	8	8	0	0	0
1938	1,630	1,650	12	74	127	275	420	439	195	90	3	9	3	0	3	0	0
1939	1,380	1,550	39	36	116	221	321	393	223	90	47	24	13	15	4	4	4
1940	1,170	1,400	20	84	82	224	434	364	120	46	14	6	0	6	0	0	0
1941	1,130	1,100	8	79	144	275	235	200	104	22	11	11	11	0	0	0	0
1942	570	630	7	18	46	117	217	121	53	28	8	9	2	2	0	0	0
1943	710	900	6	42	42	121	272	211	133	42	0	24	6	0	0	0	0
1944	920	1,050	8	26	84	114	197	202	198	93	44	31	9	22	9	4	4
1945	1,400	1,420	0	11	32	69	170	352	373	159	84	37	21	43	16	37	16
1946	1,740	1,220	19	30	78	116	240	325	217	93	47	26	11	7	0	7	4
1947	1,230	885	3	30	55	85	217	221	153	76	18	12	3	0	6	3	3
Mean percentage			0.8	3.0	5.8	12.5	23.2	26.6	16.1	6.3	2.2	1.3	0.6	0.7	0.3	0.4	0.2

EXAMPLE 2.7. A SERIES OF CATCH CURVES FOR LAKE OPEONGO LAKE TROUT. (From Ricker 1958a.)

Fry (1949) tabulated the catch of lake trout (*Salvelinus namaycush*) in Lake Opeongo by ages, based on a nearly-complete creel census and a scale sample usually of about a third of the catch (Table 2.8). Three considerations make interpretations of these catch curves difficult: (1) the lake became accessible to motorists first in 1935, so that in that year fishing effort increased sharply from a previous lower level; (2) the catch is taken almost wholly by trolling, with which kind of fishing there may be not only the slow recruitment to maximum vulnerability indicated by the table, but afterward a gradual decrease in vulnerability — perhaps because larger fish are harder to handle and not easily boated by unskillful fishermen (but see Example 8.2); and (3) Fry (p. 31) notes that the scale census was partly voluntary and therefore not completely random because of a tendency for possessors of *big* fish to bring their catch in for appraisal and approval.

Point 2 above would tend to make estimates of mortality rate too great among older fish, whereas point 3 would make them too small. As far as the latter is concerned, fish smaller than 4 kg would scarcely be exhibition pieces in a lake where those over 5 kg were fairly common, and fish of age 11 or less rarely exceeded 4 kg, so there need be little uneasiness about selective sampling of ages through 11.

Several catch curves from Table 2.8 are plotted in Fig. 2.10. All are concave, decreasing in slope at about age 12; this decrease is probably mainly the result of selection of large fish in scale sampling, but it is more pronounced during or just after periods of increasing fishing effort, as would be expected. The most useful slopes of these graphs are for ages 9–11, as indicated below:

Period	Avg effort hours	Rate of survival S	Instantaneous mortality rate Z
1936	2030	0.50	0.70
1937–39	1780	0.30	1.21
1940–42	960	0.35	1.06
1943–45	1010	0.43	0.85
1946–47	1480	0.42	0.87

The 1936 estimated mortality rate of 0.70 reflected, in part, the pre-1935 period of lighter fishing. The increase to 1.21 in 1937–39 is presumably the result of the increased exploitation, but the full effect of 2000 hours per year does not have a chance to be manifested. A residual effect of the 1936–39 years of heavy fishing remains in the samples of 1940–42, shown by the moderately large Z = 1.06, though actual fishing was least in the latter period. Considering that important recruitment extends over about 5 years, the only period where age 9–11 survival rate is approximately in balance with the observed fishing effort is 1943–45. The value Z = 0.85, or 43% survival per year, must be appropriate to a mean fishing effort of about 980 hours per year (mean of 960 and 1010). The 2 later years of greater effort, 1946–47, were sufficient to raise this only slightly.

59

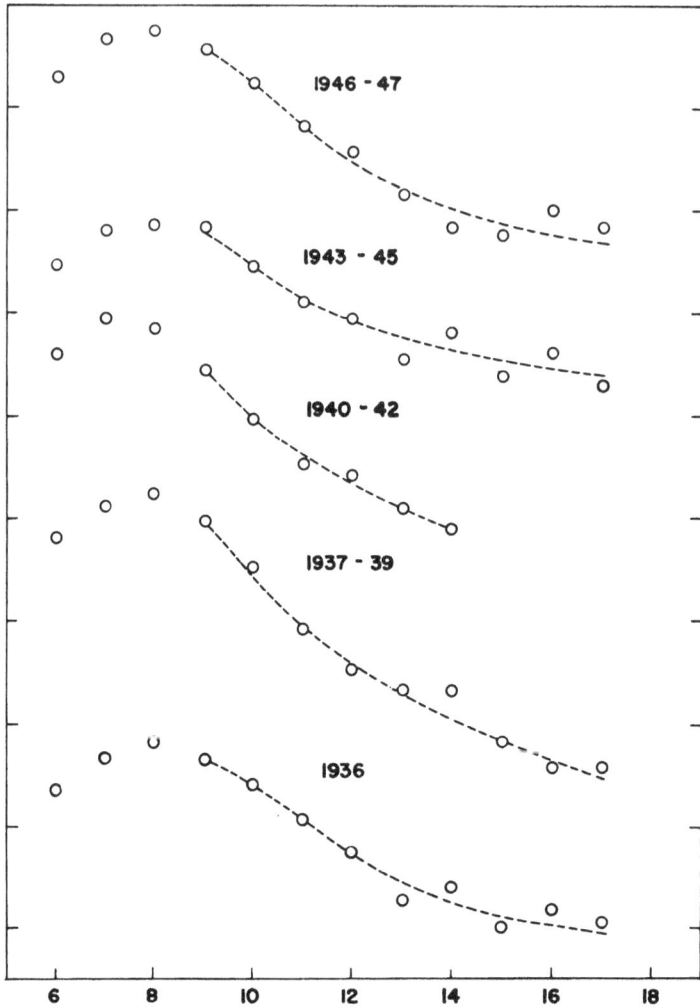

Fɪɢ. 2.10. Catch curves for Opeongo trout. Abscissa — age; ordinate — log frequency. (Data from Fry 1949.)

2.7. Catch Curves Based on Length Frequencies

It was mentioned earlier that in Heincke's and Baranov's original catch curves the logarithm of frequency per unit length interval was plotted against length, and that the relation of length to age was estimated separately. There are situations in which this method appears very attractive. For example, when assembling a representative sample of the catch from a widely scattered fishery, it may be necessary to sample so many fish that determination of the age of all of them becomes very tedious, or the scales needed for age determination may not always be forthcoming. In such a situation there would be two curves available: (A) a curve of mean length against age, based on a relatively limited body of data, and (B) a representative curve of the logarithm of frequency (log N) against length, based on all the samples available, suitably weighted. The two curves can be combined by taking the slope on each at corresponding points, i.e. at a given age on Curve A, and at that age's corresponding mean length on Curve B. The former would be represented by $dl/dt = k$ say, and the latter by $d(\log N)/dl = -Z'$, where l represents length in centimeters and t is age. Hence $d(\log N)/dt = -Z'k$, and $Z = 2.30Z'k$, according to Baranov's method of estimation described in Section 2.2.

Unfortunately, this method of computation suffers from a serious limitation: it is useful only on curves, or parts of them, where the increase in length of the fish is a constant number of centimeters per year. For this information we are again indebted to Baranov (1918), who in his fig. 12, reproduced here as Fig. 2.11, shows an artificial catch curve (A_1B_1) based on length, which was formed by adding up the contributions, to each length-interval, of a succession of overlapping age-classes which decrease in numbers by 50% per year ($Z = 0.69$). Up to age 7, the mean length of the fish is made to increase twice as fast as from age 7 onward. The result is that while the first slope of the catch curve (Z_1') obtained from ages through mean age 6, multiplied by the first rate of growth (k_1), will yield the true instantaneous mortality rate 0.69; and the increased slope (Z_2') from age 9 onward, multiplied by the slower rate of growth (k_2) for older fish, also gives the value 0.69; yet there is an interval from mean age $6\frac{1}{2}$ to mean age $8\frac{1}{2}$, approximately, in which the slope of the catch curve bears no simple relation to the survival rate.

I have constructed a similar population model in which rate of growth decreased continuously instead of changing suddenly. Without presenting the details, the annual mortality rate put into the model was $A = 0.4$, while the rates "recovered" from it at different ages by the method of the last paragraph were 0.20–0.22. As a matter of fact, when mortality rate is small and fairly steady, and rate of increase in length is decreasing at a moderate rate, the number of fish at certain intermediate sizes exceeds the number at smaller sizes nearby, as is shown by Curve CD of Fig. 2.11, and has been demonstrated for an actual fish population by Hart (1932, fig. 4). In that event $d(\log N)/dl$ becomes a positive coefficient in places, and could not possibly be used to estimate mortality rate in the manner described above.

Mortality rates estimated as above from rate of growth and length frequencies always tend to be too small, if absolute rate of increase in length is decreasing with age. Elster (1944, p. 294), for example, used a combination of length frequency distribution and rate of growth to compute a total mortality rate of 88% per year for Blaufelchen (*Coregonus wartmanni*) of commercial size in the Bodensee. Although this is a rather high rate, the method of estimation tends to make it somewhat too small, rather than too large.

At present, then, catch curves based on length frequencies are much less useful than those based on age, even when the successive ages overlap thoroughly and make a smooth curve. Their slope can be used for an unbiased estimate of survival rate only if the absolute increase in mean length of the fish between successive ages is uniform over a range of ages which, in terms of corresponding mean lengths, is somewhat greater than the range of lengths over which the slope of the graph is to be measured.

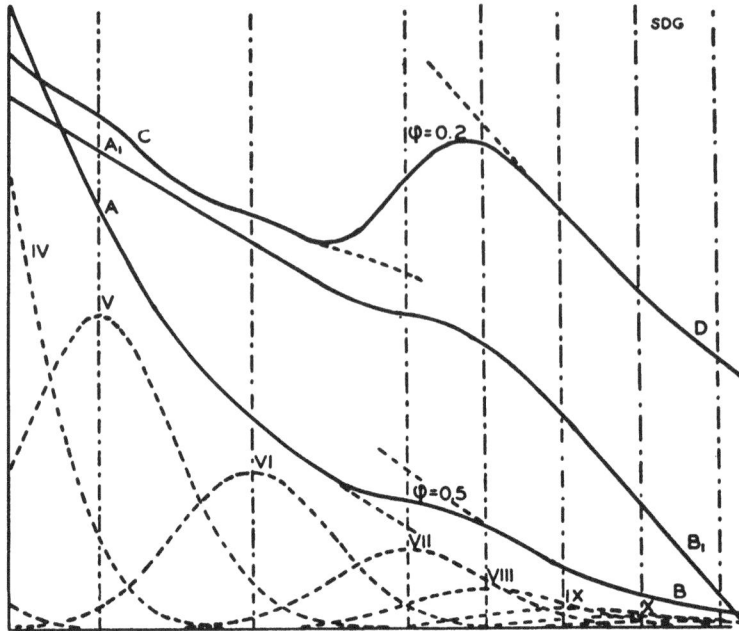

FIG. 2.11. Synthetic population curves made by summing the contributions, to successive length-classes, of several overlapping age-groups, each normally distributed as to length. The dotted bell-shaped curves are the length distributions of successive age-groups, each half as numerous as the preceding; the rate of increase in length decreases between ages 7 and 8 to half of its previous magnitude. Curve AB is the sum of the dotted curves, and shows the length frequencies of the total population. Curve A_1B_1 shows the logarithms of length frequencies of the populations, and is equivalent to a catch curve. Curve CD is a synthetic curve similar to AB, based on fish which have the same rate of growth but which decrease in numbers by only 20% per year. Abscissa — length; ordinate — frequency (log frequency for A_1B_1). (Redrawn from Baranov 1918, by S. D. Gerking.)

EXAMPLE 2.8. SURVIVAL OF PACIFIC HALIBUT: A CONCAVE CATCH CURVE BASED ON LENGTH FREQUENCY DISTRIBUTION. (From Ricker 1948, slightly modified.)

Catches of Pacific halibut (*Hippoglossus hippoglossus*) taken for tagging by Thompson and Herrington (1930), south of Cape Spencer, have been used to construct a catch curve based on length frequencies. The catches of 1925 and 1926 are combined to smooth out some of the irregularities in recruitment[4]. The catch curve (Fig. 2.12) is plotted in terms of frequency per 5-cm length interval (near the end the average for a 10-cm range has been used). Dunlop has shown that the mean length of commercially caught Goose Island halibut tended to increase by a little less than 5 cm for each year's increase in age, from age 4 to age 14; between age 9 and age 14 it is exactly 5 cm per year (Thompson and Bell 1934, p. 25). This is indicated on Fig.

---

[4] The 2 years also differ in that there are relatively more small fish in 1926 and more large ones in 1925. However, between ages 9 and 13 their curves have much the same slope. Since considerably more fish were handled in 1926, it would be somewhat better to give each year equal weight, but this has not been done here.

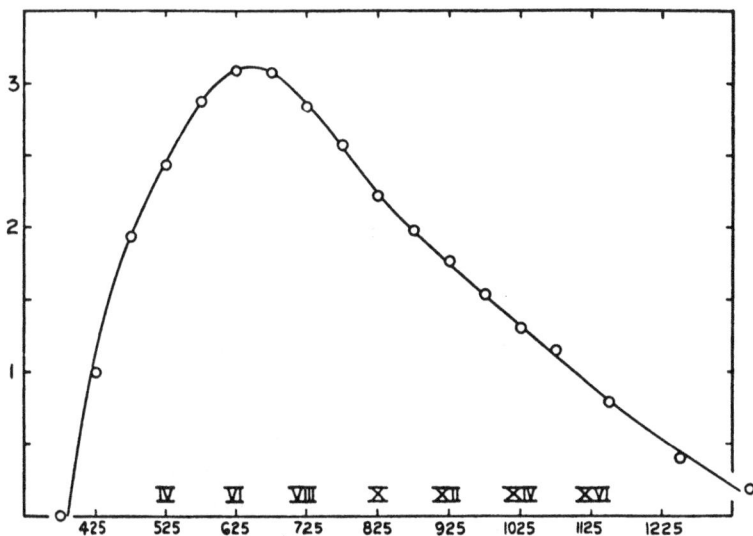

FIG. 2.12. Catch curve for the Pacific halibut population (southern grounds), from samples taken for tagging in 1925 and 1926. Abscissa — mean length (mm) of successive 5-centimeter length groups. (Ages indicated are only approximate, and at ages below IX are typical of the sample only, not of the population.) Ordinate — logarithm of the number of fish taken at each length interval. (From data of Thompson and Herrington 1930.)

2.12 by roman numerals above the approximate mean length of each age-group. Beyond age 14 there is little direct information on rate of growth; from the situation in other fishes, a decrease in rate of increase in length might be anticipated among old individuals. For estimating survival rate, the curve of Fig. 2.12 will be useful only from age 9, which is probably the first fully recruited age, to age 14, where linearity of growth may cease. Within these limits, the curve is noticeably concave, and this suggests a recent decrease in survival rate. Accordingly, the slope of the steepest part of the curve, between ages 9 and 10, will come closest to being an estimate of its current magnitude.

The revised estimates (Anon. 1962, table 7) of the halibut fishing effort in southern waters (Area 2) are shown below in terms of thousands of "skates" of gear set:

Year	Effort	Year	Effort	Year	Effort
1911	237	1917	379	1923	494
1912	340	1918	302	1924	473
1913	432	1919	325	1925	441
1914	360	1920	387	1926	478
1915	375	1921	488	1927	469
1916	265	1922	488	1928	537

If there had been a continuous increase in fishing effort and thus in total mortality rate right up to 1925–26, the curve of Fig. 2.12 would not be steep enough anywhere to represent the current rate of survival. However, there were two periods of more or less stable fishing effort, of which the more recent is 1921–27, when effort averaged 476,000 skates. This period lasted 7 years, or a year or two longer than it takes halibut to become completely vulnerable to fishing. Consequently by analogy with Curve C of Fig. 2.7 we can expect that the steepest part of the catch curve will in fact represent the survival rate which actually prevailed when the samples were taken. This steepest slope occurs between ages 9 and 10, and is –0.066 log units per centimeter, which corresponds to –0.33 log units per year. Hence $Z = 2.303 \times 0.33 = 0.76$, $A = 0.53$, and $S = 0.47^5$.

We can also make an estimate of the survival rate which obtained among fully vulnerable fish in 1916–20, when fishing effort averaged 332,000 skates per year. This will be given by the slope of the line from ages 10 to 14 inclusive, and corresponds to $S = 0.61$ and $Z = 0.49$. Obviously there is an inconsistency here. The Z values of 0.76 and 0.49, less natural mortality, should be proportional to fishing effort, whereas in fact the fishing efforts of 1921–27 and 1916–20 do not differ nearly enough for this. There are two possible explanations: the estimate of Z for 1916–20 may be in error, because it is getting into the region where the catch curves for the separate years do not agree too well; alternatively, or in addition, the efficiency of a skate of gear may have improved with time, as fishermen became better acquainted with the grounds.

2.8. CATCH CURVES FOR ANADROMOUS FISHES

Anadromous fishes may conveniently be divided into three categories: (1) those which reproduce only once and then die; (2) those which may reproduce in each of 2 or more successive years; (3) those which may reproduce more than once, but at intervals longer than 1 year. All three types usually have one feature in common, that fishing tends to be concentrated on the migrating fish which are about to mature and reproduce.

The best-known examples of the first type above are found among Pacific salmons (*Oncorhynchus* spp.). A catch curve from a sample of the migrating run of such fish is obviously of no value for estimating mortality rate, though the information may occasionally be used to estimate survival rate in another manner (Section 8.9).

Anadromous fishes of group 2, of which the Atlantic salmon (*Salmo salar*) and shad (*Alosa sapidissima*) may be taken as examples, present a somewhat different picture. Here catches taken from the spawning run can be made to give information about mortality rate, provided the maiden fish can be distinguished from those which

---

[5] There is fairly good agreement between this figure and the survival rate of 0.416 estimated by Thompson and Herrington (1930, p. 70) from recaptures during 4 years of halibut tagged in 1925. The agreement, however, is partly accidental, since halibut of all sizes tagged were used in their estimate, and those which, for at least a year after marking, were in the incompletely-vulnerable size range, were retaken relatively less frequently in the year after tagging than in later years. Since the majority of fish used were of this sort, this effect is quite important, and makes their estimate of apparent survival rate too *high*. Using completely vulnerable fish only, the tagging data yield an apparent survival rate of 0.33. Possible explanations of the discrepancy between this figure and the 0.47 obtained here are given in Example 4.4, below.

have already spawned at least once. Beginning with the first[6] age-group in which practically no maiden fish occur, the abundance of successive ages from there on will reflect the population survival rate between them, subject to the usual provisos regarding random sampling, uniformity of recruitment, and so on. However, it may be found, and this is usual among salmon, that recidivists are so rare as to constitute only a minor part of the total catch, apparently because of a heavy post-spawning or ocean mortality which is not the result of fishing. Shad, on the other hand, seem to survive in larger numbers and to greater ages (Fredin 1948).

Finally, the very interesting situation in which more than 1 year elapses between spawnings has most of the characteristics of the one just discussed. If fish are caught only in the spawning migration, the survival rate obtained from the catch curve is the (geometric) mean annual rate for all the years between one spawning migration and the next (not the overall survival for the total time elapsed between one migration and the next.) Among anadromous fishes, this behavior is best known among sturgeons (*Acipenser*); non-anadromous salmonoid fishes in some northern lakes appear to spawn only in alternate years.

EXAMPLE 2.9. CATCH CURVE FOR KURA RIVER STELLATE STURGEON. (From Ricker 1948.)

Derzhavin's (1922) comprehensive study of the sevriuga or stellate sturgeon (*Acipenser stellatus*) of the Kura River contains information on a wide variety of topics. From his table (p. 67) of the age and sex composition of this sturgeon caught

---

[6] If both the age and the number of spawnings of each fish can be determined, such comparisons can be made for all age-groups.

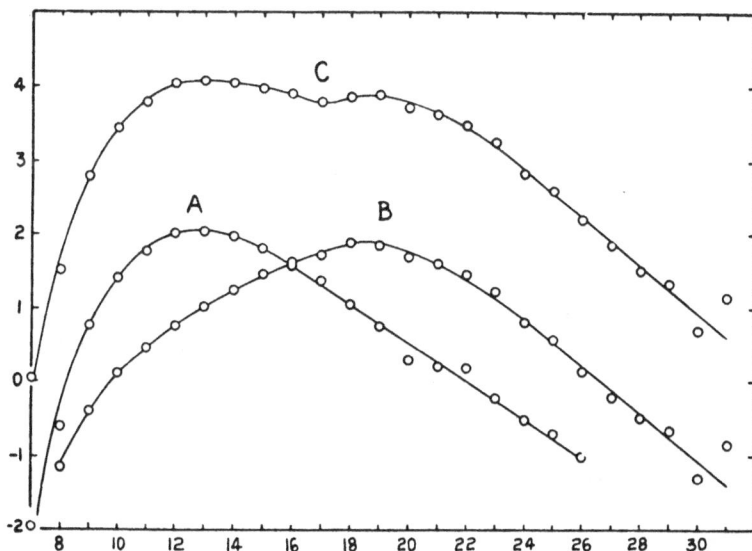

FIG. 2.13. Catch curves for the stellate sturgeon of the Kura River. (A) males; (B) females; (C) sexes combined. Abscissa — age; ordinate — logarithm of the number of fish occurring at each age, per thousand of the total sample, for curves A and B; curve C is drawn two units higher. (From data of Derzhavin 1922.)

in the Caspian Sea near the mouth of the river, the catch curves of Fig. 2.13 are plotted. When the sexes are segregated, the males are seen to occur at a much younger average age than the females. Since the fish are taken on their spawning run, this indicates that the males mature earlier, on the average. Derzhavin gives 12–15 years for males and 14–18 years for females as the principal range of ages at first maturity, though some of either sex were taken as early as 8 years. Sevriuga of both sexes spawn "at intervals of several years, possibly five," but it is not known whether the two sexes have the same average interval between spawnings. If the elapsed time were longer for the younger females, as is suggested by Roussow's (1957) work with *Acipenser fulvescens*, it would explain the longer ascending limb of their catch curve.

A point of general interest is that when vulnerability of fishing depends on maturity, differences in age at maturity of the two sexes tend to broaden the left limb and flatten the dome of the catch curve, when the sexes are not separated. In ordinary fisheries, a difference in rate of growth of the sexes will have a similar effect. However, it will probably rarely happen that the dome will actually have a dent in it, as was found for the Kura sevriuga.

2.9. STRATIFIED SAMPLING FOR AGE COMPOSITION

When an overall random sample is used to plot the catch curve and estimate survival rate in a stock, there is commonly a series of older ages which are represented by only a few individuals if the sample is of any ordinary size — say 100 to 1000 fish. The same is usually true of one or a few of the youngest ages, which ages may be of interest in examining net selectivity, etc. If good information concerning these terminal ages is desired, special effort must be expended on them.

1. A very simple plan is to take a special sample of the catch for fish above a certain size. For example, 1/1000 of the catch might be used for the general sample, and 1/100 of the large fish for the special one. Growth or survival rates computed from the special sample are used for the older ages (Ricker 1955a). However, this procedure is not particularly efficient, since some of the fish whose scales or otoliths are read in the special sample must be discarded because they belong to ages incompletely represented in the size range of that sample. Also, this consideration makes it rather unlikely that it would be profitable to use more than two different sampling fractions.

2. Ketchen (1950) suggested a different plan, which works well when a really large representative *length* sample can be obtained for the whole catch. Dividing the catches into length groups one centimeter broad, otoliths are collected for age determination from fish in the large sample, up to some fixed number in each length group or (in the terminal groups) to such smaller numbers as are available. From the percentage representation of each age in its otolith sample, an age composition for each length group of the representative length sample was determined, and the whole added by ages to build up an estimate of the age composition of that sample, thus of the catch[7].

---

[7] From a large length sample of cod, Fridriksson (1934) took a subsample for age determination and applied this computational procedure, thus decreasing the influence of sampling error and of any possible systematic bias in the subsample. However the advantage gained in this way is ordinarily small compared to what is afforded by Ketchen's procedure.

Obviously methods 1 and 2 above might advantageously be combined, when time or facilities for taking a large length sample are not available. In fact, by grafting Ketchen's procedure onto it, the method of using different sampling ratios is considerably improved: no age-determined fish need be discarded, a complete (computed) catch curve is obtained, and more than two different sampling fractions might sometimes be employed to advantage.

Both of the methods above imply that it is desirable to have more accurate information on the sparsely represented ages than what a moderate-sized single sample will supply — which is not necessarily true for all purposes, although generally so.

3. When catches from a stock are landed by many boats, at many ports, and over a considerable period of time, the assembling of a single representative length or age distribution becomes very complex — involving numerous individual samples which are eventually combined into one representative picture using a series of weighting factors. Details for particular situations have been published, but no general description would be profitable. Subsampling by length for the age determinations may be of great assistance, but sometimes age at a given length will differ significantly between different catches. Several papers describing problems and methods in use are included in Volume 140, Part I, of the Rapports et Procès-Verbaux of the International Council for the Exploration of the Sea; see especially Pope (1956) for a discussion of stratified sampling.

EXAMPLE 2.10. AGE COMPOSITION OF A LEMON SOLE CATCH OBTAINED BY KETCHEN'S STRATIFIED SUBSAMPLING METHOD. (From Ricker 1958a.)

Table 2.9, provided by Dr K. S. Ketchen, is a computation of age composition of a random catch sample (Y) using a subsample (X) stratified by length, for lemon sole (*Parophrys vetulus*). Up to 10 otoliths were read in each subsample, the age frequencies being in the left half of the Table. These are applied *pro rata* to the actual numbers in the Y-sample, in the right-hand side of the Table. Totals of these columns represent the estimate of age composition of the catch. This would be used to estimate survival rate, recruitment, etc., subject to the various considerations outlined earlier in this Chapter.

Selection of the best maximum number of fish to be included in each length-class is a matter of some importance (cf. Gulland 1955a). It depends on the breadth of the length-classes used and hence the total size of the sample to be "aged", on the number of samples actually or potentially available to represent the fishery under consideration, on the degree of difference between intrasample and intersample variability, and on the labor involved in taking additional samples.

2.10. EFFECTS OF INACCURATE AGE DETERMINATIONS ON ESTIMATES OF SURVIVAL RATE

The methods and some of the problems of age determination are discussed briefly in Chapter 9. The most accurate method may differ for different species and stocks, but all methods are subject to error of greater or less magnitude. When comparisons are made, it is commonly found that different individuals will reach different

TABLE 2.9. Age distribution in a sample of male Strait of Georgia lemon soles as determined from a stratified subsample.

Size-class (cm)	Subsample (X)	Age-groups in X						Sample (Y)	Calculated age representation in Y					
		4	5	6	7	8	9		4	5	6	7	8	9
27	6	5	1	...	...	...	...	6	5.0	1.0				
28	9	3	4	2	...	...	...	9	3.0	4.0	2.0			
29	10	4	4	1	1	...	...	30	12.0	12.0	3.0	3.0		
30	10	1	5	4	...	...	...	51	5.1	25.5	20.4			
31	10	...	8	2	...	...	...	54	...	43.2	10.8			
32	10	1	7	1	1	...	...	48	4.8	33.6	4.8	4.8		
33	10	1	3	3	2	1	...	41	4.1	12.3	12.3	8.2	4.1	
34	10	...	2	6	1	1	...	27	...	5.4	16.2	2.7	2.7	
35	10	...	1	4	3	...	2	13	...	1.3	5.2	3.9	...	2.6
36	6	...	...	1	3	2	...	6	...	...	1.0	3.0	2.0	...
37	3	...	...	1	1	1	...	3	...	...	1.0	1.0	1.0	...
38	1	...	...	...	...	1	...	1	...	...	...	...	1.0	...
Totals	95							289	34.0	138.3	76.7	26.6	10.8	2.6
Percentage									11.8	47.9	26.5	9.2	3.7	0.9

conclusions from certain scales, otoliths, etc. Usually 80–90% agreement between two individuals is considered good, and for older ages of long-lived fish it can decrease to 50% or less. How does error of this sort affect estimates of survival? Simple numerical models reveal that when true survival rate is constant between ages there is no consistent bias in either of the following situations: (1) the percentage of positive and/or negative error in age reading is the same at all ages; and (2) positive and negative errors are equal at each age, but may increase or decrease with age.

When negative error exceeds positive, and this difference increases with age, estimated survival rates will in general tend to differ from the true rates. Table 2.10 and other models show that the size of the difference can vary from a small overestimate to a large underestimate, depending on the magnitude of the errors and on the true survival rate. For Model A the error in estimated survival rate is mostly negative, but it is

TABLE 2.10. Two models of effects of error in scale reading on estimates of survival rate (S), for a population in which true S = 0.5 at all ages. Column 2 shows the true age structure for both models. *Model A*: At all ages 10% of scales are read 1 yr too high; at ages 3 through 6, 10% are read 1 yr too low; at age 7, 20%; at age 8, 30%; and so on. N' represents the age readings obtained and S' is the apparent survival rate computed from them. *Model B*: The numbers of scales shown in column 8 are misread too low, the average number of years too low increasing with age: the contributions of successively older ages to a given age are shown in columns 9–12. (For example, of the 50 misread scales of age 11, 5 are taken to be of age 10, 20 of age 9, 15 of age 8, and 10 of age 7). N' and S' are as in Model A.

1	2	3	4	5	6	7	8	9	10	11	12	13	14
			Model A							Model B			
Age	N	−	+	+	N'	S'	−	+	+	+	+	N'	S'
3	12800												
4	6400	1280	1280	320	6720	0.500	0	0	0	0	0	6400	0.500
5	3200	640	640	160	3360	0.521	0	0	0	0	0	3200	0.547
6	1600	320	320	160	1760	0.500	0	80	40	20	10	1750	0.497
7	800	240	160	120	880	0.455	80	80	40	20	10	870	0.452
8	400	160	80	80	400	0.475	120	60	30	15	8	393	0.395
9	200	100	40	50	190	0.473	120	40	20	10	5	155	0.142
10	100	60	20	30	90	0.478	100	5	8	6	3	22	0.273
11	50	35	10	18	43	0.465	50	0	1	3	2	6	0.167
12	25	20	5	10	20	0.400	25	0	0	1	0	1	0.000
13	12	11	2	5	8	0.500	12	0	0	0	0		
14	6	6	1	3	4		6	0	0	0	0		
15	3	3	1				3						

not very large in spite of what would appear to be rather large errors in ageing the fish. Furthermore, at the younger ages the error is zero or in one case positive, so that a weighted overall estimate of S would differ little from the true value 0.5. The reason is that there is at all ages some positive error, as well as negative, and the more abundant fish of the younger age of any adjacent pair add importantly to the apparent number in the older. In Model B all errors are negative (age read too low) and the error is serious at older ages. The first affected survival estimate (age 5–6) is too great, the second is almost "on," and older estimates rapidly become much too small. Again a weighted estimate of S for all ages would not be far off the mark; but if the bias in ageing were not suspected, the investigator would conclude that mortality rate increases rapidly with age.

The most effective method of checking age determinations, when it is feasible, is probably to observe the progress of a markedly dominant year-class in scale samples over a period of years. Another method is to release marked or tagged fish, recapture them after a year or more, and check their growth against their scale pattern and against the mean growth rate of the population; however, the process of capturing and the mark or tag can affect both scale structure and growth rate. In any event great care should be taken to avoid any large consistent bias in age determination.

EXAMPLE 2.11. EFFECT OF INACCURATE AGE READING ON SURVIVAL ESTIMATES OF CISCOES. (From Aass 1972.)

The usual method of determining age of ciscoes (*Coregonus albula* and allied species) has been from their scales, usually a fairly reliable method. However Aass (1972) decided to check cisco ages from Lake Mjösa in Norway by using otoliths, and obtained a very different picture (Fig. 2.14). Not only was the average age much greater, but there was a marked 3-year periodicity in the appearance of strong year-classes. This picture was substantiated by results of tagging, and particularly by the regular progression of the dominant generations.

Using mean age frequencies obtained from the scales, the estimate of survival rate between ages 4 and 5 is 0.17; the ratio of age 4 to age 3 is less (0.24), but there is no assurance that age 3 is fully recruited. The otolith ages of course produce a much gentler catch curve: its slope becomes gradually steeper from age 3 to age 12. From a freehand smoothed curve, estimated survival rate decreases from about 0.69 for age 5–6 to 0.38 for age 9–10. Since samples were taken from seines that retained fish of sizes down to and including all of age 2, probably ages 5 and older were representatively sampled.

2.11. SELECTIVITY OF FISHING GEAR

Almost all kinds of fishing gear catch fish of some sizes more readily than others, and the subject has been investigated in some detail for certain gears.

2.11.1. GILLNETS. Both theoretical and observational studies of gillnet selectivity are fairly numerous. Herring (*Clupea harengus*) have been a common object of study in salt water, beginning with Baranov (1914) and including papers by Hodgson (1933), Olsen (1959), Ishida (1962, 1964), and Holt (1963). In fresh water the lake whitefish

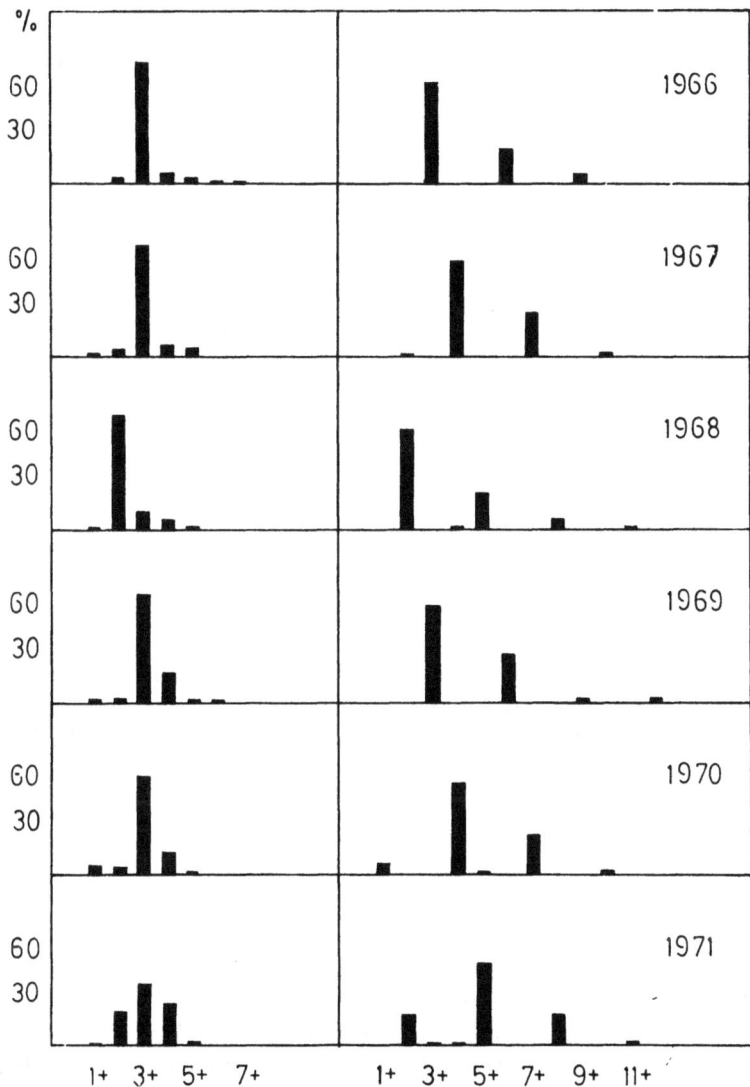

FIG. 2.14. Age frequency histograms for ciscoes of Lake Mjøsa as determined from scales (left) and from otoliths (right). (From Aass 1972, fig. 3.)

'(*Coregonus clupeaformis*) has been a favorite species (McCombie and Fry 1960, Berst 1961, Cucin and Regier 1966, Regier and Robson 1966, and Hamley 1972). Other contributions are by Baranov (1948), Andreev (1955), Gulland and Harding (1961), Hamley and Regier (1973), and Peterson (1954).

Most authors have computed only the *relative* vulnerability of each size of fish to a given size of net using a unimodal selection curve either approximately or exactly normal in shape. If two or more mesh sizes were combined into a common selectivity curve, it was assumed that all meshes caught the same fraction of the fish present at

71

their respective modal lengths of maximum vulnerability. However, experimental gill-netting over many years has shown that really small meshes catch far fewer fish than somewhat larger ones, whereas fish of the sizes best caught by small meshes must generally be more numerous than larger ones. Ricker (1949a) and Hamley (1972) demonstrated this effect quantitatively for whitefish, using the DeLury method of estimating catchabilities (Section 6.3).

The best study in this field is by Hamley and Regier (1973), who used recaptures of marked fish to estimate vulnerability of Dexter Lake walleyes (*Stizostedion vitreum*) to gillnets. Percentage recaptures increased with mesh size, and they also increased with fish size considering the gang of nets as a whole. They also found that the walleye selection curves for each mesh were bimodal, corresponding to two methods by which these fish were caught, "wedging" and "tangling." Similar effects have been observed for other toothy fishes, particularly trout of several species; for example, Ricker (1942c) found a trimodal pattern for cutthroat trout (*Salmo clarki*) and char (*Salvelinus malma*) in Cultus Lake.

2.11.2. TRAPNETS. A comprehensive study of selectivity of fairly large trapnets in three Michigan lakes was made by Latta (1959). Ten species of fish were marked by removal of a fin, and the percentage recapture of each size-class was computed. From the data presented graphically, the percentage of recaptures tended to increase throughout the whole length range for rock bass (*Ambloplites rupestris*), yellow bullheads (*Ictalurus natalis*), white suckers (*Catostomus commersoni*), and (in two out of three lakes) for bluegills (*Lepomis macrochirus*). In a third lake bluegill re-captures were almost independent of size, and this was true of largemouth bass (*Micropterus salmoides*) in one lake (if 2 years' results are averaged). For largemouth bass in another lake there was a fairly definite indication of a peak of vulnerability at an intermediate size, and this was true in a single experiment with brown bullheads (*Ictalurus nebulosus*). For all species the average size of recaptures was greater than that of the fish at tagging, though no growth had occurred meanwhile. Using a similar procedure, Hamley and Regier (1973, fig. 1) found that vulnerability of walleyes (*Stizostedion vitreum*) to trapnets increased steeply throughout their whole size range.

It is surprising that traps should be so selective, and particularly that the vulner-ability of a species sometimes increases right up to the largest individuals present. The suggestion has been made that this might reflect greater activity on the part of larger fish, but the observed behavior of large fish in captivity does not support this. Possibly they are merely more prone to seek out secluded areas, and the entrance of a trap would qualify.

2.11.3. TRAWLS AND LINES. A rate of fishing independent of age, above some minimum, would be rather likely in trawl fisheries, and Hickling (1938) in fact found that rate of return of tags from North Sea plaice (*Pleuronectes platessa*) tended to level off above a tagging length of about 25 cm.

In a line fishery for lingcod (*Ophiodon elongatus*), which begin to be caught at about 40 cm, there was no great change in rate of recapture of tagged fish in the prin-cipal size range, 66–90 cm; but among the few fish taken from that size up to 120 cm there was some decrease in rate of return — possibly resulting from an increase in natural mortality at great ages (Chatwin 1958).

Using the utilized-stock method, Fry (1949) and Fraser (1955) found that vulnerability increased in sport fisheries for lake trout (*Salvelinus namaycush*) and smallmouth bass (*Micropterus dolomieui*) over a broad range of ages. For the trout there was a suggestion of subsequent decline in vulnerability among the oldest fish, ages 10–13 (cf. Example 8.2).

2.11.4. ESTIMATION OF SURVIVAL RATE. Obviously if larger fish are increasingly vulnerable to a given gear, older fish will tend to be increasingly overrepresented and hence any estimate of survival rate from age distribution will tend to be too large. The amount of error decreases rapidly with age, however, because annual length increments are small among older fish and there is a broad range of fish sizes at each age. For example, Ricker (1949a) found that there was no consistent change in rate of removal of whitefish from a small lake by a gang of gillnets over the range 33–51 cm fork length, which included ages from 11 to 26; hence survival estimates were considered reliable over this range.

Latta (1959, table 5) shows four examples in which a simple catch curve overestimated the survival rate by 28–61%. However, in the one case where the details are given the uncorrected survival rate should have been regarded as useless anyway, because of differences in year–class strength and (possibly) increase in mortality rate with age: the ratio of age 6 to age 5 is far less than that of age 5 to age 4, and age 7 is missing completely, although a fish of age 8 did occur. This is not to discount Latta's method of estimating a weighted mean survival rate, which depends on tag recaptures (Chapter 4). In general, bias in catch–curve survival estimates from size–specific vulnerability seems unlikely to be large among the older fish in a sample, though this should not discourage efforts to assess it, or to use other more reliable methods of estimation when they are available. The general effects of changing vulnerability are discussed in Section 2.5.

CHAPTER 3. — VITAL STATISTICS FROM MARKING: SINGLE SEASON EXPERIMENTS

3.1. General Principles of Population Estimation by Marking Methods

Attaching tags to fish, or marking them by mutilating some part of the body, was first done to trace their wanderings and migrations. Toward the close of the last century, C. G. J. Petersen (1896, etc.) began the practice of using marked fish to compute, first, rate of exploitation, and, secondly, total population, of fish living in an enclosed body of water. These procedures have been widely adopted. The names usually applied are "sample censusing," "estimation by marked members," the "mark-and-recapture method," the "Petersen method," and the "Lincoln index."

The principle of this method was discovered by John Graunt and used in his "Observations on the London Bills of Mortality," first published in 1662 — a work that marks the starting point of demographic statistics (E. S. Pearson personal communication). Children born during a year were the "marked" individuals, and the ratio of births to population was ascertained from a sample. About 10 years after Petersen's first work, Knut Dahl employed the same procedure to estimate trout populations in Norwegian tarns. Applications to ocean fishes started toward the end of the first decade of the century. Sample censusing of wild birds and mammals began rather belatedly with Lincoln's (1930) estimate of abundance of ducks from band returns, while Jackson (1933) introduced the method to entomology.

The principal kinds of estimates which can be obtained from marking studies are:

1. rate of exploitation of the population
2. size of the population
3. survival rate of the population from one time interval to the next; most usefully, between times one year apart
4. rate of recruitment to the population

Of course not all mark-and-recapture experiments provide all this information; often only population size is involved. Since about 1950 there has been much activity in developing a variety of procedures for marking and recovery and, for any given procedure, there may be a variety of statistical estimates suited to different conditions. Some of the more comprehensive papers are by DeLury (1951), Chapman (1952, 1954), and Cormack (1969).

The general types of procedure involved are as follows:

1. *Single census* (Petersen type). Fish are marked only once; subsequently a single sample is taken and examined for marked fish. Whereas the *marking* should ideally be restricted to a short space of time, the subsequent sample may be taken over quite a long period.

2. *Multiple census* (Schnabel type). Fish are marked and added to the population over a considerable period, during which time (or at least during part of it) samples are taken and examined for recaptures. In this procedure samples should be replaced: otherwise the population is decreasing and the population estimate cannot refer to any definite period of time — unless, of course, the samples are a negligibly small fraction of the total population. There is some computational advantage in marking *all* fish taken in the samples, but it is not essential.

3. *Repeated censuses.* Procedures for estimating survival rate from two successive Petersen or Schnabel censuses were developed by Ricker (1942b, 1945a, b).

4. *"Point" censuses.* Samples for marking and for obtaining recoveries are made at three or more[1] periods or "points" in time, these periods being preferably short compared with the intervening periods. The first sample is for marking only, the last for recoveries only, and the intermediate one or ones for marking *and* recovery. A different mark is used each time, and subsequent sampling takes cognizance of the origin of each mark recovered. This type of census is well adapted to estimating survival rate and recruitment.

In experiments using tags, individual fish can be identified each time they are recaptured. In some insect marking experiments an individual has been given an additional mark each time it is recaptured, which serves to identify its previous recapture history. Methods for estimating population, survival rate, and recruitment from this information have been devised by Jackson (1936, etc.), Dowdeswell et al. (1940), Fisher and Ford (1947), Cox (1949), Leslie and Chitty (1951), Bailey (1951), Chapman (1951, 1952), Leslie (1952), and others. These methods vary with the kind of grouping of recaptures used, and with the mathematical model employed; they often require complicated tabulations and solving complex expressions.

With any of the above four methods, there are two or three possible procedures in taking the second or census sample.

a. *Direct census.* In direct censusing, the type usually done, the size of the sample or samples taken is fixed in advance, or is dictated by fishing success, etc.

b. *Inverse census.* In inverse censusing, the number of *recaptures* to be obtained is fixed in advance, and the experiment is stopped as soon as that number is obtained (Bailey 1951). This procedure leads to somewhat simpler statistical estimates than direct sampling. A more important consideration, possibly, is that since the size of the relative sampling error of any estimate depends mainly on the absolute number of recaptures made, fixing the number of recaptures determines the sampling accuracy of the result within fairly narrow limits. Inverse censusing is likely to be most useful with single censuses, but it can also be applied to multiple censusing (Chapman 1952).

In practice, sampling can be and probably usually is somewhat intermediate between direct and inverse. An experimenter may have time for up to two weeks of census sampling, for example, but would be glad to stop earlier if a reasonable number of recaptures has been taken. However, if he decides to finish at the end of a certain day, rather than at exactly the time the nth recapture is made, the procedure is most akin to direct sampling.

---

[1]If only two points are used, this method is indistinguishable from the Petersen type.

c. *Modified inverse sampling.* A procedure described by Chapman (1952) works toward a predetermined number of *unmarked* fish in the sample, but here the only advantage appears to be statistical convenience.

d. *Sequential censuses.* If the problem is to find whether a population is greater or less than some fixed number, sampling can be done by stages and terminated whenever this point is settled, at any desired degree of confidence. Suitable formulae are given by Chapman (1952).

Only the better-known, easier, or more practical of the above procedures will be presented here. The simple Petersen situation is described first, followed by a review of possible systematic errors, then a description of other procedures.

3.2. PETERSEN METHOD (SINGLE CENSUS)

3.2.1. SIMPLE PETERSEN ESTIMATES. A number of tagged or marked fish are put into a body of water. Record is then kept of the total number of fish caught out of it during a year or other interval, and of the number of marked ones among them. We have:

M   number of fish marked
C   catch or sample taken for census
R   number of recaptured marks in the sample

We wish to know:

u   rate of exploitation of the population
N   size of population at time of marking

An estimate of rate of exploitation of the population is given by:

$$u = \frac{R}{M} \tag{3.1}$$

Leslie (1952) shows that this is an unbiased maximum likelihood estimate. Assuming random mixing of marked and unmarked fish, its variance is found from the binomial distribution to be:

$$\frac{C}{MN}\left(1 - \frac{M}{N}\right)$$

With large numbers of recoveries, R/C can be used as an approximation for the unknown M/N, giving:

$$V(u) = \frac{R(C - R)}{M^2 C} \tag{3.2}$$

Similarly, an unbiased estimate of the reciprocal of population abundance is, by direct proportion:

$$\frac{1}{N} = \frac{u}{C} = \frac{R}{MC} \tag{3.3}$$

77

The large-sample sampling variance of (3.3) is:

$$V(1/N) = \frac{R(C-R)}{M^2C^3} \tag{3.4}$$

The reciprocal of (3.3) is a consistent estimate of N; that is,

$$N = \frac{MC}{R} = \frac{C}{u} \tag{3.5}$$

with a sampling variance of:

$$V(N) = \frac{M^2C(C-R)}{R^3} \tag{3.6}$$

This is expression (2.6) of Bailey (1951). However, values of MC/R are not symmetrically distributed, whereas those of R/MC are; thus if the normal curve of error is used to calculate limits of confidence, it is best to calculate them for $1/N$ using (3.4), and then invert them to obtain limits for N.

Confidence limits can be obtained more simply, however, by treating R as a Poisson or binomial variable (whichever is appropriate), obtaining limits for it directly from a chart or table (Appendix II), and substituting these in (3.5).

3.2.2. ADJUSTED PETERSEN ESTIMATE. Although expression (3.5) is a consistent estimate of N, in that it tends to the correct value as sample size is increased, it is not quite the best estimate[2]. This is true whether sampling is direct or inverse. Bailey (1951) and Chapman (1951) have shown that with ordinary "direct" sampling (3.5) tends to overestimate the true population. They proposed modified formulae which give an unbiased estimate in most situations. Chapman's version is as follows (omitting –1, which is of no practical significance):

$$N^* = \frac{(M+1)(C+1)}{R+1} \tag{3.7}$$

It is usually worthwhile to use (3.7) in place of (3.5) in direct sampling, even though with large values of R there is little difference.

The large-sample sampling variance for N^* in (3.7) is given by Chapman as approximately equal to:

$$V(N^*) = \frac{(M+1)^2(C+1)(C-R)}{(R+1)^2(R+2)} = \frac{N^2(C-R)}{(C+1)(R+2)} \tag{3.8}$$

Again, however, it is better to obtain approximate confidence intervals from charts or tables appropriate to the binomial or Poisson distributions, using R as the entering variable (cf. Example 3.1).

---

[2]That a best estimate does not remain a best estimate when inverted is one of the uncomfortable facts of statistical life. The same is true between a statistic and *any* function of it, other than a linear one. For analogous examples see Sections 2.1 and 11.4.2.

Expressions (3.3)–(3.8) are applicable whether the fish captured are removed from the population or whether they are returned to it (Chapman 1952, p. 300).

Bailey's (1951) expression corresponding to (3.7) differs slightly:

$$N = \frac{M(C + 1)}{R + 1} \tag{3.9}$$

and his expression for the variance is similarly adjusted, but practically these are indistinguishable from Chapman's formulae.

For "inverse" sampling — which ceases when a predetermined R has been taken — (3.5) is close to being an unbiased estimate of N. Nevertheless, a modified formula is slightly better (Bailey, p. 298):

$$N = \frac{C(M + 1)}{R} - 1 \tag{3.10}$$

3.2.3. STATISTICAL BIAS IN PETERSEN ESTIMATES. Expression (3.7) provides an unbiased estimate of N if $(M + C) > N$, so that there is no chance that R might be zero because of sampling variability (Chapman 1951, Robson and Regier 1964). If this condition is not met the estimate N^* has negative bias. Provided $N > 100$ this bias is close to:

$$-Ne^{-MC/N} \tag{3.11}$$

For $MC/N = 3$ the exponential in (3.11) is 0.050, and for $MC/N = 4$ it is 0.018. Therefore, in practice, a less stringent condition can be used: that MC be greater than four times the true population N, in which event the probability of bias will be less than 2% (Robson and Regier 1964).

Since true N is unknown it is more convenient to have a rule based on an observed statistic, the number of recaptures (R). For the Poisson situation (i.e. when M/N is small) the lower confidence limits in Appendix II will indicate the probability of $R = 0$ for any observed R, and thus whether systematic bias of this type is likely. For 95% confidence, true R will not be less than 1 if observed $R = 3$ or more; and for 99% confidence this is true when observed $R = 4$ or more. If M/N is not small, these limits are somewhat more severe than is necessary. Thus the probability of statistical bias can be ignored if recaptures number 3–4 or more.

Similar statistical bias exists when very small numbers of recaptures are made with other kinds of estimates of population, survival rate, and rate of exploitation, described in the sections and chapters to follow.

3.2.4. SAMPLING ERROR AND SAMPLE SIZE. Sampling errors for Petersen estimates are most easily obtained from tables or charts of fiducial limits for the binomial, Poisson, or normal approximations to the hypergeometric distribution. Suitable charts have been published by Clopper and Pearson (1934), Adams (1951), and Davis (1964), while Ricker (1937 and Appendix II here) tabulates limits for the Poisson distribution: the latter can be used as an approximation for the others, since they will never give too favorable a picture. The observed number of recaptures R is entered in the x column of Appendix II, and the 95% or 99% confidence limits read off. The latter are then substituted for R in (3.5) or (3.7), and corresponding

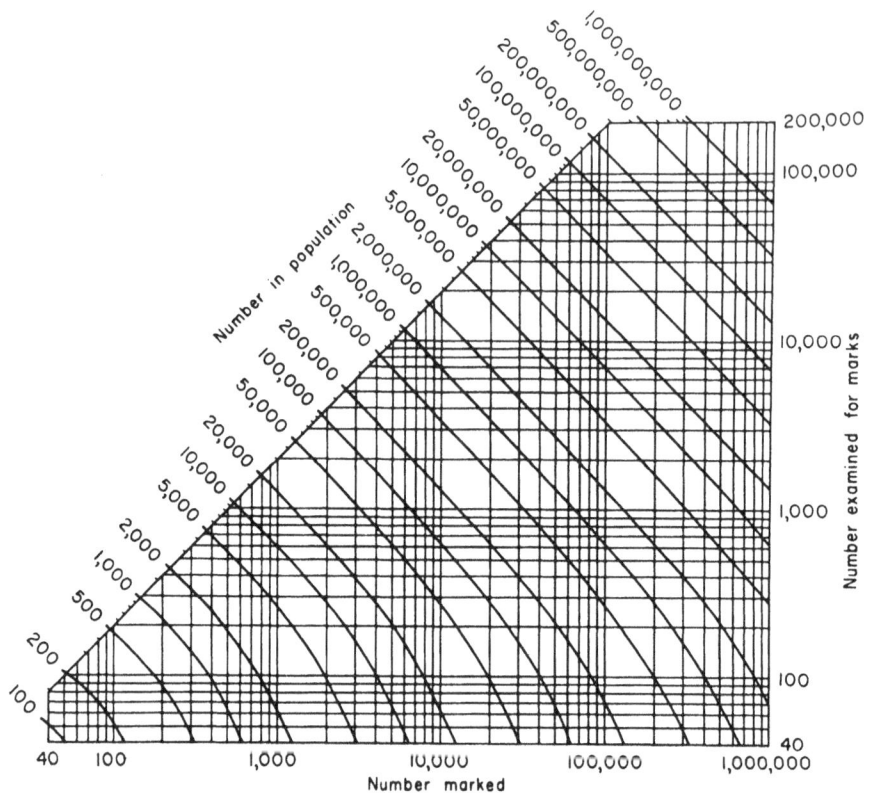

FIG. 3.1. Combinations of number of marks (M) and number subsequently examined for marks (C) for a series of population sizes (N — curved lines), which will provide Petersen estimates of N that in 95% of trials will deviate no more than 25% from the true value of N. (From Robson and Regier 1964, fig. 5.)

limits for N are obtained. Provided $R > 4$, so that the statistical bias of Section 3.2.3 is improbable, these represent the likely limits of error for the population estimate.

Robson and Regier (1964) have combined statistical bias with sampling error in a series of charts that show the combinations of M and C (fish marked and fish examined for marks) that will provide estimates with error no greater than 50%, 25%, or 10% of the true value 19 times in 20. The most useful of these, that for large populations and 25% accuracy, is reproduced here as Fig. 3.1. The expected number of recaptures when obtaining this degree of accuracy, with 95% confidence, varies from approximately 25 to 75 for populations from 10^2 to 10^9.

Robson and Regier (1964) also discuss in detail the optimum allocation of resources in Petersen experiments to increase accuracy as much as possible.

EXAMPLE 3.1. TROUT IN UPPER RÖDLI TARN: A SIMPLE PETERSEN EXPERIMENT. (From Ricker 1948, slightly modified.)

An early application of Petersen's method was made by Knut Dahl, beginning in 1912. He wished to estimate the population of brown trout (*Salmo trutta*) of some small Norwegian tarns, as a guide to the amount of fishing they should have. From 100 to 200 trout were caught by seining, marked by removing a fin, and distributed in systematic fashion around the tarn so that they would quickly become randomly mixed with the unmarked trout. Shortly afterward, more seining was done, and the fraction of marked fish in the catch determined. In the account which I have (Dahl 1919, 1943), the actual numbers of fish marked and recaptured are not given, but from the resulting estimates for the 1912 experiment in Upper Rödli tarn, the following table is prepared, in which these figures are of the right general magnitude:

	Total number of trout	Number of marked trout	Ratio
In the sample—			
Actual number	177(C)	57(R)	0.322
Limits of 95% confidence	....	46–71	0.26–0.40
In the tarn—			
Actual number	337(N)	109(M)	
Limits of 95% confidence	417–272	....	
Ratio of catch to population	....	0.52	
Limits of 95% confidence	....	0.42–0.65	

The steps in preparing this schedule are as follows: The ratio of marked to total trout in the sample is first estimated as $57/177 = 0.322$, and by reference to Clopper and Pearson's (1934) chart the 95% limits of confidence of this ratio are $0.26 – 0.40$. Multiplying these by 177, the limits of confidence for the actual number of recaptures are 46–71. The best estimate of the number of fish in the population is now calculated from (3.7) as:

$$N^* = \frac{(M+1)(C+1)}{(R+1)} = \frac{110 \times 178}{58} = 337$$

By substituting 46 and 71 for R in the above, the confidence limits for N* are 417 and 272.

In this experiment the product MC = 19,300, which is much more than 4 times any possible population size; thus bias of the type discussed in Section 3.2.3 is completely negligible.

Finally, rate of exploitation is $u = R/M = 57/109 = 0.52$; its range for 95% confidence is $46/109 = 0.42$ to $71/109 = 0.65$. In Dahl's experiment the rate of exploitation played an important part, for he undertook to fish the tarn until about half of its fish were removed, as estimated from recovery of marked ones.

3.3. EFFECT OF RECRUITMENT

A straightforward application of formulae 3.1–3.10 is justified only if a number of conditions are met, chief among which are the following:

1. The marked fish suffer the same natural mortality as the unmarked.

2. The marked fish are as vulnerable to the fishing being carried on as are the unmarked ones.

3.   The marked fish do not lose their mark.

4.   The marked fish become randomly mixed with the unmarked; *or* the distribution of fishing effort (in subsequent sampling) is proportional to the number of fish present in different parts of the body of water.

5.   All marks are recognized and reported on recovery.

6.   There is only a negligible amount of recruitment to the catchable population during the time the recoveries are being made.

All of these conditions are of general applicability to experiments of this type, and are discussed in more detail below. Number 6 is essential to the estimate of population, but not to estimating rate of exploitation. Notice that natural mortality will not interfere with the accuracy of the results, as long as it is the same for both marked and unmarked groups. The population estimate obtained applies to the time at which the marked fish were released.

Of the requirements above, the condition that recruitment be negligible is one that often will not be met. Where it is not, the estimate of population is too great. A correction for this effect can be applied by one of several methods.

1. If the population being estimated is divided into age-groups which overlap only a little in length, then by choosing the lower limit of size of fish to be marked at the trough between two age-groups, a boundary can be established whose position will advance as the season progresses and the fish grow larger. In this way there will be little or no recruitment into the marked size range, and C and R should remain in strict proportion throughout the time recoveries are obtained; always provided that the marked fish grow as much as the unmarked, and that they suffer the same mortality.

2. If the age-groups in the fishery overlap so thoroughly that no such point of demarcation can be found, the growth rate of the fish throughout the season can sometimes still be estimated by scale-reading. Suppose, for example, that we wish an estimate of the fish 200 mm long or longer as of July 1st. Assume for the moment that a sufficient number of fish can be marked immediately prior to July 1 to give adequate recoveries later. Take the scales from a sample of fish caught near July 1 and ascertain the mean growth increment, from the time of the last annulus, of fish of the two age-classes whose mean length lies nearest to 200 mm. From time to time throughout the fishing season take additional samples and determine the increment of these same age-classes. By applying these increments proportionately, the average seasonal growth of fish which on July 1 were 200 mm long can be determined with fair accuracy. Now by including only fish greater than this size in the daily catches (C), the effect of recruitment is avoided, and the population estimate consequently will be a true one.

3. When information on growth rate is not obtainable in the detail necessary for the method just outlined, an approximate correction, which is far better than none at all, can often be made. First calculate the per annum rate of growth of fish

of the appropriate size, using scales from a single group of fish taken at any time (though consideration must be given to possible effects of selective sampling, cf. Section 9.1.3). Then divide by the fraction of the growing season that is concerned, i.e. from July 1 to the successive days of the fishing season on which fishing is done. Add these successive values to 200 mm and proceed as above.

The fact that recoveries are being made over a considerable period of time, rather than on a single day or other short interval, is in itself no obstacle to the accurate estimation of population, after the effects of recruitment have been excluded.

If it were necessary to mark fish for a considerable period prior to July 1 in order to get a sufficient number, the same procedure as described above could be extended backward. That is, fish less than 200 mm could be marked in May and June, the exact minimum size in successive weeks to be determined by an examination of rate of growth prior to July 1. It is not *essential* that such smaller fish be used, provided the total mortality rate remains substantially the same over the length range in question, but it will provide more fish for marking than would otherwise be available. In either event there is a disadvantage in extending the marking period too far backward, for natural mortality will remove some of the marked fish before July 1 and make subsequent population estimates too great. If necessary, approximate corrections can be made for this by deducting the estimated mortality for the fraction of the growing season concerned.

4. A method that does not involve age or growth estimates has been described by Parker (1955). After a marking, addition of new fish to the catchable population "dilutes" the marks, and the ratio of recaptures to total sample, R_t/C_t, tends to fall off with time, t. If this fraction is plotted against time and a line fitted, the intercept at $t = 0$ is an estimate of R_t/C_t at time of marking, which can be divided into the number marked, M, to get an estimate of initial population. It may be preferable to use some transformation of R_t/C_t in the graph: the logarithm may be convenient, or the arcsin of its square root as used by Parker.

This method is most useful when the experiment extends over a sufficient period of time for recruitment to be quite pronounced. An estimate of error in the transformed R_t/C_t can be made by calculating the standard deviation from the regression line and then the standard error of the intercept at $t = 0$ (see Snedecor 1946, section 6.9). Transformed back to original units and converted to population by dividing into M, these limits will generally be wider than those based on Poisson or hypergeometric theory. They will also be more realistic, since the variation about the regression line may be greater than expected because of non-random distribution and sampling.

EXAMPLE 3.2. BLUEGILLS IN MUSKELLUNGE LAKE: A PETERSEN EXPERIMENT WITH RECRUITMENT ELIMINATED BY MEANS OF LENGTH ANALYSIS. (From Ricker 1948, slightly modified.)

Figure 3.2 shows the length distribution of bluegills (*Lepomis macrochirus*) handled in a marking experiment on Muskellunge Lake, Indiana (Ricker 1945a).

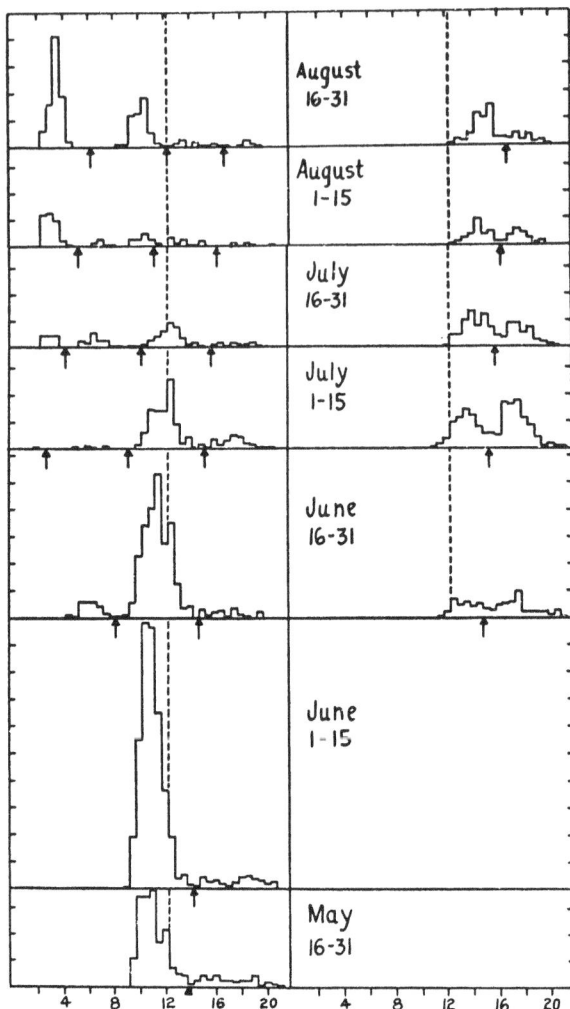

FIG. 3.2. Length frequency distribution of bluegills caught in traps (left) and by fishermen (right) in Muskellunge Lake, Ind., 1942, by semimonthly periods. Each ordinate division represents 20 fish. The vertical broken line represents the minimum size of fish marked, and the minimum size which could legally be taken by fishermen. Ordinate — frequency; abscissa — length (cm).

The population was sampled with two kinds of traps, which took small fish and larger fish, respectively; although unfortunately the intermediate length range, 60 to 90 mm, was poorly sampled. From length frequencies and scale-reading the stock could be divided into age-groups fairly well, as shown by the arrows in Fig. 3.2. Fish of 123 mm and longer were marked. Recaptures were obtained in traps and from fishermen's catches from June 16 to September 7. From the figure, the legal-sized population (125 mm group and up) at the beginning of this period contained a majority of 3-year-old

84

and older fish, but by the end of summer the 2-year-old group had almost completely grown into the fishery, and the older ones contributed only a minor share of the catch. The point of division moves from between the 135- and 140-mm groups in May to between the 165- and 170-mm groups in the latter part of August, advancing 5 mm each half-month. The fact that the marked fish grew as rapidly as the unmarked was shown by the increase in the minimum size of marked fish recaptured of about 5 mm each half-month following June 15. (In a later experiment different marks were used for fish greater and less than 142.5 mm in early June, with the same result.)

The data of the experiment are summarized in Table 3.1. Considering first the fish of age 3 and older, the ratio of marked to unmarked is about the same in traps and in fishermen's catches, so the combined estimate of $28/727 = 0.0385$ gives the mean fraction of marked ones in the population. The estimated population as of the first half of June is therefore $140/0.0385 = 3640$; or better, from (3.7), $N^* = 141 \times 728/29 = 3540$. Since MC is much greater than 4N, there is no appreciable statistical bias, though there is a fairly large random sampling error.

The estimate 3540 is doubtless slightly high because no account is taken of natural mortality during the short period marking was in progress. An approximate correction for this could be made, but it would be unlikely to exceed about 5%.

Rate of exploitation by fishermen is estimated very simply from Table 3.1 as $u = 23/140 = 16\%$. The correction just mentioned would slightly increase this estimate, as would an allowance for fish caught by the few boats whose catches were not checked.

TABLE 3.1. Bluegills marked prior to June 16, 1942, in Muskellunge Lake; number of recaptures; and the catch from which recaptures were taken.

Half-month period	6-II	7-I	7-II	8-I	8-II	9-I	Total
A. Age 3 and older fish: 140 marked							
Traps							
Recaptures..................	3	0	1	0	1	...	5
Total catch.................	35	50	21	10	12	...	128
Fishermen							
Recaptures..................	3	9	8	2	1	0	23
Total catch.................	120	230	165	39	36	9	599
B. Age 2 fish: 90 marked, of legal size in early June							
Traps							
Recaptures..................	2	0	0	0	0	...	2
Total catch (legal in early June)..	77	25	10	5	8	...	125
Total catch (whole age-group)...	487	187	80	21	20	...	795
Fishermen							
Recaptures..................	1	5	6	1	4	0	17
Total catch (legal in early June)..	44	96	92	44	80	19	375

Turning now to the age 2 fish of Table 3.1, we observe that the ratio of marked to unmarked "legal" fish is smaller in trap recaptures than in fishermen's, but not significantly so. Combining the two, the best estimate of population in early June, from (3.7), is $N^* = 91 \times 501/20 = 2280$. The rate of exploitation by fishermen is $u = 17/90 = 19\%$, not significantly different from that for larger fish.

We can also try to estimate the size of the whole of the age 2 group of fish from the trap records, by assuming the marked and unmarked portions to be equally vulnerable to trapping. From the table, the whole age-group should be $795/125 = 6.36$ times as numerous as is the part of it which was of legal size in early June (compare the relative sizes of the parts of the age-group in June 1-15 on either side of the dotted line in Fig. 3.2). The whole age 2 brood is therefore estimated as $6.36 \times 2280 = 14,500$ fish.

3.4. EFFECTS OF MARKING AND TAGGING

3.4.1. DIFFERENTIAL MORTALITY. A frequent effect of marking is extra mortality among marked fish, either as a direct result of the mark or tag, or indirectly from the exertion and handling incidental to marking operations. In either event recoveries will be too few to be representative; thus population estimates made from them will be too great and rates of exploitation will be too small. For example, Foerster (1936) found that yearling sockeye salmon (*Oncorhynchus nerka*) marked by removal of the ventral fins survived to maturity only about 38% as often as did unmarked ones. Foerster's method of estimating and correcting for this error depended on special circumstances of the migratory behavior of the salmon, so it is usually necessary to look to other methods. One approach is to compare returns from different kinds of tags or marks. If one method of marking obviously involves more mutilation of the fish than another, yet both marks are recaptured with equal frequency, then neither is likely to be producing any significant mortality. The opposite result, however, while suggesting that mortality is caused by the more severe procedure, would not necessarily exonerate the milder one. Neither result would shed light on effects of capture and handling, as distinct from the marking proper. When fish are being tagged, and are more or less obviously bruised or abraded in the process of capture, it is possible and useful to keep a record of the degree of injury and apparent vigor for each fish separately. When recaptures come in, these can be checked against the record to see if the less vigorous fish are less frequently retaken.

Both of the above checks were made in an experiment on Shoe Lake, Indiana (Ricker 1942b). Half the bluegill and other sunfishes (*Lepomis* spp.) were marked by removing the two pelvic fins; the other half were given a jaw tag in addition to the mark. It turned out that tagged fish were retaken as frequently as untagged in traps, but in anglers' catches they were much less numerous than untagged ones; this situation lasted through the second summer of the experiment. Among tagged fish, there was no association between rate of recapture and an estimate of trap damage based chiefly on the extent to which the tail was split. Because the tag produced a rather serious and prolonged lesion, while the fin scars and tail membranes healed quickly, it was concluded that trapping, handling, removing the fins, and even the presence of the tag all resulted in very little or no mortality; but that the tag, presumably by interfering with feeding, vitiated estimates of population made from recoveries of line-caught fish. On large-mouthed fishes, however, the jaw tag interferes much less with normal feeding.

Another disadvantage of jaw tags, doubtless related to the above, was that they reduced the growth rate markedly in all species of fish on which they were used. This is not too important, perhaps, since the number on the tag makes it possible to identify the size class to which the fish belonged when tagged. Fortunately, when medium-sized fish are marked by removing a fin or fins, no such retardation of growth occurs (Example 3.2; Ricker 1949b).

3.4.2. DIFFERENCES IN VULNERABILITY OF MARKED AND UNMARKED FISH. A more insidious source of error is a tendency for marked or tagged fish to be either more, or less, vulnerable to fishing than are native wild fish. This may result from several causes.

1. If the fish used were not originally part of the population being estimated, they may obviously behave differently, whether or not they are marked or tagged. This consideration usually makes hatchery-reared fish, or wild fish from strange waters, useless for estimating native populations.

2. When tags are used, the tag itself may make a fish more, or less, vulnerable to fishing. The jaw-tagged bluegills mentioned above are a case in point: the tagged ones were much less vulnerable to angling. Another example is of salmon tagged with two disks joined by a wire passing through the body. Though excellent from several standpoints, these "Petersen disks" made the fish more vulnerable to gillnets than untagged fish, because the twine caught under the disk.

3. Of more general applicability are differences in behavior as a result of tagging or marking. Capturing and marking a fish subjects it to physiological stress (Black 1957 and many subsequent authors), and possibly psychological disturbance as well. Thus it is not surprising to find them behaving differently afterward, for a longer or shorter period. For example, marked centrarchids, when first released, usually swim down and burrow into the weeds. The same tendency, if it persists, might make them more apt to enter a trap funnel than an untouched fish. Any fish, after marking, may be "off its feed", and thus less likely to be caught by methods involving baited hooks. If marking makes a fish less inclined to move about, it will be less apt to be caught in fixed gear like traps or gillnets, but it may be more likely to be caught in moving gear like seines or otter trawls. With other fish a tag may be a stimulus resulting in increased or more erratic movement for some days or weeks. For example, Dannevig (1953, fig. 3) found that tagged cod were retaken by gillnets with rapidly decreasing frequency over the first 15–20 days after tagging, but during the same period recaptures from hook gear remained steady (1948) or actually increased (1949).

Effects of these sorts will generally be hard to detect, and hard to distinguish from actual mortality due to tagging. Rate of recapture in successive weeks or months after tagging may provide suggestive information. So may comparisons of recaptures by different methods of fishing, for vulnerability to one kind of gear may be affected, but not to another, as in the case of the jaw-tagged sunfish or the cod mentioned above. What makes the use of these criteria difficult is that ordinarily recaptures are none too numerous, and their limits of sampling error may be so wide that significant systematic errors are hard to demonstrate.

EXAMPLE 3.3. CORRECTION FOR EFFECTS OF TAGGING ON VULNERABILITY OF CHUM SALMON IN JOHNSTONE STRAIT, B.C. (From Ricker 1958a, after Chatwin 1953.)

Chum salmon (*Oncorhynchus keta*) were tagged at two sites along their migration route from Queen Charlotte Sound through narrow 100-mile-long Johnstone Strait into the Strait of Georgia (Table 3.2). The fish moved from Area 12 (upper Johnstone Strait) through Area 13 (lower Johnstone Strait), and were tagged about midway along each Area.

TABLE 3.2. Chum salmon tagged and recovered in Upper Johnstone Strait and Lower Johnstone Strait, with estimated percentage returns for fish *entering* the strait.

Tagging locality	No. tagged	Percentage recovery by localities				
		Area 12	Area 13	Other	Unknown	Total
Area 12.....................	1733	15.98	10.09	11.74	1.73	39.54
Area 13....................	1952	0.15	14.65	14.81	1.33	30.94
Entrance of 12 (computed)....	...	13.10	10.44	12.81	1.45	37.80

Recaptures of Area 12 fish were expected to be about twice as great in Area 13 as in 12, since they were exposed to only half of the Area 12 fishery and there was about the same amount of fishing in each Area. In fact, however, more were caught in 12 than in 13 (15.98% and 10.09%, respectively). This fact, plus a consideration of times of tag recoveries, indicated that the tag or the tagging procedure delayed the fish's movement by a few days. (Similar effects have been observed in river tagging; see Killick 1955.)

For estimating rate of exploitation, the data of the chum experiment have two defects: (1) there is the extra vulnerability due to the tagged salmon's delay in resuming migration; and (2) it would be desirable to refer the results to a (hypothetical) tagging point for fish *as they first enter the fishery* at the upper end of Area 12. Chatwin made both these adjustments in a single operation, by assuming that fish tagged in Area 13 were delayed to the same degree as those tagged in Area 12. The rate of recovery of tagged fish *entering* Area 13 is, from Table 3.2, $10.09/(1 - 0.1598) = 12.01\%$; as compared with 14.65% recovery of those tagged *in* Area 13. If the same relation applies in Area 12, where 15.98% of local tags were retaken, the corrected rate of exploitation in Area 12, applicable to untagged fish entering the Area, is:

$$12.01 \times 15.98/14.65 = 13.10\%$$

Of the 86.90% which remain after traversing Area 12, 12.01% are taken in Area 13, or 10.44% of the original arrivals to the fishery. In a similar way the recaptures below Area 13, of fish entering Area 12, were estimated as 12.81%. These three percentages are then added, and increased by the small percentage of "unknown" recaptures, to obtain a final representative rate of exploitation of 37.8%. However, there were a few other complications in the situation, one being the possibility of incomplete reporting of tags recaptured.

In this experiment only the rate of exploitation could be estimated, and not the total population, because in the lower Strait of Georgia the Johnstone Strait chums became mixed with others, and the catch statistics cannot distinguish them by origin.

3.5. LOSS OF MARKS OR TAGS

Another source of error in population estimates concerns the tags or marks themselves. Tags have been placed, at one time or another, on many different parts of a fish. The conventional strap tag is usually attached either at the base of the tail fin, on the gill cover, or around the lower or upper jaw. Tags attached with wires are usually run through the flesh near or beneath the dorsal fin. Visceral tags are inserted into the body cavity. Whatever tag or tagging site is used, it is important that the attachment be reasonably permanent, if results of the experiment are to be used to estimate population abundance. Evidence of nonpermanent attachment can sometimes be had by examining a sample of the catch closely, to detect scars left by shed tags.

When fish are marked, rather than tagged, a similar loss of the mark may occur. An early method of marking, used by Petersen on plaice, was to punch holes in the dorsal fin. For more normally-shaped fish the usual method, in fresh water at least, is to remove one or more fins. Many fishes possess considerable power of regeneration of fins, especially when they are cut not too close to the base. I have seen regenerated pectoral fins of large crappies (*Pomoxis annularis*) which were perfect except for a certain waviness of the rays; these had been clipped about one-fifth way from the base a year earlier. Experience in Indiana with post-fingerling largemouth bass (*Micropterus salmoides*), black crappies (*Pomoxis nigromaculatus*), and a variety of sunfishes (*Lepomis*), bullheads (*Ictalurus*), and yellow perch (*Perca flavescens*) showed that the pectoral fins did not regenerate at all, and the pelvic fins usually did not, when cut as closely as possible to the base. At most, the pelvic fins regenerated imperfectly, so they could be distinguished by even a quick inspection, and very rarely did both fins of a pair regenerate significantly.

For really young fish, results have been more variable. Young Indiana bass, 50 to 75 mm long when clipped, exhibited at most a very imperfect regeneration of pectoral or pelvic fins over a period of two or three months in ponds, or up to eight months in aquaria (Ricker 1949b). However, Meehan (1940) reported that young largemouth bass marked in Florida usually regenerated closely-clipped pectoral and ventral fins perfectly within a few weeks. Possibly this is associated with more rapid growth in southern waters. The anal and soft dorsal fins of even large centrarchids, however, regenerated quickly and often practically perfectly, no matter how closely cut.

In salmonid fishes regeneration is apparently less easy, and dorsal, anal, and adipose fins, as well as the paired fins, have all been used with good results. Some regeneration may occur, particularly of the adipose, but it is practically always imperfect, unless the cutting is done when the fish are very small. It is comparatively easy to check on the extent of fin regeneration by keeping a number in captivity, or by sampling wild marked stock at frequent intervals, or by using two unassociated fins for the mark.

A source of error similar to regeneration is the natural absence of fins from wild fish. Among Pacific salmon their frequency evidently varies from stock to stock and from year to year (Foerster 1935; Davidson 1940; Ricker 1972, p. 54). They *can* be numerous enough to complicate interpretation of single-fin experiments when recoveries are obtained at low incidence over a wide area. Other kinds of fish have been less studied from this point of view, but I have never seen naturally-missing fins among spiny-rayed fishes in fresh water.

3.6. NON-RANDOM DISTRIBUTION OF MARKS AND OF FISHING EFFORT

To make a marking experiment representative, either the marked fish, or the total fishing effort, must be randomly distributed over the population being sampled. Consider a population consisting of 10,000 fish in each of two halves of a lake, 20,000 in all. Twice as many traps are set in one half as in the other, so that, both for marking and for recoveries, one end is sampled twice as efficiently as the other. In an experiment of the Petersen type, 1/5 of the fish at one end are marked, and 1/10 of those at the other. Similarly, after mixing of the marked fish into the unmarked, 1/5 and 1/10, respectively, are taken and the marked fish among them recorded. Eliminating sampling error, the result is as follows:

	First half	Second half	Total
Actual population (N)..	10,000	10,000	20,000
Number marked (M)....	2,000	1,000	3,000
Sample taken (C).........	2,000	1,000	3,000
Recoveries (R).............	400	100	500

If the data are treated as a whole, the estimated population is $3000 \times 3000/500 = 18,000$, which is 10% low. This error can be avoided, however, by considering the two halves of the lake separately and calculating the population of each. When there is any reason to suspect unequal fishing effort in two or more parts of a lake, it will be valuable to divide the experiment into parts in this way, as was done for example by Lagler and Ricker (1942). This type of error always tends to make the result of a common calculation less than the sum of the separate calculations.

C. H. N. Jackson seems to have been the first to point out that if either the marking *or* the subsequent sampling is done randomly,[3] the estimate obtained is not biased.

---

[3]The randomness is relative to the population structure; it need not necessarily exist in any geographical sense.

For example, if after the non-random marking in the illustration above a random sample were taken, say of one-fourth, the total number of fish in it would be 5000, and the number of marked fish 750, giving a population estimate of $3000 \times 5000/750 = 20,000$, the correct figure.

To play safe, it is well to try to make both marking and subsequent sampling random, even though either one singly would suffice. Proceeding in this way, it was not difficult to obtain a representative picture of the populations of most of the spiny-rayed fishes of small Indiana lakes (Ricker 1942b, 1945a, 1955a; Gerking 1953a). Other information concerning the randomness of the procedure can be obtained by comparing the ratio of marked to unmarked fish caught by different types of gear, or gear set in different situations, provided the gear does not tend to select marked from unmarked fish, or vice versa. Schumacher and Eschmeyer (1943) tested the randomness of distribution of their marked fish in a pond of 28 hectares, by draining it and recovering a large part of the total fish present. They found the ratio of marked to unmarked fish, of several species, to be little different from what they had previously computed from their trap samples, but bullheads (*Ictalurus* spp.), carp (*Cyprinus carpio*) and bigmouth buffalo (*Ictiobus cyprinellus*) showed significant or near-significant differences. This they attribute to the fact that a large part of the pond was too shallow for their nets, the fish in question being presumably insufficiently active to attain a random distribution during the two weeks of their experiment. Similarly Lagler and Ricker (1942) found little mixture of the fish populations of two ends of a long narrow pond, over a two-months' period. Additional tests have been reported by Carlander and Lewis (1948), Fredin (1950), and others.

A salutary measure, when it is feasible, is to take the sample in which recaptures are sought by using an entirely different kind of gear from that used to catch fish for marking. For example, if fish for marking are taken in traps, and recoveries are obtained by angling, there is little likelihood of *similar* bias being present in both gears.

Large lakes, river systems, and ocean banks present even more difficult problems. Many ocean fisheries cover so wide an area that representative tagging of the whole population is impossible, while fishing effort may vary greatly from bank to bank. This makes it necessary to select smaller units for examination, in which event the problem of wandering may be troublesome.

River fish also are amenable to enumeration by Petersen's method, if they are not of a roving disposition; and as a matter of fact their populations often prove to be surprisingly stable (Scott 1949; Gerking 1953b). Adjustments for a small amount of movement were made by these authors, this being determined by sampling at sites above and below the section under consideration.

The first report of an application of the Petersen method to a migrating fish was apparently by Pritchard and Neave (1942). Coho salmon (*Oncorhynchus kisutch*) were tagged at Skutz Falls on the Cowichan River, British Columbia, and recoveries were made in tributaries of Cowichan Lake, many miles upstream. Close agreement of the tagged:untagged ratio in widely-separated tributaries provided evidence that tagging had been random with respect to the destination of the fish and to their expectation of recovery. Howard (1948) described a more extensive study with sockeye salmon (*O. nerka*) at Cultus Lake, British Columbia, noting various kinds of heterogeneity in the data and the procedure necessary for a reasonably reliable result.

3.7. Unequal Vulnerability of Fish of Different Sizes

Unequal vulnerability of different sizes of fish to fishing gear is a source of systematic error in population estimates similar to that just discussed. It can be illustrated by the same numerical data as used in Section 3.6, putting, in place of the two halves of the lake, two size groups of fish, one twice as vulnerable to fishing as the other. Detection of possible unequal vulnerability can be accomplished by comparing the rate of recapture of marked fish of different sizes, when enough recaptures are made to minimize the effects of sampling error. However, differential mortality, or different behavior of marked fish as compared with unmarked, might give a similar picture if it affected, for example, småll fish more than large ones.

In general, it is likely that variation in vulnerability with size, though a common enough phenomenon, will not usually be a serious problem. For one thing its effects can be minimized by excluding from consideration fish near the limits of vulnerability to any given type of fishing gear, or by using less selective types of gear for experiments of this sort, or by dividing the fish into two or more size groups. Even in the example of Section 3.6, which probably represents a rather extreme situation, the bias in population estimate was only 10%. Cooper and Lagler (1956) found that the efficiency of an electric shocker varied from about 7% for 3-inch trout up to 40% for 11-inch ones; even so, a Petersen estimate made for the whole population was only 30% low. Similarly, for Seber–Jolly population estimates (Section 5.4), Gilbert (1973, fig. 3, 4) illustrated the negative bias obtained for a number of mixtures of fish having different catchabilities.

What should always be avoided is the combining of data concerning two or more *species* to make a common estimate. There may sometimes be a temptation to do this, when data are available for two or more species of similar kind and size, with only a few recaptures for each; but obviously different species may differ greatly in vulnerability over the whole size range of both, and consequently such a combined estimate can be much too low. Thus in small Indiana lakes the redear sunfish (*Lepomis microlophus*) is about 10 times as vulnerable to trapping as is the similar bluegill (*L. macrochirus*), while its abundance is usually about a fifth of that of the bluegill (Ricker 1945a, 1955a). In an experiment based wholly on trap data, the number of redears marked would be twice the number of bluegills, and the number of marked redears recaptured would be 20 times the number of marked bluegills. A calculation similar to that of the last section will show that if the two species were to be treated as a unit, the resulting population estimate would be less than the combined population of the two species by 64%. An example is provided by Krumholz (1944), who found that the sum of the estimates of the population of bass (*Micropterus*), bluegills, and pumpkinseeds (*Lepomis gibbosus*) in a small lake, when calculated separately, was 19,080, whereas the figure obtained from an estimate made by lumping all species together was 9700.

EXAMPLE 3.4. PLAICE PLANTED IN THISTED-BREDNING: A PETERSEN EXPERIMENT WITH UNEQUAL VULNERABILITY BY SIZE. (From Ricker 1948.)

Petersen (1896, p. 12) marked 10,900 out of 82,580 plaice transported into Thisted-Bredning, one of the expansions of the Limfjord, by punching a hole in the dorsal fin. These fish were of almost commercial size and were available to fishermen the same year. Two samples of plaice from the fishery were examined, 1000 in all, of which 193 had the mark. Now this is a curious result, for the fraction of marked fish in the sample (0.193) is greater than in the original number transported (0.132); whereas, if any native fish at all were present in Thisted-Bredning, we should expect the fraction of marked ones in the sample to be smaller.

To see if the difference is greater than could be ascribed to sampling error, we proceed as follows:

	Transported	In the sample	Limits of 95% confidence
Total number............................	82,580	1000	
Number of marked ones........	10,900	193	168–222
Ratio..	0.132	0.193	0.168–0.222

Only once in about 40 times, on the average, would a similar sample have a fraction of marked ones as low as 0.168, whereas the actual fraction put in was 0.132. We may accordingly conclude, as did Petersen, that the experiment does not wholly meet the requirements of random sampling. A possible disturbing factor would be, for example, a tendency for markers to select larger fish for marking, combined with a tendency for larger fish to be more quickly caught by fishermen than smaller ones. Though there is thus an element of uncertainty in the actual determination, there is no reason to question Petersen's conclusion that the Thisted-Bredning plaice were almost all of imported origin.

Notice that the rate of commercial exploitation cannot be calculated in this example without knowing either the total number of fish, or the total number of marked fish, which were removed from the broad. Petersen did make estimates of rate of exploitation, but for this he used tagged fish.

EXAMPLE 3.5. A PETERSEN ESTIMATE OF THE LEMON SOLES OF HECATE STRAIT: ADJUSTMENTS FOR SIZE DIFFERENCE IN VULNERABILITY, AND FOR MIGRATION. (From Ricker 1958a, after Ketchen 1953.)

Ketchen (p. 468) tagged and released 3003 English soles (*Parophrys vetulus*) into a population being actively fished in Hecate Strait, British Columbia. Recaptures were made by commercial boats. However, the average length of the commercial catch was somewhat greater than that of the group tagged. To obtain an estimate of the stock of commercial sizes, the number of tags released was reduced by an approximate factor obtained by superimposing the two frequency distributions (Fig. 3.3). The lined area of the graph includes 23.9 "per cent units," so the number of tags put out was reduced by this percentage, to 2285. (Of these, 30 had been retaken before the start of the period shown in Table 3.3.)

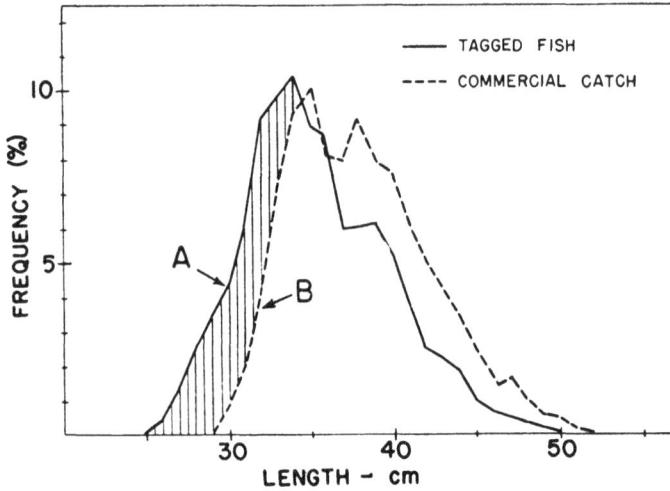

Fɪɢ. 3.3. Length frequency distributions of lemon soles taken by the commercial fishery, and of those tagged and released, as percentage. The lined region comprises 23.9% of the area of either polygon, and represents the percentage by which the number of tags must be reduced to obtain the number "effectively" tagged for this fishery. (From Ketchen 1953, fig. 3.)

Two factors affected the representativeness of the recoveries. First, the stock was moving gradually northward, so that new fish were entering the fishing area and old ones (including tagged ones) were moving out. Secondly, tagging was done from a single boat and the tagged fish, whether from their position or their behavior, were temporarily less catchable than the untagged ones. The latter effect was indicated by disproportionately few recaptures made in the first few days after tagging. Both effects tend to make for too large an estimate (of the population present at time of tagging), but the first increases in importance with time, whereas the second decreases. Consequently, from a computation of population at two-day intervals (Table 3.3),

TABLE 3.3. Petersen estimates of a lemon sole population, from recaptures made at 2-day intervals.

Interval	Tags recap-tured	Total fish caught	"Effective" no. of tags at large	Population estimate (from expression 3.7)
	R	C	M	N
	pieces	pieces	pieces	millions
1	19	81,000	2255	9.1
2	19	46,400	2236	5.2
3	27	67,900	2217	5.4
4	41	132,100	2190	6.9
5	74	173,600	2149	5.0
6	45	102,500	2075	4.6
7	50	118,800	2030	4.7
8	60	146,300	1980	4.7
9	47	127,600	1920	5.1

it is possible to select the low point as the best available estimate of the stock on the grounds when tagged. This can be taken as 4.7 million fish of commercial size (average, 0.937 lb), or 4.4 million pounds — an estimate which is still probably somewhat high. For a different estimate of this population, see Example 6.3.

3.8.   INCOMPLETE CHECKING OF MARKS

It need hardly be added that incomplete discovery or return of tags or marks can lead to serious error. When fish are examined by observers employed especially for the purpose, or by efficient mechanical devices for detecting metal tags, this danger is minimized. Often, however, reliance must be placed on commercial or sport fishermen to turn in records. Experience shows that this is almost certain to give incomplete returns — varying a great deal, of course, with local interest, publicity given to the experiment, the amount of handling the fish get, the type of tag or mark used, and the size of the reward offered if any. Cash rewards are undoubtedly a great help, but they tend to be expensive and have been utilized chiefly in commercial fisheries. The same principle has been applied to sport fisheries by using returned tags as tickets in a sweepstakes, with the prizes donated by local merchants or sportsmen's organizations. Whatever type of inducement is used to encourage non-professional reporting, it will always be desirable to have a substantial part of the catch examined by trained observers, if this is at all possible.

3.9.   MULTIPLE CENSUSES

3.9.1.   GENERAL CONSIDERATIONS. During the mid-1930's David H. Thompson in Illinois and Chancey Juday in Wisconsin began making population estimates from experiments in which marking and recapture were done concurrently. Neither published his results, but Dr Juday interested Zoe Schnabel (1938) in a study of the theory of the method.

Strictly speaking, the method requires that population be constant, with no recruitment and no mortality during the experiment; but it is often useful even if these conditions are only approximately satisfied. The following information is available:

M_t total marked fish at large at the start of the tth day (or other interval), i.e. the number previously marked less any accidentally killed at previous recaptures.

M ΣM_t, total number marked.

C_t total sample taken on day t.

R_t number of recaptures in the sample C_t.

R ΣR_t, total recaptures during the experiment.

The theory of this situation has been discussed by Schumacher and Eschmeyer, De Lury, Chapman, and others. We wish to estimate N, the population present throughout the experiment.

95

3.9.2. MEAN OF PETERSEN ESTIMATES. The simplest approach is to use the results for each day (or other short interval) for a Petersen estimate. (The minimum number of recaptures should preferably be 3 or 4, so as to avoid the statistical bias described in Section 3.2.3.). The mean of these Petersen estimates is taken as the estimate of N, and the differences between each day's estimate and the mean will provide an estimate of the standard deviation and standard error of the mean. The principal disadvantage of this procedure is that a series of estimates of generally increasing reliability is treated as though they were uniformly reliable. The advantage lies partly in the fact that the method provides an estimate of error based on observed variability (rather than a theoretical figure based on the assumption of random mixing); Underhill (1940) used it for this purpose, even though he rightly felt that expression (3.15) below provided a better estimate of the population. Another possible advantage of this approach is that the series of estimates can reveal trends that might indicate departure from the basic postulates above.

3.9.3. SCHUMACHER AND ESCHMEYER'S ESTIMATE. Consider a line fitted to values of R_t/C_t plotted against M_t, with the restriction that it go through the origin; the slope of this line is an estimate of $1/N$. The best estimate of $1/N$ is obtained if each point is weighted as C_t, and this leads to the estimate:

$$\frac{1}{N} = \frac{\Sigma(M_t R_t)}{\Sigma(C_t M_t^2)} \tag{3.12}$$

The reciprocal of (3.12) is an estimate of N. For the variance of (3.12), the basic datum is the mean of the squares of deviations from the line of R_t/C_t against M_t, given by Schumacher and Eschmeyer as:

$$s^2 = \frac{\Sigma(R_t^2/C_t) - (\Sigma R_t M_t)^2/\Sigma(C_t M_t^2)}{m - 1} \tag{3.13}$$

where m is the number of catches examined. However, instead of computing confidence limits directly for N, as Schumacher and Eschmeyer do, it is better to compute them for the more symmetrically distributed $1/N$ (DeLury 1958). The variance of $1/N$, which is a regression coefficient, is:

$$\frac{s^2}{\Sigma C_t M_t^2} \tag{3.14}$$

For computing limits of confidence for $1/N$ from (3.14), t-values are used corresponding to $m - 1$ degrees of freedom. Limits of confidence for N are found by inverting those obtained for $1/N$.

3.9.4. SCHNABEL'S METHOD. An approximation to the maximum likelihood estimate of N from multiple censuses is given by Schnabel's (1938) short formula:

$$N = \frac{\Sigma(C_t M_t)}{\Sigma R_t} = \frac{\Sigma(C_t M_t)}{R} \tag{3.15}$$

This estimate, like (3.5), is asymmetrically distributed. Limits of confidence (based on the assumption of random mixing) can be computed by treating R as a Poisson variable and using the table in Appendix II, particularly when R is small. For medium to large R, advantage can be taken of the fact that $1/N$ is distributed nearly normally with variance:

$$V(1/N) = \frac{R}{(\Sigma C_t M_t)^2} \tag{3.16}$$

From the estimated standard error (the square root of 3.16) limits of confidence can be calculated for $1/N$ using t-values for the normal curve. These limits are then inverted to give a confidence range for N.

Chapman (1952, 1954) points out that inverting an estimate of $1/N$ does not give quite the best estimate of N itself. For (3.15) a simple adjustment is available that gives a better result:

$$N = \frac{\Sigma(C_t M_t)}{R + 1} \tag{3.17}$$

Approximate limits of confidence for (3.17) can be obtained by considering R as a Poisson variable (Appendix II).

DeLury (1951) described an iterative method of obtaining the true maximum likelihood estimate of N, but later (1958) he abandoned it in favor of (3.12) above, on the grounds that the iterative procedure depended too heavily on the postulate of random mixing.

3.9.5. STATISTICAL BIAS. Like Petersen estimates, estimates from multiple censuses are subject to negative bias when the combination of number of fish marked and number examined falls too low. I have not found this discussed specifically but it seems likely that, as in Section 3.2.3, this bias can be ignored whenever the number of recaptures is 4 or more.

EXAMPLE 3.6. SCHNABEL AND SCHUMACHER ESTIMATES OF REDEAR SUNFISH IN GORDY LAKE, INDIANA. (From Ricker 1958a, after Gerking 1953a.)

Gerking (1953a) compared different estimates of populations of various sunfishes in a small lake. Our Table 3.4 reproduces part of his table 3 for part of the stock of redear sunfish (*Lepomis microlophus*). As often happens, a few marked fish died from effects of trapping or from other causes; these are deducted from the number marked on the day in question, and therefore from the number at large next day (M_t).

Columns 2 and 5 of Table 3.4 provide the Schnabel-type estimates. The short Schnabel formula (3.15) gives $N = 10740/24 = 448$; the modified Schnabel (3.17) is $N = 10740/25 = 430$.

TABLE 3.4. Computations for Schnabel and Schumacher estimates for age 3 redear sunfish in Gordy Lake, Indiana, from trap recaptures. (Data from Gerking 1953a, table 3, using only the June 2–15 data.)

1	2	3	4	5	6	7	8
Number caught C_t	Recap- tures R_t	Number marked (less removals)	Marked fish at large M_t	C_tM_t	M_tR_t	$C_tM_t^2$	$R_t^2C_t$
10	0	10	0	0	0	0	0
27	0	27	10	270	0	2,700	0
17	0	17	37	629	0	23,273	0
7	0	7	54	378	0	20,412	0
1	0	1	61	61	0	3,721	0
5	0	5	62	310	0	19,220	0
6	2	4	67	402	134	26,934	0.6677
15	1	14	71	1,065	71	75,615	0.0667
9	5	4	85	765	425	65,025	2.7778
18	5	13	89	1,602	445	142,578	1.3889
16	4	10	102	1,632	408	166,464	1.0000
5	2	3	112	560	224	62,720	0.8000
7	2	4	115	805	230	92,575	0.5714
19	3	...	119	2,261	357	269,059	0.4737
162	24	119	984	10,740	2,294	970,296	7.7452

Columns 6–8 contain the products needed for the Schumacher estimate and its standard error. The estimate of $1/N$ is $2294/970296 = 0.0023642$; hence N = 423. Variance from the regression line is, from (3.13):

$$s^2 = \frac{7.7452 - (2294)^2/970296}{14 - 1} = 0.17851$$

$$s = 0.42250$$

From (3.14):

$$s_{1/N} = \sqrt{\frac{0.17851}{970296}} = 0.00042892$$

Since $t = 2.160$ for 13 degrees of freedom (Snedecor 1946, table 3.8), the 95% confidence range for $1/N$ is 2.16 times the above, or ±0.0009265. Confidence limits for $1/N$ are 0.0023642 ± 0.0009265 or 0.0014377 and 0.0032907, and the reciprocals give limits for N.

98

These estimates and their estimated confidence ranges are summarized below:

Kind of estimate	N	95% range
Original Schnabel (from 3.15)	448	320–746 (from 3.16)
Modified Schnabel (3.17)	430	302–697 (Poisson)
Schumacher (from 3.12)	423	304–696 (from 3.14)
DeLury's weighting formula	440	

Gerking (1953a) computed an estimate using DeLury's weighting formula; using only the data of Table 3.4, it is shown as the last item above. In this and similar comparisons, differences among estimates are small compared to the range of the confidence limits.

3.10. Systematic Errors in Multiple Censuses

In population estimates from multiple censuses, systematic errors can assume complex forms, and to examine their effects theoretically would be a protracted task. In general, all the sources of error discussed earlier in the chapter must be considered here too. Three which are of greater importance in this method are:

1. *Error due to recruitment.* This can sometimes be avoided by the method discussed earlier, of making allowance for fish growth and confining the marking (or the calculation) to a single age-class or some otherwise-restricted segment of the population. For examples see Wohlschlag and Woodhull (1953). Another method is to plot the trend of successive population estimates, and extrapolate back to time zero (Example 3.7).

2. *Error due to natural mortality.* In the absence of recruitment, the effect of natural mortality, affecting marked and unmarked fish equally, is to make a Schnabel estimate less than the initial population size, though greater than the final population size. If natural mortality is exactly balanced by recruitment, the Schnabel estimate becomes greater than the population size since the replacements will not have marked fish among them.

3. *Error due to fishing mortality.* This differs from the last in that it is usually possible to obtain a record or estimate of marked fish removed in this way, and if so this number can be subtracted from the number of marked fish at large in the lake. Fishermen's records are also an additional source of data for the population estimate. However, unless recruitment exactly balances the loss to fishermen, the population estimate will not be equal to the initial population present, nor even exactly equal to the average population present.

The probable effects of these and other errors should be examined in each experiment separately. Other things being equal, the shorter the time in which recoveries are made, the better the estimates obtained by Schnabel's method; and this provides an incentive to more intensive work. (If the experiment extends over a long period, it can be broken up for analysis by the "point-census" method if numbered tags are

99

used, or if the marks used are changed at intervals.) However, *too* short a period makes it difficult to attain a random distribution of the marked fish.

3.11. ESTIMATION OF NATURAL LOSSES AND ADDITIONS TO THE STOCK

If natural mortality or emigration from a stock occurs during a multiple census experiment, but additions are excluded, a Schnabel estimate tends to be less than a Petersen estimate, the former being affected by the losses whereas the latter is not. DeLury (1951) points out that the difference between these two estimates can be used to estimate the magnitude of the rate of loss during the course of the experiment. Expression 1.19 on page 292 of his paper can be used for an approximate direct estimate.

Alternatively, trial values of the rate of loss could be introduced into the Schnabel computation until one is obtained which makes the final estimate equal to a previously-obtained Petersen (or other unbiased) estimate of initial population.

Schnabel estimates of $1/N$ made on successive days during an experiment tend to increase with time when there are losses from the population but no additions to it. Thus another possible criterion for the best trial estimate of natural mortality rate would be that which eliminates this trend from successive *daily* estimates of $1/N$ (*not* the cumulative estimates).

If both mortality and recruitment (or emigration and immigration) can occur, DeLury (1958) shows that estimates of rates of mortality and recruitment can be obtained by a multiple regression procedure.

Unfortunately, the sampling errors of all these estimates tend to be large, and DeLury's bead-drawing trials suggest that it would rarely be possible to obtain useful values for rate of accession or loss from the Schnabel situation.

EXAMPLE 3.7. SCHNABEL AND TIME-ADJUSTED ESTIMATES OF CRAPPIES OF FOOTS POND. (Modified from Ricker 1948.)

Lagler and Ricker (1942) give estimates of the numbers of various species of fishes of Foots Pond, Indiana, using Schnabel's method of estimation. Fish were caught in traps near shore, but were released in open water in an attempt to get them mixed into the unmarked population. Recoveries extended over a period of 7 weeks during the summer. All recoveries were from the same traps as used to catch fish for marking, since line fishing during this time was negligible. Table 3.5 gives the data for white crappies (*Pomoxis annularis*) of the northern part of the pond, accumulated by 10-day periods so as to have at least 3 recaptures per period and so minimize statistical bias of the type described in Section 3.2.3. The first 5 days are omitted because of probable incomplete mixing or aberrant behavior of recently-marked fish.

The direct unweighted Schnabel estimate, using (3.17), is $65,050/22 = 2960$ fish. The estimates tend to increase throughout the experiment. This may be merely sampling variability, but it could result either from recruitment to the population or from different mortality or behavior of marked fish. There is no good way of deciding between the alternatives, which of course do not exclude each other. An adjustment

TABLE 3.5 Estimation of the number of crappies in the north half of Foot's Pond, Indiana. (Data from Lagler and Ricker 1942.)

Mid-period time (days)	$C_t M_t$	Recaptures R_t	Individual estimate $C_t M_t/(R_t+1)$	$\dfrac{R_t \times 10^6}{C_t M_t}$
10.5	5,560	3	1390	540
20.5	13,250	5	2210	373
28.0	14,880	6	2130	403
40.5	24,520	4	4900	163
48.0	6,840	3	1710	439
Total	65,050	21	2960	–

for them can be made similar to Parker's correction for recruitment in Petersen experiments (Section 3.3). Although these data are perhaps too scanty to make it worthwhile, the method is illustrated below.

Successive Petersen estimates of $1/N$ are shown in the last column of Table 3.5. These are regressed against the mid-times of successive 10-day periods of the experiment (in the 3rd and last periods trapping was done only during the first 5 days). The regression coefficient is -0.0475 and the intercept for time 0 is 523.6×10^{-6}. Inverting the latter gives 1910 as an estimate of the initial population. This is 1050 less than the Schnabel estimate, suggesting that the time effect is important in this situation.

3.12. SCHAEFER METHOD FOR STRATIFIED POPULATIONS

In work with migratory or diadromous fishes, it often happens that the fish can be sampled and marked at one point along their migration route, and recovered later at a different place. In effect, the population is divided into a series of units, each partially distinct from adjacent units. This is an example of *stratification*, which has been considered at length by Chapman and Junge (1954). Stratification may also exist in respect to space, for non-migratory fishes.

We noticed earlier that if *either* the marking sample *or* the recovery sample is random, an unbiased (consistent) estimate of the total population can be obtained by the Petersen method. But if both the original marking and the sampling for recoveries are selective, the Petersen estimate may be biased. If both marking and recovery favor the same portion of the population, the Petersen estimate tends to be too small. For the estimate proposed by Schaefer (1951a,b), time of marking is divided into periods here designated by i, and time of recovery into periods designated by j. We have:

M_i number of fish marked in the ith period of marking (T_α of Schaefer)

M ΣM_i, total number marked

C_j number of fish caught and examined in the jth period of recovery (C_i or c_i of Schaefer)

C ΣC_j, total number examined

R_{ij} number of fish marked in the ith marking period which are recaptured in the jth recovery period (m_{ai} of Schaefer)

R_i total recaptures of fish tagged in the ith period ($m_{a.}$ of Schaefer)

R_j total recaptures during the jth period ($m_{i.}$ of Schaefer)

These data are arranged in a table of double entry, as shown in Table 3.6 of example 3.8. For each cell of the table, an estimate is made of the portion of the population available for marking in period i *and* available for recovery in period j; and the sum of these for all cells is the total population:

$$N = \Sigma N_{ij} = \Sigma \left(R_{ij} \cdot \frac{M_i}{R_i} \cdot \frac{C_j}{R_j} \right) \qquad (3.18)$$

This is expression (32) of Schaefer (1951a, b).

Chapman and Junge's analysis indicates that (3.18) gives a maximum likelihood estimate only under the same conditions that the Petersen estimate does: that is, when either tagging or subsequent sampling for recoveries is done without bias. However, (3.18) will frequently give a better estimate than (3.5). This is because (3.18) is consistent, and (3.5) is not, in the limiting situation where the successive "strata" tagged maintain their separate identity and can be treated as separate populations. In that event only the diagonal cells of a table like Table 3.6 would be occupied, and the formula (3.18) becomes the sum of a number of independent Petersen estimates (since $R_{ij} = R_i = R_j$ in that event).

In many practical situations there will be a considerable degree of distinctness maintained among successive groups of fish tagged, along with some intermingling between groups (for examples see Killick 1955). This intermediate situation is less favorable for estimation of population than is either complete separation or completely random mixing at tagging or recovery; nevertheless (3.18) performs rather well in such circumstances. Another advantage of the Schaefer treatment is that it can provide estimates of the population present in successive time intervals, both at point of tagging and at point of recovery.

Chapman and Junge (1954) proposed another possible estimate of N for stratified populations (their estimate N_3), but it is rather cumbersome and, to be consistent, it needs the same assumptions as (3.18). In fact, Chapman and Junge demonstrate that no consistent estimate of N is possible if neither tagging nor subsequent sampling takes a constant fraction of the successive strata.

EXAMPLE 3.8. ESTIMATION OF A RUN OF SOCKEYE SALMON, USING STRATIFIED TAGGING AND RECOVERY. (From Ricker 1958a, after Schaefer 1951a.)

The Birkenhead run of sockeye salmon (*Oncorhynchus nerka*) was tagged near Harrison Mills, British Columbia, and recoveries of tags were made on the spawning grounds, about 200 miles upriver. The distribution of tagging and recoveries is shown in Table 3.6. Stock estimates from formula (3.18) are given in Table 3.7. In the latter, the last row shows the approximate abundance of fish going past the tagging point in successive weeks, while the last column shows the approximate number reaching the spawning stream in successive weeks.

TABLE 3.6. Recoveries from sockeye salmon tagged in successive weeks at Harrison Mills, divided according to week of recovery upstream; together with the total number tagged each week (M_i), and the number recovered and examined for tags (C_j). (Data from Schaefer 1951a, table 3.)

	Week of tagging (i)								Tagged fish re-covered	Total fish re-covered	C_j/R_j
	1	2	3	4	5	6	7	8			
Week of recovery (j):									R_j	C_j	
1	1	1	1	...	...	...	...	...	3	19	6.33
2	...	3	11	5	...	...	...	...	19	132	6.95
3	2	7	33	29	11	...	...	...	82	800	9.76
4	...	...	24	79	67	14	...	...	184	2,848	15.48
5	...	...	5	52	77	25	...	...	159	3,476	21.86
6	...	...	1	3	2	3	...	...	9	644	71.56
7	...	...	...	2	16	10	1	1	30	1,247	41.57
8	...	...	1	7	7	6	5	...	26	930	35.77
9	...	...	...	3	3	2	...	...	8	376	47.00
Tagged fish recovered (R_i)	3	11	76	180	183	60	6	1	520		
Total fish tagged (M_i)	15	59	410	695	773	335	59	5			
M_i/R_i	5.00	5.36	5.39	3.86	4.22	5.58	9.83	5.00			

TABLE 3.7. Computed estimates of sockeye salmon passing Harrison Mills, using Schaefer's method. (From Schaefer 1951a, table 4.)

	Week of tagging (i)								Total
	1	2	3	4	5	6	7	8	
Week of recovery (j):									
1	32	34	34	...	...	...	...	...	100
2	...	112	412	134	...	...	...	...	658
3	98	366	1,736	1,093	453	...	...	...	3,746
4	...	...	2,002	4,720	4,377	1,209	...	...	12,308
5	...	...	589	4,388	7,103	3,049	...	...	15,129
6	...	...	386	829	604	1,198	...	...	3,017
7	...	...	...	321	2,807	2,320	409	208	6,065
8	...	...	193	967	1,057	1,198	1,758	...	5,173
9	...	...	...	544	595	525	...	...	1,664
Total	130	512	5,352	12,996	16,996	9,499	2,167	208	47,860

Schaefer notes that since the values of M_i/R_i in the last row of Table 3.6 do not vary greatly, a simple Petersen estimate should be fairly close to the result in Table 3.7. The sum of the C_j column is $\Sigma C_j = 10,472$, and the sum of the M_i row is $\Sigma M_i = 2351$. The Petersen estimate is therefore (from expression 3.5):

$$N = 10,472 \times 2351/520 = 47,340$$

as compared with 47,860 from Table 3.7. Such close agreement would be infrequent.

3.13. CONTRIBUTIONS OF SEPARATE STOCKS TO A COMMON FISHERY

The marking technique can be used, if various conditions are satisfied, to estimate the contribution of each of a number of river races of salmon to a common oceanic fishery. Marking is done on young fish before they leave the river. Subsequently the ratio of marked to unmarked is observed (a) in the fishery concerned, and (b) in the various rivers where the migrants were tagged. Junge and Bayliff (1955) have outlined the conditions necessary for an unbiased estimate, and these are sufficiently formidable that the authors have no example of an experiment satisfactory from this point of view.

CHAPTER 4. — POPULATION STATISTICS FROM MARKING EXPERIMENTS EXTENDING THROUGH TWO OR MORE TIME INTERVALS, WITH CONSTANT SURVIVAL RATE

4.1. MARKING DONE PRIOR TO THE FIRST FISHING SEASON

4.1.1. TWO YEARS OF RECAPTURES. When marking experiments are done in two or more successive years, or when a single year's experiment is divided into two or more parts, it becomes possible to estimate rate of survival in the population, in addition to population size and rate of exploitation. This is easiest if the survival rate does not vary between the periods being examined (though it may exhibit parallel seasonal fluctuations *within* each period). This chapter describes procedures used when survival rate is constant, while the next deals with estimates when survival rate changes.

Suppose that fish are marked during a short period of time at the start of a year. During that year and later years they are susceptible to the same fishing and natural mortality rate as are unmarked fish, which rates do not vary appreciably over a period of years. We are given:

M   number of fish marked
R_1 recaptures in year of marking
R_2, R_3, etc. recaptures in later years

We want to know:

S   survival rate between years
N_1 population at the start of year 1
u   rate of exploitation

Table 4.1 shows the expected values of fishing mortality, natural mortality, and total mortality. The situation is similar to that for estimating survival from age composition, and the various formulae of Chapter 2 can be used, substituting the series of recaptures R_1, R_2, etc. for the series of ages N_1, N_2, etc. However, the potential sources of systematic error are rather different. Here there is no need to worry about between-years differences in recruitment, because the recaptures of successive years all stem from a single known marking. On the other hand, there can be problems of loss of marks, of incomplete reporting of marks, and of differences in behavior between marked and unmarked fish, as discussed in Sections 4.3 and 4.4

By analogy with expressions (2.1) and (2.2), we may state, for any two successive years of recaptures:

$$S = \frac{R_2}{R_1} \qquad (4.1)$$

$$Z = -(\log_e R_2 - \log_e R_1) \qquad (4.2)$$

TABLE 4.1. Expected mortality and survival in a stock of M fish marked at the beginning of year 1, in which rate of exploitation (u), expectation of natural death (v), and consequently total mortality rate (A) and survival rate (S) are all constant over a period of 5 yr. The successive entries in the "Recoveries" row are the expected values of the recaptures R_1, R_2, etc.

Year	1	2	3	4	5
Initial stock of marked fish	M	MS	MS^2	MS^3	MS^4
Recoveries	Mu	MuS	MuS^2	MuS^3	MuS^4
Natural deaths	Mv	MvS	MvS^2	MvS^3	MvS^4
Total mortality	MA	MAS	MAS^2	MAS^3	MAS^4

Here S is approximately the geometric mean of the survival rates of years 1 and 2, and Z is approximately the arithmetic mean of the two instantaneous mortality rates. When mortality rates are the same in the two years, then Z is an unbiased estimate of both.

4.1.2. RECAPTURES FROM A SERIES OF YEARS. When more than two years of recaptures are available, a first step is to plot a graph of logarithms of recaptures against time. If the points fall in a straight line, it suggests that survival rate has been uniform over the period in question. In that event S and Z can be estimated from the slope of the line. The points can be weighted as the number of recaptures each represents, but if some variation in F (hence also Z) between years is possible, it may be better to give all points the same weight.

Given such an estimate of S, the stock of marked fish at the beginning of successive years is estimated as:

$$M, \ MS, \ MS^2, \ \text{etc.} \tag{4.3}$$

These can be summed over the whole of the experiment and divided into the total recoveries, yielding a weighted estimate of mean rate of exploitation:

$$u = \frac{R_1 + R_2 + \ldots + R_n}{M(1 + S + S^2 + \ldots + S^{n-1})} \tag{4.4}$$

If data concerning recoveries in the year of marking are lacking or imperfect, it is still possible to estimate S and u by extrapolating back. Then (4.4) becomes:

$$u = \frac{R_2 + R_3 + \ldots + R_n}{SM(1 + S + S^2 + \ldots + S^{n-2})} \tag{4.5}$$

In using (4.5), it is important to make sure that whatever influences have made the first year's data unusable have not affected total mortality rate of marked fish in that

106

year. Such effects, however, would not affect the estimate of S, except to increase its sampling error by decreasing the number of recaptures on which it is based.

Because it is difficult to mark large numbers of fish in a short time, the device has sometimes been used of computing, for each recaptured fish, its exact "time out" in days, and dividing up recaptures by weeks, months, or years on that basis (e.g. Hickling 1938). This works especially well with fisheries that are prosecuted on a year-round basis, so that there is no serious seasonal variation in expectation of recovery of tags. It also works better if the spread of tagging dates is not *too* protracted. With a seasonal fishery, however, expectation of recapture of a tagged fish varies with the time of year it is released, and any broad mixture of tagging dates introduces an additional effect into the interpretation of recoveries.

EXAMPLE 4.1. SURVIVAL ESTIMATE WHEN FISH ARE MARKED PRIOR TO THE FISHING SEASON IN SUCCESSIVE YEARS. (From Ricker 1948.)

Data for a hypothetical marking experiment are as follows. Five thousand fish were marked just before the first fishing season, well distributed over the fishing area. Recoveries were: 1st year, 2583; 2nd year, 594; 3rd year, 175; 4th year, 40; 5th year, 7; 6th year, 0; these representing a complete canvas of the fishery.

The most obvious piece of information from these data is that the estimate of rate of exploitation, from data of the first year, is $u = 2583/5000 = 0.517$. To obtain survival rate, logarithms of recoveries are plotted and a predictive regression line is fitted (Fig. 4.1). The line has a slope of –0.608 log-units per year, corresponding to a survival rate of antilog $\bar{1}.392 = 24.7\%$ per year (expression (2.5) also gives 24.7%).

Having obtained S, a schedule can be constructed (Table 4.2), similar to Table 4.1, on the basis of 5000 fish marked. The mean rate of exploitation, from (4.4), is

TABLE 4.2. Mortality and survival in a population, based on the indicated recoveries from an initial stock of 5000 fish, and the assumption of a constant rate of exploitation and total mortality rate.

Year	1	2	3	4	5	Total
Initial stock of marked fish	5000	1235	305	76	18	6634
Recoveries	2583	594	175	40	7	3399
Natural deaths	1182	336	54	18	6	1596

$u = 3399/6634 = 0.512$. Since total mortality is $A = 1 - 0.247 = 0.753$, it follows that annual expectation of natural death is $v = 0.753 - 0.512 = 0.241$. From Appendix I, $Z = 1.398$ $F = uZ/A = 0.95$, and $M = Z - F = 0.45$.

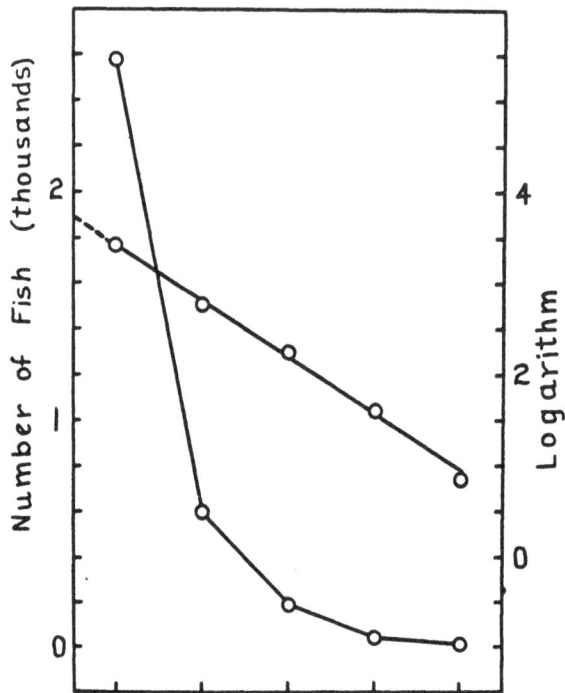

FIG. 4.1. Number of recoveries (curved line) and their logarithms (straight line) in successive years of the experiment of Example 4.1.

EXAMPLE 4.2. SURVIVAL OF NORTH SEA PLAICE ESTIMATED FROM TAGGING EXPERIMENTS. (From Ricker 1958a.)

Hickling (1938) reviewed the extensive English plaice-tagging experiment of 1929–32. Individual experiments were done over periods usually no longer than a month, and in any event returns are tabulated according to actual number of days elapsed from the day of tagging, grouped in sequences of 365 days. In the published data, recaptures are separated into two groups: those of the first year and those of all subsequent years.

Data for plaice marked off Heligoland in May, 1951, are shown in Table 4.3. The rate of first-year recovery increases with increase in fish size, from 4% to 45–50%. According to Hickling's figure 27, showing recaptures for all experiments, the first-year rate of recapture of plaice commonly reaches a plateau at 25–26 cm. Plaice of this size and larger can reasonably be considered fully recruited to the fishery, even though it may be possible that a part of the poorer returns from smaller fish is a result of greater tagging mortality or loss of tags among them. The figures 45–50% are, however, called here the "apparent" rate of exploitation of the fully-vulnerable

108

TABLE 4.3. Plaice marked off Heligoland in May, 1931, and recaptures made, arranged by 5-cm length-classes. (Data from Hickling 1938, table 16.)

1	2	3	4	5	6	7	8	9
Length-class	No. marked	1st-yr recaptures	Apparent rate of exploitation	Later recaptures	Apparent survival rate	Apparent survival rate (interpolated)	Apparent total mortality rate	Apparent expectation of natural death
cm	M	R_1	u'	$R_2+..$	$\bar{S}'$	S'	A'	v'
15–19	249	9	0.036	10		...	...	...
					0.527			
20–24	300	66	0.220	21		...	...	...
					0.241			
25–29	342	154	0.450	43		0.230	0.770	0.320
					0.218			
30–39	112	56	0.500	11		0.189	0.811	0.311
					0.164			

fish, because it would be desirable to examine possible systematic errors before accepting them wholeheartedly (see below).

On the assumption that the survival rate of *fully-vulnerable* tagged fish[1] is constant from one year to the next, its numerical value is estimated in column 6, using expression (2.4). Survival figures are estimated principally from the ratio of first-year to second-year recoveries: thus they pertain to a period of time when the fish are, on the average, at least half a year older than when the corresponding rate of exploitation was estimated (from recoveries in the 12 months immediately after marking). This is indicated in Table 4.3 by setting the primary survival estimates in the spaces between the exploitation estimates (column 6). The ratio of later recaptures to first-year recaptures of course decreases with size at marking; however, it should be completely stabilized for the *marking* size which is less than 25 cm by half a year's growth (about 2 cm), and it should be *nearly* stable for sizes 2 or 3 cm less. Consequently, the estimated apparent survival rate of 0.241 from the 20–24 cm tagging class is probably very little biased, while estimates from the two larger classes should not be biased at all (by incomplete vulnerability). The 15–19 cm group, however, yields an erroneous (too high) estimate of survival — although, since exploitation is less, we could reasonably expect the survival rate for the small fish to be appreciably greater than for larger fish. In the last two columns of Table 4.3 values of v' (apparent expectation of natural death) are obtained by subtraction ($= A' - u'$). The corresponding apparent instantaneous rates of natural mortality, M', are, from (1.4) and (1.13), 0.61 and 0.64.

---

[1] Hickling, following Thompson and Herrington (1930), estimated survival from the returns of all tagged fish regardless of size, and consequently obtained a composite figure which does not apply to any particular part of the stock, nor yet to the stock as a whole.

These estimates are considerably higher than natural mortality figures obtained by other methods, and suggest that in these experiments there may be systematic error of one or more of the types described in Sections 4.3 and 4.4 below. Some of the possibilities could be examined using month-by-month and year-by-year recoveries (cf. Example 4.3), but the plaice data have not been published in sufficient detail for this.

4.2. MARKING DONE THROUGHOUT THE FIRST FISHING SEASON, WITH RECOVERIES IN AT LEAST TWO SEASONS

With large-scale Type 2 fisheries in big bodies of water it is difficult or impossible to capture and mark a large number of fish in a short time, distributing them more or less evenly over the population under review, and the seasonal character of the fishing may make it undesirable to divide up recaptures according to the number of "days out." It is necessary therefore to see whether estimates of survival, etc., can be made when, for practical reasons, marking is carried on *during* instead of *before* a fishing season.

If the experiment is to estimate rate of fishing, it is important that marking be done in some rather definite manner, in relation to incidence of mortality in the population. Ideally, the natural and fishing mortality rates in the population would both be distributed evenly over the whole year, in which case it is best that marking be done at a uniform absolute rate throughout the year. Such a group of marked fish would be analogous to a year-class of recruits entering a fishery at a uniform absolute rate over a year's time. If the fishery is more seasonal it will be best, and often easiest, to mark fish at a rate more or less proportional to the industry's weekly landings, which would correspond exactly to the situation above if natural mortality were negligible, or were similarly distributed. Generally little or nothing will be known of the seasonal distribution of natural mortality, so that our ideal situation will often be as good an assumption as any. However, if the fishery is sharply limited seasonally, corrections could be introduced on the basis of natural mortality occurring throughout the year or throughout the growing season.

4.2.1. SEVERAL YEARS' RECAPTURES. The discussion here will concern only the simpler situation postulated above. In Section 1.5.6 it was shown that if Z and A represent, respectively, instantaneous and yearly rates of total mortality in a Type 2 fishery, the mortality among fish recruited or marked at a uniform absolute rate would be $(Z - A)/Z$. Of these the fraction F/Z or u/A would be killed by capture, and M/Z or v/A would die from natural causes. From this a schedule can be constructed (Table 4.4) showing catch and total mortality in all years.

An interesting and somewhat unexpected feature of this tabulation concerns the recoveries. In order to plot a catch curve involving the year of marking, to show survival rate directly, it will be necessary to adjust the number of first-year recoveries. A first impulse would be to double their number, since if the fish are marked at a uniform rate through a season, it might seem that on the average they would be

TABLE 4.4. Mortality and survival in a stock of fish marked throughout year 1. In all years, the annual mortality rate $= A$; survival rate $= S$. In year 1, the actual mortality is $(Z - A)/Z$, and survival is A/Z (see the text). All entries are fractions of the number of fish marked during year 1.

Year	1	2	3	4	5
Marked fish at large at start of year	...	$\dfrac{A}{Z}$	$\dfrac{SA}{Z}$	$\dfrac{S^2A}{Z}$	$\dfrac{S^3A}{Z}$
Recoveries	$\dfrac{u(Z-A)}{AZ}$	$\dfrac{uA}{Z}$	$\dfrac{uSA}{Z}$	$\dfrac{uS^2A}{Z}$	$\dfrac{uS^3A}{Z}$
	$\dfrac{F(Z-A)}{Z^2}$	$\dfrac{FA^2}{Z^2}$	$\dfrac{FA^2S}{Z^2}$	$\dfrac{FA^2S^2}{Z^2}$	$\dfrac{FA^2S^3}{Z^2}$
Natural deaths	$\dfrac{v(Z-A)}{AZ}$	$\dfrac{vA}{Z}$	$\dfrac{vSA}{Z}$	$\dfrac{vS^2A}{Z}$	$\dfrac{vS^3A}{Z}$
Total deaths	$\dfrac{Z-A}{Z}$	$\dfrac{A^2}{Z}$	$\dfrac{SA^2}{Z}$	$\dfrac{S^2A^2}{Z}$	$\dfrac{S^3A^2}{Z}$

subject to only half the mortality of those present from the start of the season. However, for the number of recoveries in the first year (R_1) to be a member of the geometric series of later years $(R_2, R_3,$ etc.), it should equal $M'uA/SZ$ (M' is the number marked). As R_1 actually equals $M'u(Z-A)/AZ$, the factor by which R_1 must be multiplied to get a value that fits into the series is:

$$\frac{AZ}{M'u(Z-A)} \times \frac{M'uA}{SZ} = \frac{A^2}{S(Z-A)} \tag{4.6}$$

Accordingly, before plotting the catch curve, recoveries of the first year must be multiplied by $A^2/S(Z-A)$.

We are then comparing the two quantities:

$$\frac{R_1A^2}{S(Z-A)} \tag{4.7}$$

$$\frac{R_2 + R_3 + \ldots}{S + S^2 + \ldots} \tag{4.8}$$

Both of these represent the number of first-year recoveries that would be expected *if* tagging had been done at the start of the first year instead of throughout it. Expression

111

(4.7) is computed from the actual number of first-year recoveries, whereas (4.8) is computed from recoveries of later years — using a uniform instantaneous mortality rate in both cases.

If this comparison shows that first-year recoveries do not agree with those for later years, it suggests the presence of error of Type C, described in Section 4.4 below. If, however, the adjusted point for year 1 lies close to the line established by later years, we are faced with the problem of getting the best combined estimate of rate of exploitation. The simplest procedure is to combine two separate estimates. From Table 4.4, we have for year 1:

$$u = \frac{R_1 AZ}{M'(Z - A)} \tag{4.9}$$

And thus:

$$F = \frac{R_1 Z^2}{M'(Z - A)} \tag{4.10}$$

For recaptures of later years, we can use a modification of formula (4.5), which can readily be deduced by comparing Table 4.4 with Table 4.1:

$$u = \frac{Z(R_2 + R_3 + \ldots + R_n)}{AM'(1 + S + S^2 + \ldots + S^{n-2})} \tag{4.11}$$

$$F = \frac{uZ}{A} = \frac{Z^2(R_2 + R_3 + \ldots + R_n)}{A^2M'(1 + S + S^2 + \ldots + S^{n-2})} \tag{4.12}$$

The two estimates of F, (4.10) and (4.12), can be averaged arithmetically, weighting each as the total number of recaptures involved: viz., R_1 and $(R_2 + R_3 + \ldots + R_n)$, respectively.

4.2.2. TWO YEARS' RECAPTURES. When marking is done throughout a year, and recaptures for only 2 years are obtained (the year of marking and the following one), computation of a survival rate becomes hazardous, because there is no check on its constancy. However, if the latter be assumed, letting M' be the number of fish marked in year 1, and R_1 and R_2 be the number of these recaptured during year 1 and year 2, respectively, the data available are, from Table 4.4:

$$\frac{R_1}{M'} = \frac{u(Z - A)}{AZ} = \frac{F(Z - A)}{Z^2} \tag{4.13}$$

$$\frac{R_2}{M'} = \frac{uA}{Z} = \frac{FA^2}{Z^2} \tag{4.14}$$

112

Dividing (4.13) into (4.14):

$$\frac{R_2}{R_1} = \frac{A^2}{Z - A} \qquad (4.15)$$

The right-hand member of this equation is a simple function of S or Z, which can be taken directly from Appendix I.

Because of the uncertainty of this method, it is desirable to do a marking experiment in 2 successive years, whenever only 1 year following the year of marking can be expected to yield a substantial number of recoveries.

4.3. SYSTEMATIC ERRORS: TYPES A AND B

The general discussion of various kinds of systematic errors in Chapter 3 is applicable also to experiments in which survival rate is being estimated. Some types of error are of special interest and importance when recoveries extend over a long period. They can be classified according to their effects on the various statistics being estimated:

4.3.1. TYPE A ERRORS. There are sources of error which affect the estimate of rate of fishing, but not the estimate of total mortality and survival. In this category can be placed (1) the death of any considerable number of fish, or the loss of their tags shortly after marking or tagging; and (2) incomplete reporting of marks or tags taken by fishermen (assuming the reporting to be equally efficient or inefficient during all the years of the experiment). Errors of this sort scarcely require further comment. If fish die just after tagging, the apparent rate of exploitation obtained will be less than the true rate; the true rate is equal to the apparent rate divided by the ratio of the number of fish which survive the effects of tagging to the total number put out. Or if reporting is incomplete, the true rate of exploitation will be equal to the apparent rate divided by the fraction reported. That the estimates of total mortality and survival will remain unaffected by either of these is obvious from the fact that in formulae (4.1) and (4.2) the number of fish marked does not appear. Special efforts must be made to discover possible errors of these two kinds, since the data of the experiment give no clue to them. For example, to check on marking mortality or immediate loss of tags, fish of different degrees of apparent vigor, or fish tagged in different ways, can be used; or the fish can be held under observation. To check on efficiency of reporting of tags by fishermen, or of their recovery by mechanical devices, part of the catch can be examined by special observers; this is always a desirable procedure anyway. An elaborate series of corrections of this kind has been made for sardine tagging experiments (see Clark and Janssen 1945a, b; Janssen and Aplin 1945).

4.3.2. TYPE B ERRORS. A second group of errors includes those which affect the estimate of total mortality, but not the estimate of rate of fishing. Here belong

(1) any loss of tags from fish which occurs at a steady instantaneous rate throughout the whole period of the experiment; (2) extra mortality among tagged or marked fish, similarly distributed in time; and (3) emigration of fish from the fishing area, similarly distributed in time. The effects of any of these three are in most ways comparable to ordinary natural mortality. Suppose the loss takes place at instantaneous rate U, making the total instantaneous rate of disappearance $F + M + U = Z'$, compared with the true mortality rate $F + M = Z$. The annual disappearance rate corresponding to Z' will be A', a larger quantity than the mortality rate A. The apparent rate of exploitation (i.e. rate of recovery of tagged fish with tags still attached) is, say, $u' = FA'/Z'$. Obviously the rate of fishing F is equal to $u'Z'/A'$, just as much as uZ/A; and since u', Z', and A' are all available from the data of the experiment, an unbiased estimate of F can be had.

Often an independent estimate of Z and A for the population will be available, made from an analysis of ages of fish in the catch. Given a satisfactory estimate of Z from this source, and of F from a marking experiment, a complete and unbiased description of mortality in the population becomes possible.

Another variant of Type B error occurs when tags or marks are continuously lost or disappear but the rate of loss is variable. For example, the rate of loss of tags might accelerate with time as tags worked loose from fish, so that very few fish with tags would remain after 2 or 3 years, even though many actually lived (without their tags) much longer. Such a situation would be reflected in a nonlinear recapture curve; that is, the graph of the logarithm of recaptures against time would be convex upward. Alternatively, the more loosely applied tags might come off rapidly at first, so that there would be a deceleration of the rate of loss of tags in general, resulting in an upwardly concave recapture curve. If in such cases the rate of acceleration or deceleration is constant, the differences between successive logarithms of recaptures should be in a linear sequence when plotted against time, and this second derivative line could be used as a basis for an unbiased estimate of rate of fishing. The latter could be computed along lines analogous to those just described, or, more simply but less accurately, by using one of the graphical methods described in Section 4.5. Similarly, any empirical relationship derived from the observed trend of the recaptures might be used, though perhaps with less assurance than when the formula describes an easily-grasped theoretical position. For example, Graham (1938a) fitted a straight line to the logarithms of the logarithms of number of recaptures, and extrapolated back along it.

A comparison of the apparent total mortality rate obtained from a fin-clipping or tagging experiment with the value obtained from a catch curve is probably the best method of discovering any variety of Type B error. If this is impossible, it will be useful to compare survival rates estimated from different types of marks or tags, to see if any differences appear. It may also be helpful to examine a large number of fish to see if holes left by lost tags can be found, though since such scars often heal up quickly, no quantitative estimate of the loss will usually be obtained in this way.

EXAMPLE 4.3. FISH MARKED PRIOR TO THE FISHING SEASON, WITH INCOMPLETE TAG RECOVERY. (From Ricker 1948.)

The tag recovery data of Example 4.1 can also be used to illustrate a situation where the search for tags among the fish caught is incomplete. Suppose, for example, that the tags in question are internal iron tags, and are recovered with something less than complete efficiency by magnets installed in processing plants. Trial runs with these magnets showed their efficiency to have been, in successive years, 0.88, 0.70, 0.92, 0.90, and 0.82. A similar situation would arise if the recoveries were made not from the commercial catch, but from experimental catches in which the fish were not killed; the series of figures just given would then represent the relative sizes of these catches in successive years. In either event a correction must be made for variations in the size of the catch effectively examined; that is, each year's recoveries must be reduced to the basis of 100% efficiency, or to the basis of some standard size of catch (cf. Jackson 1939).

In the present example, the adjusted number of recoveries for the first year is $2583/0.88 = 2930$; later years yield, in the same manner, 848, 190, 44, and 8, respectively. These adjusted figures can now be applied in expression (2.5), giving:

$$S = \frac{1332}{4020 + 1332 - 1} = 0.249$$

A somewhat better estimate would be obtained by fitting a straight line to the logarithms of the adjusted values, weighting each point as the *unadjusted* number of recaptures on which it is based. The rate of exploitation is found as in Example 4.1, but using the adjusted figures.

EXAMPLE 4.4. TYPE B ERROR IN HALIBUT TAGGING EXPERIMENTS. (Modified from Ricker 1948.)

Thompson and Herrington's (1930) tagging experiments are used in Examples 5.4 and 5.5 of the next chapter to estimate total rate of disappearance of tags from, and rate of fishing for, fully-vulnerable halibut on grounds south of Cape Spencer. In Example 2.8 survival rate was estimated from the catch curve of the halibut taken for tagging. These estimates are compared below:

			Instantaneous rate of		
Year	Method	Apparent survival rate	Total loss	Fishing	Natural mortality and other losses
		S'	Z'	F	M + U
1926	Tagging	0.331	1.11	0.57	0.54
1927	Tagging	0.320	1.14	0.51	0.63
		Survival rate, S	Mortality rate, Z		
	Catch curve	0.47	0.76		

115

The survival figure 47% was obtained from size distribution of fish used for tagging in 1925 and 1926, and almost the same value can be computed from relative numbers of halibut used by Dunlop for age determination (Thompson and Bell 1934, p. 25). As shown in Example 2.8, total fishing effort was remarkably steady during the period 1921–27, so the 47% survival obtained from age distribution should be entirely comparable to the 32% or 33% obtained from tag recoveries.

Even without this cross-check, the values for $M + U$ shown above, which are impossibly high for the true natural mortality rate of a long-lived fish, would indicate that something besides natural mortality contributes to the disappearance of the tagged fish. The only obvious possibility is that there has been Type B error throughout the time of the experiment, resulting from a continuous loss of tags from fish or from a movement of tagged fish out of the fishery. On the assumption that the 0.76 mortality rate from age-frequencies is the true one, the necessary instantaneous rates of loss can be computed by difference: they are $1.11 - 0.76 = 0.35$ for 1926, and $1.14 - 0.76 = 0.38$ for 1927. Shedding of tags is believed to be infrequent in these experiments, so that movement of tagged fish away from the fishing grounds probably accounts for most of this. The latter is a likely-enough possibility, because halibut move a lot and are found, sparsely, over a much greater area of the sea bottom than the grounds customarily fished. In nature, wandering away from the fishing grounds is presumably balanced by return movements back onto the grounds, but in the first year or two after tagging, the outward movement of tagged fish would exceed their return, and it is these years which mainly determine the survival rates of Examples 5.4 and 5.5.

4.4. Systematic Errors: Type C

A third group of errors includes those which make the first year's recoveries not directly usable in estimating either total mortality or rate of fishing, but which do not prejudice the estimation of either of these from the data of later years. Here may be mentioned (1) abnormal behavior of marked or tagged fish during the season of their marking, and (2) non-random distribution of marked fish in the general population during the year of marking, combined with (possibly only temporary) non-random distribution of fishing effort. In either event the marked fish may be either more or less vulnerable to capture during the year of marking than during later years; but they are assumed to have regained their usual behavior by the beginning of the year following marking, and in the latter year either fishing effort or the fish marked must be randomly distributed.

Type C errors can be serious when few recoveries are made beyond the year of marking. If, however, fair numbers of marked fish are obtained for at least two years after the year of marking, Type C error is merely troublesome: that is, it complicates estimation of rate of fishing but does not distort the result.

4.4.1. Fish marked just before the start of a fishing season. A model is shown in Table 4.5. The first year's rate of fishing for marked fish (F_1) is either greater or less than that for later years (F_2). Consequently, rate of exploitation (u_1), expectation

TABLE 4.5. Mortality and survival in a population of marked fish where rate of exploitation (u_1), natural deaths (v_1), and total mortality (A_1) are different in the first year from later years (u_2, v_2, A_2), but are identical among those later years. A unit number of fish was marked just prior to year 1.

Year	1	2	3	4
Initial population	1	S_1	S_1S_2	$S_1S_2^2$
Recoveries	u_1	u_2S_1	$u_2S_1S_2$	$u_2S_1S_2^2$
	$\dfrac{F_1A_1}{Z_1}$	$\dfrac{F_2A_2S_1}{Z_2}$	$\dfrac{F_2A_2S_1S_2}{Z_2}$	$\dfrac{F_2A_2S_1S_2^2}{Z_2}$
Natural deaths	v_1	v_2S_1	$v_2S_1S_2$	$v_2S_1S_2^2$
Total deaths	A_1	A_2S_1	$A_2S_1S_2$	$A_2S_1S_2^2$

of natural death (v_1), and total mortality rate (A_1) for the first year all differ from corresponding statistics for later years. The estimate of mortality rate (A_2) made from recaptures of marked fish after the first year should reflect the real mortality rate of the population. To calculate rate of fishing we proceed from the assumption that instantaneous rate of natural mortality (M) is the same in the first year as in later years. However, no direct equation can be set up because of the exponential relation between A and Z, and it is necessary to proceed by successive trials.

In Table 4.5, the fraction of recoveries in year 2, which is represented by u_2S_1, is available in the data of the experiment as R_2/M' where M' is the number marked; thus:

$$u_2 = \frac{R_2}{S_1M'}$$

If all the later years' data are used, this expression becomes, by analogy with (4.5):

$$u_2 = \frac{R_2 + R_3 + R_4 + \ldots + R_n}{S_1M'(1 + S_2 + S_2^2 + \ldots + S_2^{n-2})} \tag{4.16}$$

In these expressions everything is known except S_1.

Another estimate can be made somewhat less directly. The necessary data are available to evaluate:

$$u_1 = \frac{R_1}{M'}$$

From (1.13):

$$F_1 = \frac{u_1Z_1}{A_1} = \frac{R_1Z_1}{M'A_1}$$

117

Now the natural mortality rate M is equal to $Z_1 - F_1$, and M being the same in the second as in the first year, we have also $F_2 = Z_2 - M$. From (1.13), and substituting:

$$u_2 = \frac{F_2 A_2}{Z_2} = \frac{A_2}{Z_2}\left(Z_2 - M\right)$$

$$= \frac{A_2}{Z_2}\left(Z_2 - Z_1 + F_1\right)$$

$$= \frac{A_2}{Z_2}\left(Z_2 - Z_1 + \frac{R_1 Z_1}{M' A_1}\right) \tag{4.17}$$

In this expression A_2 and Z_2, R_1 and M', are all available directly from the experiment, while Z_1 and A_1 are directly related to the unknown S_1 of equation (4.16). Thus for any trial value of Z_1 (or A_1 or S_1), u_2 can be calculated from both (4.16) and (4.17), and successive trials will yield a best value which makes the two estimates equal.

4.4.2. FISH MARKED THROUGHOUT THE FISHING SEASON. When marking is done *throughout* the first year, as in Table 4.4, some modification of the above procedure will be necessary. Here F_1 will be the first year's instantaneous rate of fishing mortality, but it will be directly applicable only to fish marked at the very beginning of the season. The total instantaneous mortality rate, applicable to such fish, is $F_1 + M = Z_1$. From Table 4.4, the total first-year mortality among the marked group as a whole is $(Z_1 - A_1)/Z_1$, and the survivors are A_1/Z_1. A possible estimate of u_2 is therefore:

$$u_2 = \frac{R_2 Z_1}{M' A_1}$$

or, if all recoveries beyond year 1 be used, then by analogy with (4.11):

$$u_2 = \frac{Z_1(R_2 + R_3 + R_4 + \ldots + R_n)}{A_1 M'(1 + S_2 + S_2^2 + \ldots + S_2^{n-2})} \tag{4.18}$$

The first-year recoveries, as a fraction of the total fish marked, will be:

$$\frac{R_1}{M'} = \frac{F_1}{Z_1}\left(\frac{Z_1 - A_1}{Z_1}\right)$$

Evaluating F_1 from this, and proceeding as in the development of (4.17):

$$u_2 = \frac{A_2}{Z_2}\left(Z_2 - Z_1 + \frac{R_1 Z_1^2}{M'(Z_1 - A_1)}\right) \tag{4.19}$$

118

The rate of exploitation u_2 can now be evaluated as before. (It saves time to know that if (4.18) turns out greater than (4.19), the trial value of Z_1 is too *great*.)

4.4.3. DETECTION OF TYPE C ERRORS. Type C errors are easy to detect. If marking is done just before the fishing season, it will show up once on a graph of logarithms of recoveries in successive years, as a displacement of the point for the first year above or below the straight line drawn through the points for later years. If marking is done during the fishing season, recaptures of year 1 should first be multiplied by $A_2^2/S_2(Z_2 - A_2)$ before taking the logarithm and plotting (Section 4.2.1).

If Type B error (continuous loss of tags, etc.) is present as well as Type C, then it is the *apparent* survival and mortality rates which should be used throughout the calculations above, rather than the true rates, to obtain an unbiased estimate of F_2.

EXAMPLE 4.5. FISH MARKED THROUGHOUT THE FISHING SEASON, WITH UNREPRE-SENTATIVE FIRST-SEASON MORTALITY. (From Ricker 1948.)

In an hypothetical population, 1500 fish were marked throughout a fishing season. Recoveries were: same year, 450; 2nd year, 312; 3rd year, 125; 4th year, 50; 5th year, 20. Presumably if later years' data were available they would produce additional recoveries; therefore expression (2.5) cannot be used to estimate survival rate, but (2.7) or a regression line may be used instead. The survival rate after the first year is $S_2 = 195/487 = 0.400$, $A_2 = 0.600$, and $Z_2 = 0.916$. Fish were tagged from and returned to schools which were being actively fished, so there is reason to suspect the first year's mortality may be too great to be representative of the population as a whole. To test this, we evaluate $450A_2^2/S_2(Z_2 - A_2) = 1282$, and finding it greater than $312/S_2 = 780$, conclude that our suspicions are justified (cf. Fig. 4.2). Consequently, it is necessary to depend on recaptures in the second and later years to obtain an estimate of rate of fishing.

Using equations (4.18) and (4.19), we select trial values of Z_1, and obtain the following:

Trial value of Z_1................	1.00	1.20	1.22	1.24
u_2, from (4.18)......................	0.329	0.357	0.360	0.363
u_2, from (4.19)......................	0.479	0.379	0.369	0.358

Graphical interpolation between the last two gives 1.233 as the best estimate of Z_1, and 0.362 as the best estimate of u_2. From the latter, the rate of fishing is $F_2 = 0.362 \times 0.916/0.600 = 0.553$; and $M_2 = 0.916 - 0.553 = 0.363$.

It is possible to check the value of u_2 obtained above by an approximate calculation. If there were no natural mortality at all in the first year, the survivors at the start of the second year would number 1500 - 450 = 1050; hence a minimal estimate of u_2 is $312/1050 = 0.297$, and a maximal value of v^2 is $0.600 - 0.297 = 0.303$. But the fraction of fish marked in the first year which die naturally in the same year should be about half this, since they enter the population in uniform numbers throughout the season. Thus a fairly good estimate of the first year's expectation of natural death among all fish marked will be $0.303/2 = 0.152$, and the actual deaths will be $0.152 \times 1500 = 228$. A better estimate of u_2 will therefore be $312/(1500-450-228) = 0.38$, which is close to the 0.362 obtained in the last paragraph. This approximate computation should be fairly good as long as v_2 is, say, less than 0.4.

119

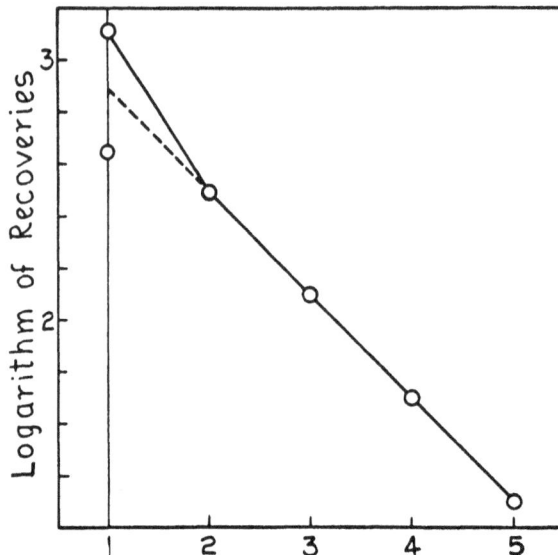

FIG. 4.2. Logarithms of recoveries of marked fish in successive years of the experiment of Example 4.5. The lower point
for year 1 represents the actual logarithm of the number of
recoveries; the higher point is the logarithm of $A^2/S(Z-A)$
times the recoveries. The latter should lie on the line established by recoveries in later years, if there were no type C
error.

If desired, a schedule similar to Table 4.5 can now be constructed, showing
population and natural mortality among the 1500 fish tagged in successive years:

Year	1	2	3	4	5
Initial population	...	862	345	138	55
Catch	450	312	125	50	20
Natural mortality	188	205	82	33	13
Total mortality	638	517	207	83	33

Another cause of increased (or decreased) returns in the first year would be a
failure to get the fish marked at a rate proportional to that at which the fishery is
making landings. For example, if relatively more marking were done near the beginning of the fishing season, then the rate of exploitation of the marked fish during
the year of marking would be greater than $F(Z-A)/Z^2$. In its effect upon recoveries,
this state of affairs would resemble the situation just considered, and could be treated
in the same fashion, except that the first year's natural mortality could be increased
somewhat to compensate for the longer average time the marked fish are at large.

120

4.5. GRAPHICAL METHODS OF ESTIMATING EFFECTIVENESS OF FISHING

Various graphical methods of estimating the effectiveness of fishing have been proposed, for example by Thompson and Herrington (1930), Graham (1938a), and Clark and Janssen (1945a), of which a comparison and critique were given by Ricker (1948, Section 26). Here we illustrate the principles and limitations of graphical extrapolation for two typical situations.

4.5.1. MARKING DONE AT THE START OF A YEAR. Figure 4.1 illustrates a possible procedure. The straight line fitted to logarithms of successive catches can be produced back to the beginning of year 1, the time of marking. It could be argued that this intercept represents the logarithm of the number of fish which would be retaken if recaptures were to be continuously made at the rate established immediately after tagging (before natural mortality had a chance to reduce their number), and that hence the antilogarithm of the intercept, divided by N, should be the rate of fishing, F.

Closer consideration shows that such an estimate of F will be approximate only. Turning to Table 4.1, if the "recoveries" of each year be considered as pertaining to the middle point of the year,[2] we have the series:

$$M'u, M'uS, M'uS^2, M'uS^3, \text{etc.}$$

separated by unit time intervals. These constitute a geometric series with common ratio S. The point we are interested in (the beginning of the first year) lies one-half of a unit time interval to the left of $M'u$, and must therefore be $M'uS^{-0.5}$ or $M'u/\sqrt{S}$. Dividing by the number of fish marked, M', gives the expression:

$$\frac{u}{\sqrt{S}} \tag{4.20}$$

which we originally proposed to identify as rate of fishing, F. Comparing with the true formula $F = uZ/A$, we find that (4.20) differs in so far as $1/\sqrt{S}$ differs from Z/A. From Appendix I it is evident that as $Z\rightarrow0$, these two expressions become the same; for larger values of Z we have:

Z	Z/A	$1/\sqrt{S}$
0.5	1.27	1.28
1.0	1.58	1.65
1.5	1.93	2.12
2.0	2.31	2.72

[2] It is obvious that it is this assumption which is incorrect, in both this argument and that of method 2 below. The mean date of recapture of tags is in advance of the middle of the fishing of each year, and when mortality rate is moderate-to-large the difference is important.

121

Thus over a considerable part of the range of Z values likely to be encountered, (4.20) could be used for F without serious error; but when Z exceeds 1.0 the error becomes considerable (i.e. $u/\sqrt{S}$ is greater than F). In that event it will be worthwhile to calculate F from the u obtained by expression (4.5) or, what amounts to the same thing, to interpret the intercept obtained by graphical extrapolation in terms of (4.20).

4.5.2. MARKING DONE THROUGHOUT A YEAR. When marking is done throughout the first year a similar possibility of extrapolating exists, which can be illustrated from Fig. 4.2. We might argue that the marking, which actually was spread evenly through the first year, could be considered as having been concentrated at its middle. Similarly the recaptures, which are spread through the succeeding years, could be considered as concentrated at the middle of each. Then, in Fig. 4.2, the intercept of the straight line (logarithms of recoveries) with the ordinate for the middle of year 1 should represent the logarithm of the number of fish which would be recaptured in year 1 if the fishery were compressed into a short space of time immediately following the marking at the middle of the year, without allowing time for natural mortality to take effect. Such an intercept, divided by the number of fish marked, would seem to be an estimate of the rate of fishing, F.

Considering the "Recoveries" row of Table 4.4, from year 2 onward, it is evident that it constitutes a geometric series with common ratio S, and that the point at year 1 which fits into the series will be uA/ZS. Substituting $R_2/M' = uA/Z$, the true year 1 intercept becomes $R_2/M'S$; or, if all data for later years be included, this intercept is:

$$\frac{R_2 + R_3 + \ldots + R_n}{SM(1 + S + S^2 + \ldots + S^{n-2})} \tag{4.21}$$

Comparing this with the true F which can be estimated from recoveries after the first year, shown in (4.12), it is evident that they differ in that (4.12) has Z^2/A^2 where (4.21) has $1/S$. As may be seen by comparing A^2/Z^2 and S in Appendix I, the two latter expressions do not differ greatly over a part of the range of Z values likely to be encountered in work of this sort; but when Z becomes larger than, say, 0.8, the error is considerable, making (4.21) greater than F.

CHAPTER 5. — POPULATION STATISTICS FROM MARKING EXPERIMENTS EXTENDING THROUGH TWO OR MORE TIME INTERVALS, WITH VARIABLE SURVIVAL RATE

5.1. POPULATION AND SURVIVAL RATE WHEN MARKING IS DONE AT THE START OF FISHING IN TWO CONSECUTIVE YEARS — RICKER'S METHOD

5.1.1. NATURAL MORTALITY THE SAME AT ALL AGES. The most direct approach to an estimate of survival by marking is to run similar marking experiments in two successive years (or other interval), using different marks for each. When marking is all done right at the start of the fishing season, we have the following:

M_1 number of fish marked at the start of the first year
M_2 number of fish marked at the start of the second year
R_{11} recaptures of first-year marks in the first year
R_{12} recaptures of first-year marks in the second year
R_{22} recaptures of second-year marks in the second year

We wish to know:

S_1 survival rate during year 1 (from time of marking in year 1 to time of marking in year 2)

We may reason as follows: the number of fish, M_2, marked at the start of the second year, yields R_{22} recaptures that year; thus the rate of exploitation in year 2 is $u_2 = R_{22}/M_2$. Of the M_1 fish marked in year 1, R_{12} are caught in year 2. The number of first-year marked fish still at large at the start of year 2 should be R_{12}/u_2, or $R_{12}M_2/R_{22}$. The latter number must be compared with the number of marked fish at large at the start of year 1, M_1, to obtain the survival rate over that period:

$$S_1 = \frac{R_{12}M_2}{M_1R_{22}} \tag{5.1}$$

This is the large-sample formula of Ricker (1945a, 1948). Notice that it is not essential that all the marked fish recaptured be reported. It is only necessary that, during year 2, the marks put on at the start of year 1 be reported as completely as those put on at the start of year 2.

The large-sample variance of (5.1) can be obtained by substituting the estimates in Seber's (1972) more general expression (see Section 5.1.3), as follows:

$$V(S_1) = S_1^2\left(\frac{1}{R_{12}} + \frac{1}{R_{22}} - \frac{1}{M_1} - \frac{1}{M_2}\right) \tag{5.2}$$

Expression (5.1) can be modified for small numbers of recaptures by analogy with expressions (3.7) and (5.12):

$$S_1 = \frac{R_{12}M_2}{M_1(R_{22} + 1)} \tag{5.3}$$

Expression (5.2) can be used for the variance of this estimate also.

The estimate of S_1 from (5.1), rather than (5.3), has the advantage that it can be transformed directly to A_1 and Z_1; hence F_1 and M can be computed using (5.28) provided the recaptured marks are reported completely. The value of F_2 is also available, if acceptable data of the type shown by equation (5.27) are at hand.

5.1.2. NATURAL MORTALITY VARIES WITH AGE. The method above can also be made to take care of any changes in natural mortality rate associated with age which may occur among the fish. If such are important, the fish marked in the second year, M_2, should have a minimum size greater than that of the fish marked in the first year, M_1, by one year's growth (Ricker 1945a). Still better, the computation can be made to apply to one or more definite year-classes or length-groups of fish in two successive years of their existence by using different marks, or merely by advancing the boundary between the groups as the fish increase in size, as in Example 3.2 above.

EXAMPLE 5.1. SURVIVAL RATE OF BLUEGILLS IN MUSKELLUNGE LAKE, FROM MARKINGS DONE AT THE START OF TWO CONSECUTIVE YEARS. (From Ricker 1948.)

The procedure of Section 5.1 was the principal one used during the 1940's to estimate vital statistics of fish populations in small Indiana lakes. An example concerning bluegills (*Lepomis macrochirus*) of Muskellunge Lake will be described; the data are from Ricker (1945a, p. 383–384).

Of $M_1 = 230$ bluegills marked before the start of the 1942 fishing season, $R_{12} = 13$ were captured in 1943. Of $M_2 = 93$ marked before the start of the 1943 fishing season, $R_{22} = 13$ were recaptured in 1943. The survival rate in the first year is therefore, from (5.3):

$$S_1 = \frac{13 \times 93}{230 \times 14} = 0.37546$$

From (5.2), the variance of S_1 is:

$$V(S_1) = 0.37546^2\left(\frac{1}{13} + \frac{1}{13} - \frac{1}{230} - \frac{1}{93}\right)$$

$$= 0.14097 \times 0.13874 = 0.01956$$

The standard error is the square root of this, or 0.1398.

The attractive simplicity of this procedure is unfortunately often marred by doubts occasioned by a possible lack of homogeneity among the fish being handled, or by within-season variations in mortality rate which are not the same for all age-groups. A discussion of some of these considerations can be found in the paper cited above, particularly the section on pumpkinseeds (*Lepomis gibbosus*) (p. 385–386).

5.1.3. ESTIMATION OF SURVIVAL RATE FROM A SERIES OF MARKINGS — ROBSON–SEBER METHOD. If markings are continued for more than two years, the best procedure will usually be to make a series of estimates of survival rate from expression (5.1) or (5.3). This is because survival rate may change as fish get older, and hence the inclusion of recaptures of third and later years may distort the estimate obtained; also, if tags are used, there is danger they may become increasingly lost in later years. However, if survival rates are assumed not to vary with age, and tags are permanent, it is possible to use all available recaptures to estimate each year's survival rate. The necessary computations were developed by Robson (1963), and in a more general form by Seber (1972).

Suppose M_1, M_2, M_3, etc. fish are marked at the start of three successive years, and recoveries are made over four years. The schedule below gives the categories of marks and recoveries, either or both of which could continue for an indefinite number of years:

Year	No. of fish marked	Recaptures made and reported in year				Totals
		1	2	3	4	
1	M_1	R_{11}	R_{12}	R_{13}	R_{14}	R_1
2	M_2	....	R_{22}	R_{23}	R_{24}	R_2
3	M_3	....	....	R_{33}	R_{34}	R_3
Totals	....	m_1	m_2	m_3	m_4	....

A second schedule is made by adding the above primary recapture entries cumulatively from the right. For example, $b_{14} = R_{14}$; $b_{13} = R_{14} + R_{13}$; $b_{12} = R_{14} + R_{13} + R_{12}$; and so on:

Year of marking	Cumulative recaptures			
	1	2	3	4
1	b_{11}	b_{12}	b_{13}	b_{14}
2		b_{22}	b_{23}	b_{24}
3			b_{33}	b_{34}
Totals	T_1	T_2	T_3	T_4

Totals of the columns of the second schedule, and totals of both rows and columns of the first, are used to obtain the estimates below. For any year i of the experiment:

Survival rate in year i:

$$S_i = \frac{R_i M_{i+1}(T_{i+1} - R_{i+1})}{R_{i+1} M_i T_i} \tag{5.3a}$$

125

Rate of exploitation in year i:

$$u_i = \frac{m_i R_i}{T_i M_i} \tag{5.3b}$$

From the above the values of Z, F, and M can be calculated in the usual manner (Chapter 1). Seber (1972, p. 315) gives an expression for the large-sample variance of an estimate of S_i (our symbols S_i and M_i are his ϕ_i and a_i, respectively).

For (5.3b) to be valid it is necessary that all marked fish be reported when caught; otherwise u and F will be too small, and M too great. However, the estimate of survival rate of expression (5.3a), like (5.1), is valid if reporting is incomplete and even if the percentage of recaptures reported varies from year to year.

5.2. SURVIVAL RATE WHEN MARKING IS DONE THROUGHOUT THE YEAR

Suppose M_1 fish are marked during year 1, of which R_{11} are retaken that year and R_{12} in year 2; also M_2 fish are marked during year 2, of which R_{22} are retaken the same year. From Table 4.4, suitably modified, three equations can be taken:

$$\frac{R_{11}}{M_1} = \frac{F_1(Z_1 - A_1)}{Z_1^2} \tag{5.4}$$

$$\frac{R_{12}}{M_1} = \frac{F_2 A_2 A_1}{Z_2 Z_1} \tag{5.5}$$

$$\frac{R_{22}}{M_2} = \frac{F_2(Z_2 - A_2)}{Z_2^2} \tag{5.6}$$

Dividing (5.6) into (5.5) gives:

$$\frac{R_{12} M_2}{R_{22} M_1} = \frac{A_1 A_2 Z_2}{Z_1(Z_2 - A_2)} = \frac{A_2^2}{(Z_2 - A_2)} \times \frac{A_1 Z_2}{A_2 Z_1} \tag{5.7}$$

The simplest situation is where $Z_1 = Z_2$, so that:

$$\frac{R_{12} M_2}{R_{22} M_1} = \frac{A_2^2}{Z_2 - A_2} \tag{5.8}$$

Then S or Z can be obtained directly from the corresponding entry in Appendix I.

Generally Z_2 is not the same as Z_1. In that event a more accurate estimate can be made if the natural mortality rate, M, can be considered the same in both years. The correction term $A_1 Z_2 / A_2 Z_1$ in (5.7) can be evaluated by two-stage iteration: (1) Take the approximate values of Z_2, F_2, and $Z_2 - F_2 = M$ obtained above as first estimates. (2) Select a reasonable trial value of F_1, add M to get a trial of Z_1, and calculate the right-hand side (RHS) of (5.4); repeat until the Z_1 is obtained which

126

makes the RHS equal the LHS. (3) Using this Z_1, calculate the correction term A_1Z_2/A_2Z_1 in (5.7) and compute $A_2^2/(Z_2 - A_2)$; the latter will correspond to a new estimate of Z_2 which can conveniently be obtained from Appendix I. (4) Using this new Z_2, calculate F_2 from (5.6) and get M by subtraction. These improved estimates of Z_2, F_2, and M are used to start again at stage (2) above, and the process continues until there is no further improvement.

EXAMPLE 5.2. SURVIVAL RATE OF MUSKELLUNGE LAKE BLUEGILLS, FROM MARKING DONE THROUGHOUT TWO CONSECUTIVE YEARS. (Modified from Ricker 1948.)

The marking of bluegills during the 1942 fishing season included 100 fully-vulnerable age 3 individuals, of which 7 were recaptured that year, so that $R_{11}/M_1 = 0.07$. The total number of legal fish marked that year was 400, of which 41 were retaken by fishermen in 1943, so that $R_{12}/M_1 = 0.1025$. Finally, 131 age 3 individuals were marked during the 1943 fishing season, with 14 recaptures the same year, giving $R_{22}/M_2 = 0.1068$ (Ricker 1945a).

From the approximate relationship (5.8):

$$\frac{A_2^2}{Z_2 - A_2} = \frac{R_{12}M_2}{M_1R_{22}} = \frac{0.1025}{0.1068} = 0.958$$

From Appendix I, $Z_2 = 1.23$; and from (5.6), $F_2 = 0.1068 \times 1.5129/0.5223 = 0.309$, and $M = 1.23 - 0.31 = 0.92$.

This is as good a result as these data are apt to provide, considering the small number of recaptures in categories R_{11} and R_{22}. However, in order to illustrate the complete method we will proceed. Take a trial $Z_1 = Z_2 = 1.23$; $F_1 = F_2 = 0.31$. Using the tabulated values of $Z^2/(Z - A)$ in Appendix I, the RHS of (5.4) becomes $0.31/2.897 = 0.1068$, as compared with the actual 0.07. Varying F_1, with M constant at 0.92, gives the additional values below:

F_1	M	Z_1	$Z_1^2/(Z_1 - A_1)$	RHS of (5.4)
0.31	0.92	1.23	2.897	0.1068
0.25	0.92	1.17	2.850	0.0877
0.20	0.92	1.12	2.811	0.0715
0.19	0.92	1.11	2.803	0.0677

Interpolation between the last two gives $F_1 = 0.194$ as the best value; thus $Z_1 = 0.194 + 0.92 = 1.114$. Again using Appendix I, the correction term in (5.7) is evaluated as:

$$A_1Z_2/A_2Z_1 = 0.6031/0.5754 = 1.048$$

and the adjusted $A_2^2/(Z_2 - A_2)$ will be $R_{12}M_2/R_2M_1$ divided by this, that is, $0.958/1.048 = 0.914$. This corresponds to $Z_2 = 1.32$; hence from (5.6), $F_2 = 0.1068 \times 1.742/0.587 = 0.314$. Thus the estimate of F_2 is not much changed, but $M = 1.32 - 0.31 = 1.01$, which is appreciably greater than 0.92.

127

Again solving (5.4) by iteration, with trial F_1 values:

F_1	M	Z_1	$Z_1^2/(Z_1 - A_1)$	RHS of (5.4)
0.19	1.01	1.20	2.873	0.0662
0.20	1.01	1.21	2.881	0.0692
0.21	1.01	1.22	2.889	0.0726

The interpolated value of F_1 is now 0.202, and $Z_1 = 1.01 + 0.20 = 1.21$. A second estimate of the correction term in (5.7) is:

$$A_1 Z_2 / A_2 Z_1 = 0.5800/0.5552 = 1.045$$

This is practically the same as the 1.048 obtained on the previous trial, so the definitive estimates can be taken to be $Z_2 = 1.32$, $Z_1 = 1.21$, $M = 1.01$, $F_2 = 0.31$, $F_1 = 0.20$.

Corresponding to $Z_1 = 1.21$ is $S_1 = 0.30$, and this may be compared with the estimate $S_1 = 0.375$ obtained for the same population in Example 5.1. The difference is less than the standard error of the latter (0.140).

5.3. POPULATION, SURVIVAL, AND RECRUITMENT FROM A TRIPLE-CATCH EXPERIMENT — BAILEY'S METHOD

5.3.1. GENERAL. Estimation of insect populations by marking, begun by Jackson in 1933, has led to an extensive literature of statistical estimation based on a series of three or more "point" samples, usually separated by a rather short period of time (a week, for example). Different methods of grouping the recaptures have been examined, and both deterministic and stochastic models have been used to develop appropriate estimators (Dowdeswell et al. 1940; Bailey 1951; Leslie 1952; Moran 1952; Craig 1953; Goodman 1953; Hamersley 1953; Darroch 1958, 1959, 1961; Seber 1962, 1965; and Jolly 1963, 1965).

These methods can be applied to fish, and the interval between markings can be as long as a year. Because two or more estimates of the percentage of marks in the population are obtained, it is possible to estimate additions to the population as well as losses. Of the various models proposed, we will consider only two that have explicit rather than iterative solutions.

Those who work with insect populations rarely deal with a population that is definitely bounded in space, so that immigration is often as important or more important than recruitment from the local stock, while emigration may be more significant than mortality. Thus they usually refer merely to gains and losses. The same can apply to some fish populations, but in what follows it is convenient to speak simply of recruitment and mortality.

5.3.2. BAILEY'S DETERMINISTIC MODEL. Three catches or samples are taken. On the first occasion (Time 1) the fish are marked; at Time 2 the recaptures are noted and returned to the water, and unmarked fish are given a different mark; at Time 3 the previous marks of both categories are listed (as well as the unmarked individuals).

Loss of some fish by accidental death due to the fishing procedure affects the result only by reducing the population to that extent. However, if a previously-marked fish is accidentally killed at the second sampling, it should be replaced by a new one similarly marked. Indeed it may be an advantage to kill all the recaptures and replace them by fresh fish of the same size given the same mark, as this tends to reduce bias from capture-proneness.

The categories of individuals in the 3 samplings are as shown in Table 5.1. We wish to know:

N_1, N_2, N_3 the population present at each sampling (N_2 = Bailey's x)

S_{12}, S_{23} survival rates between Times 1 and 2, and 2 and 3, respectively ($S \leq 1$). (S_{12} = Bailey's λ)

r_{12}, r_{23} rates of accession of new recruits between the same times: these rates are strictly analogous to survival rates; they represent initial stock plus *all* new recruits during the period, divided by the initial abundance at the start of the period ($r \geq 1$). (r_{23} = Bailey's μ)

TABLE 5.1. Categories of fish newly marked, examined, and recaptured, in Bailey's triple-catch method. Shown in brackets below our symbols are those used by Bailey 1951 (at left), and by Wohlschlag 1954 (at right).

Time	Fish newly marked	Fish examined for marks	Recaptures from 1st marking	Recaptures from 2nd marking
1 (1,0)	M_1 (s_1, R_0)	...	...	...
2 (2,1)	M_2 (s_2, R_1)	C_2 (n_2, F_1)	R_{12} (n_{21}, m_{01})	...
3 (3,2)	...	C_3 (n_3, F_2)	R_{13} (n_{31}, m_{02})	R_{23} (n_{32}, m_{12})

Values of S and r can be used to calculate instantaneous rates of mortality (Z) and recruitment (z), respectively, each in terms of its own time period:

$$Z_{12} = -\log_e S_{12} \tag{5.9}$$

$$z_{23} = +\log_e r_{23} \tag{5.10}$$

(Z/t_{12} = Bailey's γ; z/t_{23} = Bailey's β)

Bailey's small-sample formulae for direct estimates of N_2, S_{12}, and r_{23}, are:

$$N_2 = \frac{M_2(C_2 + 1)(R_{13})}{(R_{12} + 1)(R_{23} + 1)} \tag{5.11}$$

$$S_{12} = \frac{M_2 R_{13}}{M_1(R_{23} + 1)} \tag{5.12}$$

129

$$r_{23} = \frac{R_{12}(C_3 + 1)}{C_2(R_{13} + 1)} \qquad (5.13)$$

With large numbers of recaptures the "+1" can be omitted in each case, if desired. Approximate variances for these expressions are also given by Bailey:

$$V(N_2) = N_2^2 - \frac{M_2^2(C_2 + 1)(C_2 + 2)R_{13}(R_{13} - 1)}{(R_{12} + 1)(R_{12} + 2)(R_{23} + 1)(R_{23} + 2)} \qquad (5.14)$$

$$V(S_{12}) = S_{12}^2 - \frac{M_2^2 R_{13}(R_{13} - 1)}{M_1^2(R_{23} + 1)(R_{23} + 2)} \qquad (5.15)$$

$$V(r_{23}) = r_{23}^2 - \frac{R_{12}(R_{12} - 1)C_3(C_3 + 2)}{C_2(C_2 - 1)(R_{13} + 1)(R_{13} + 2)} \qquad (5.16)$$

Assuming that mortality rate is the same in interval t_{23} as in interval t_{12}, the value of N_3 can be calculated, as well as the actual number of recruits in t_{23}. For this the instantaneous rate of mortality in (5.9) must be adjusted to the same time interval as the instantaneous rate of recruitment in (5.10), i.e. to $Z_{12}t_{23}/t_{12}$. Then the instantaneous rate of increase in N from Time 2 to Time 3 becomes $z_{23} - Z_{12}t_{23}/t_{12}$, and the actual increase or decrease is the exponential of this quantity (found in column 2 or 12 of Appendix I).

A similar computation can be made for N_1, but the assumption required — that rate of recruitment is the same in the two time intervals — will nearly always be unrealistic.

5.3.3. ACCURACY. The accuracy of a population estimate from (5.11) depends principally upon the magnitudes of the three R-items, of which R_{13}, in the numerator, will normally tend to be the smallest. Good design in such an experiment should aim at having R_{12}, R_{13}, and R_{23} all about the same size, and this is likely to be accomplished if M_1 is made larger than M_2, and C_3 larger than C_2. But if R_{13} is in fact small, it is advisable to explore the applicability of (3.7) for estimates of N_1 and N_2, using one of the devices discussed in Section 3.3 to remove the effects of recruitment. In the notation used here, the accuracy of an estimate of N_1 from (3.7) would depend mainly on the magnitude of R_{12}, while the accuracy of an estimate of N_2 would depend on R_{23}.

5.3.4. CONTINUOUS MARKING AND RECOVERY. Although ideally the three samples above should be taken at "points" or very short intervals of time, Wohlschlag (1954) used the method in a continuous experiment, by dividing the experiment into 3 equal periods and considering all sampling and marking as though it had been concentrated at the middle of each period. This seems accurate enough for any experiment on the usual scale, but with very large numbers of recaptures, and a high rate of recapture, a correction for the non-central expectation of average time of recapture might be introduced.

5.3.5. REPEATED EXPERIMENTS. If there are three markings and four "Times" available in an experiment of this type, Bailey's 3-point analysis can be used twice, simply by moving the whole procedure one interval ahead; and similarly for any additional markings. Bailey gives expressions whereby common estimates of S and r can be obtained from an indefinite number of markings. However, the tabulations are complex and the solutions are iterative, while the assumption that r will remain constant throughout an experiment of this type will usually be unrealistic.

5.3.6. COMPARISONS. The survival formula (5.12) is identical with (5.3) of Section 5.1, though the symbols are different: C_2, R_{12}, and R_{22} of Section (5.1) are equivalent to C_3, R_{13}, and R_{23} here. The identity of (5.3) and (5.12) is a consequence of the fact that, in making Petersen estimates, expectation of the recovery ratio R/C is unchanged by any natural mortality affecting marked and unmarked equally; thus recoveries can be made over a protracted period. Similarly expression (5.15) is equivalent to (5.2), in a different form.

The instantaneous rate of recruitment, z, has not usually been computed for fish populations because it does not have the same direct biological meaning that a mortality rate does. However, there is no objection to computing it for descriptive purposes.

5.3.7. ILLUSTRATION OF BAILEY'S PROCEDURE. Table 5.2 shows a hypothetical experiment in which the initial population is 2000 and the survival rate is S = 0.6 during both intervals, while a quarter of the stock is caught and examined at Time 2 and half of it at Time 3. The recaptures have been given their expected values (to the nearest integer), so the large-sample form of formulae (5.11)–(5.13) should give exact estimates:

$$N_2 = \frac{250 \times 450 \times 56}{60 \times 58} = 1800$$

$$S_{12} = \frac{250 \times 56}{400 \times 58} = 0.600$$

$$r_{23} = \frac{60 \times 1050}{450 \times 56} = 2.50$$

TABLE 5.2. Marks put out and recaptures made in the triple-catch experiment of Section 5.3.7.

Time	Population	Fish newly marked	Fish examined for marks	Recaptures marked at Time 1	Recaptures marked at Time 2
1	$N_1 = 2000$	$M_1 = 400$	....	....	....
2	$N_2 = 1800$	$M_2 = 250$	$C_2 = 450$	$R_{12} = 60$	....
3	$N_3 = 2100$	....	$C_3 = 1050$	$R_{13} = 56$	$R_{23} = 58$

Suppose that the first interval, t_{12}, was 20 days long, while t_{23} was 30 days. The instantaneous rate of mortality for t_{12} is $Z_{12} = -\log_e 0.6 = 0.5108$, or 0.025540/day. The instantaneous rate of recruitment for t_{23} is $z_{23} = \log_e 2.50 = 0.9163$, or 0.030543/day. Thus if the mortality rate during t_{12} applies also to t_{23}, the population was increasing at an instantaneous rate of $0.030543 - 0.025540 = 0.005003$/day during t_{23}.

From the above and the previously-obtained estimate of N_2, an estimate of N_3 can be computed. The 30-day net instantaneous rate of increase during t_{23} is $30 \times 0.005003 = 0.1501$ and $e^{0.1501} = 1.163$. Hence $N_3 = 1.163 \times 1800 = 2093$, which agrees with the tabulated value 2100 within the limits of accuracy of the calculation.

5.4. POPULATION, SURVIVAL, AND RECRUITMENT FROM A 4-CATCH OR LONGER EXPERIMENT — SEBER–JOLLY METHOD

5.4.1. STOCHASTIC MODEL OF JOLLY (1965) AND SEBER (1965). Four or more catches or samples are taken. On the first occasion (Time 1) the fish are marked. At Time 2 the total sample and the marked fish recaptured in it are enumerated. Part or all of the sample, including all the recaptures, are given a new mark and are released; but the double-marked fish are preferably to be considered as unmarked at any future recapture[1]. At Time 3 and subsequently, the same procedure is followed, the recaptures from each marking being tabulated separately. On the last occasion, of course, no fish need be marked. Table 5.3 shows the categories involved.

TABLE 5.3. Categories of marked and recaptured fish in a Seber–Jolly experiment with five operation times. Where Jolly's (1965) notation is different from ours, his is shown in brackets. Values of K_i are the sum of all recaptures made later than Time i of fish marked before Time i.

Time	Fish newly marked	Fish examined for marks	Recaptures of fish marked at Time 1	Time 2	Time 3	Time 4	Total	$K_i(Z_i)$
1	$M_1(s_1)$	....	....	....	....	....	....	....
2	$M_2(s_2)$	$C_2(n_2)$	R_{12}	....	....	....	m_2	$K_2 = R_{13} + R_{14} + R_{15}$
3	$M_3(s_3)$	$C_3(n_3)$	R_{13}	R_{23}	....	....	m_3	$K_3 = R_{14} + R_{15} + R_{24} + R_{25}$
4	$M_4(s_4)$	$C_4(n_4)$	R_{14}	R_{24}	R_{34}	....	m_4	$K_4 = R_{15} + R_{25} + R_{35}$
5	....	$C_5(n_5)$	R_{15}	R_{25}	R_{35}	R_{45}	m_5	....
Total	....	....	R_1	R_2	R_3	R_4	....	....

---

[1]If recaptures are only a small fraction of the total population it is simpler just to kill them. Some recommend that the remarked fish be included in the count of those marked and released at Time 2, any subsequent recovery being counted as belonging to the Time 2 marking. The major disadvantage of this is that it magnifies the bias arising from any capture-proneness of some members of the population; also, carrying more than one mark may stress the fish unduly.

Two sets of ratios are computed from the quantities in Table 5.3. The first is simply m_i/C_i (Jolly's $\hat{a}_i$), being an estimate of the fraction of marked fish in the population at the time of capturing the ith sample. The second ratio comes less directly. Let the quantity β_i (Jolly's M_i) represent the number of marked fish in the population just prior to capturing the ith sample. Then, paraphrasing Jolly, immediately *after* Time i there are two groups of marked animals in the population, the $\beta_i - m_i$ old marks that were not captured at Time i, and the M_i new marks that have just been released. Of the former, K_i are subsequently caught (see Table 5.3 for the definition of K_i); and of the latter, R_i are subsequently caught. Since the chances of recapture are assumed to be the same for both groups, one would expect the ratios $K_i/(\beta_i - m_i)$ and R_i/M_i to be approximately equal, and this equality can be rearranged as an estimate of β_i:

$$\beta_i = \frac{M_i K_i}{R_i} + m_i \tag{5.17}$$

From the above, estimates of the three parameters of interest are as follows. (1) The total number of fish in the population (marked and unmarked) just before the ith sample was taken is obtained from the Petersen formula:

$$N_i = \frac{\beta_i C_i}{m_i} \tag{5.18}$$

(2) The survival rate of fish from Time i to Time $i + 1$ (Jolly's ϕ_i) is:

$$S_i = \frac{\beta_{i+1}}{\beta_i - m_i + M_i} \tag{5.19}$$

This does not include the effect of any accidental or purposeful mortality during sampling.

(3) The number of fish that joined the population between Times i and $i + 1$ *and* survived to Time $i + 1$ is:

$$B_i = N_{i+1} - S_i(N_i - C_i + M_i) \tag{5.20}$$

Inspection of Table 5.3 shows that this method provides direct estimates of N for all sampling times except the first and last, while S and B are estimated for all time *intervals* except the first and last. Thus a minimum of four operation times is required, of which the first involves marking only, the last recovery only, and the others both.

5.4.2. MODIFIED SEBER–JOLLY ESTIMATES. Expressions (5.18) – (5.20) are consistent estimates, suitable for large numbers of recaptures. For smaller numbers of recaptures modified expressions are proposed by Seber (1965), analogous to the modified Petersen estimate of expression (3.7). In our notation these are:

$$\beta_i^* = \frac{(M_i + 1)K_i}{R_i + 1} + m_i + 1 \tag{5.21}$$

$$N_i^* = \frac{\beta_i^*(C_i + 1)}{M_i + 1} \tag{5.22}$$

$$S_i^* = \frac{\beta_{i+1}}{\beta_i^* - m_i + M_i} \tag{5.23}$$

$$B_i^* = N_{i+1}^* - S_i^*(N_i^* - C_i + M_i) \tag{5.24}$$

5.4.3. STATISTICAL BIAS AND SAMPLING ERROR. As usual, the above estimates are subject to bias when expectation of recaptures is small, in either of the categories R_1, R_2, etc. or m_2, m_3, etc. of Table 5.3. Gilbert (1973) illustrates this bias graphically for the situation in which the population is constant and equal to 50. Based on *observed* recaptures, the rule that each R or m should be larger than 3 or 4 seems good enough.

Jolly (1965) was unable to develop any general formula for the variance of Seber–Jolly estimates. With moderate-to-large numbers of recaptures in categories R_1, R_2, etc. and m_2, m_3, etc. of Table 5.3, a useful approximation is to assign each R_i and m_i a variance equal to itself. Then the variance of the estimates of N, S, and B can be calculated using the law of propagation of error for the normal distribution (see Section 5.4.6).

5.4.4. ESTIMATION OF FIRST-YEAR SURVIVAL. The Seber–Jolly method does not provide an estimate of survival rate during the first year. However such an estimate can be obtained from data of a Seber–Jolly experiment by the method of Section 5.1. In Table 5.3, the M_i fish marked at Time 1 become M_iS_i by Time 2. At that time they are reduced by the R_{12} recaptures, and the remainder suffer the mortality rate S_2 so that $(M_1S_1 - R_{12})S_2$ survive at Time 3. At Time 3 the fraction of the population sampled is estimated by R_{23}/M_2S_2. This fraction should apply approximately to the M_1 series also, that is:

$$\frac{R_{13}}{(M_1S_1 - R_{12})S_2} \sim \frac{R_{23}}{M_2S_2} \tag{5.25}$$

Thus the estimate of S_1 becomes:

$$S_1 = \frac{R_{12} + R_{13}M_2/R_{23}}{M_1} \tag{5.26}$$

5.4.5. ILLUSTRATION OF THE SEBER–JOLLY METHOD. Table 5.4 shows values of B, N, C, M, R, and K for a constructed population in which S = 0.5 in all intervals, and the fraction of the population sampled is 0.2 at Time 2, 0.3 at Time 3, and 0.25 at Time 4. The initial value of N is 100,000, and the values of C_i, B_i, and M_i are as

TABLE 5.4 Constructed population, markings, and recaptures for Seber–Jolly estimates.

| Time | Recruits | Population | Sample | Newly marked | Recaptures from markings at | | | | |
					Time 1	Time 2	Time 3	Total	K_i
1	....	$N_1 = 100{,}000$	....	$M_1 = 10{,}000$	....	....	....	....	....
2	$B_1 = 40{,}000$	$N_2 = 90{,}000$	$C_2 = 18{,}000$	$M_2 = 16{,}000$	$R_{12} = 1000$	....	....	$m_2 = 1000$	$K_2 = 775$
3	$B_2 = 65{,}000$	$N_3 = 109{,}000$	$C_3 = 32{,}700$	$M_3 = 20{,}000$	$R_{13} = 600$	$R_{23} = 2400$	....	$m_3 = 3000$	$K_3 = 875$
4	$B_3 = 55{,}000$	$N_4 = 103{,}150$	$C_4 = 25{,}788$	....	$R_{14} = 175$	$R_{24} = 700$	$R_{34} = 2500$	$m_4 = 3375$	....
Total	....	....	....	....	$R_1 = 1775$	$R_2 = 3100$	$R_3 = 2500$	....	....

shown. Numbers of recaptures have their exact expected values. Items which would not be known in an experiment are in italics. The estimates of marked fish in the population are, from (5.17):

$$\beta_2 = \frac{16{,}000 \times 775}{3100} + 1000 = 5000$$

$$\beta_3 = \frac{20{,}000 \times 875}{2500} + 3000 = 10{,}000$$

The primary population statistics are, from (5.18)–(5.20):

$$N_2 = 5000 \times 18{,}000/1000 = 90{,}000$$

$$N_3 = 10{,}000 \times 32{,}700/3000 = 109{,}000$$

$$S_2 = \frac{10{,}000}{5000 - 1000 + 16{,}000} = 0.5$$

$$B_2 = 109{,}000 - 0.5(90{,}000 - 18{,}000 + 16{,}000) = 65{,}000$$

The value of S_1 by the Ricker method is, from (5.26):

$$S_1 = \frac{100 + 600 \times 16{,}000/2400}{10{,}000} = 0.5$$

Thus all the "estimates" agree with the values postulated. In this illustration no additional values of N can be computed because the number of recruits varies unpredictably between intervals.

5.4.6. ESTIMATION OF APPROXIMATE SAMPLING ERRORS. The "estimates" above have no sampling error because the recaptures were all given their expected values. However, for the population parameters shown we will illustrate the method of calculating approximate variances by postulating that the recaptures are in fact Poisson variables, each with a variance equal to itself, and by using the large-sample method of estimating errors of sums and products. Let m_i and s_i stand for the means and standard deviations of the primary estimates, and let M and S be the mean and standard deviation of the combination. Then:

If $M = m_1 \pm m_2 \pm \ldots$, $\qquad S^2 = s_1^2 + s_2^2 + \ldots$,

If $M = K m_1^{p_1} m_2^{p_2} \ldots$, $\qquad S^2 = M^2(p_1^2 s_1^2/m_1^2 + p_2^2 s_2^2/m_2^2 + \ldots)$

K is a quantity known without error, and p_1, p_2, etc. can have any value — positive or negative, integral or fractional (Tuttle and Satterly 1925, p. 219).

The Poisson assumption is correct for samples in which N is much larger than C (100 times or more), which is the usual situation in fishery work. In the illustration, C_1/N_1 was made equal to 0.2,

0.3, and 0.25. respectively. Thus the use of the Poisson variances would be conservative: they are 1.25 to 1.43 times the true binomial variances. Using the computed population estimates N_i, the binomial variances could be estimated as $(N_i - C_i)/N_i$ times the actual numbers of recaptures in each category, for all samples but the last.

The first requirement is the variance of β_1 and β_2. In expression (5.17) M_i is without error, and the variances of K_i, R_i and m_i are the respective sums of the variances of their component R-values. For β_2 this means $V(K_2) = 775$, $V(R_2) = 3100$, and $V(m_2) = 1000$. The first term of the expression for β_2 is 4000, and its variance is:

$$4000^2\left(\frac{775}{775^2} + \frac{3100}{3100^2}\right) = 25,808$$

The variance of m_2 is 1000; thus the total variance of the estimate $\beta_2 = 5000$ is 25,808 + 1000 = 26,808. Similarly the variance of the estimate $\beta_3 = 10,000$ is 78,600.

For the variance of N_2, in expression (5.18) C_i is without error, and β_i has the variance estimated above, and m_i has a variance equal to itself. The combined variance of an estimated $N_2 = 90,000$ is:

$$90,000^2\left(\frac{26,808}{5000^2} + \frac{1000}{1000^2}\right) = 16,786,000$$

The standard error of N_2 is therefore approximately 4100. Similarly the variance of $N_3 = 109,000$ is 13,298,800.

In the denominator of (5.19) M_i is without error, so its variance is the sum of those of the first two terms; for $i = 2$ this is 25,808 + 1000 = 26,808. The approximate variance of $S_2 = 0.5$ is therefore:

$$0.5^2\left(\frac{78,600}{10,000^2} + \frac{26,808}{20,000^2}\right) = 0.00036405$$

and its standard deviation is about 0.0191.

In expression (5.20), C_i and M_i are without error. For $i = 2$ the part of the expression in brackets is equal to 88,000 and its variance is 16,786,000, as found above. The variance of S_2 times 88,000 = 44,000 is:

$$44,000^2\left(\frac{0.00036405}{0.5^2} + \frac{16,786,000}{88,000^2}\right) = 7,038,100$$

The variance of $B_2 = 65,000$ is therefore 13,298,800 + 7,038,100 = 20,336,900, and its standard error is 4510.

The variance of the estimate of S_1 in (5.26) can be obtained using the same procedure. The variance of the right hand term of the numerator is:

$$4000^2\left(\frac{600}{600^2} + \frac{2400}{2400^2}\right) = 33,333$$

137

Thus the whole numerator (5000) has a variance of 34,333. The survival rate $S_1 = 0.5$ therefore has an approximate variance of $34,333/10,000^2 = 0.00034333$, and its standard deviation is 0.0185. The latter is much the same as the 0.0191 found for S_2, though this is accidental. Notice that S_2 also could be estimated by the procedure of (5.26), and the sampling error of this estimate is somewhat less — 0.0153 — because larger numbers of recaptures are used.

5.5. SURVIVAL ESTIMATED FROM MARKING IN ONE SEASON, IN CONJUNCTION WITH FISHING EFFORT DATA

Consider a change in rate of fishing that results from a change in fishing effort from one year to the next. Suppose that F_1, M, Z_1, A_1, etc. are statistics describing the first year of an experiment, while F_2, M, Z_2, A_2, etc. describe the second year, only M being common to both. Of M_1 fully-vulnerable fish marked at the start of the first year, R_{11} are recaptured that year, and R_{12} the next year. To estimate survival rate (S_1), another piece of information is necessary. In default of a second year's marking, this may be provided by data on fishing effort (f) in the two years; which data, if they really represent effective effort as the fish encounter it, will be proportional to rate of fishing, F. We have:

$$\frac{f_1}{f_2} = \frac{F_1}{F_2} \qquad (5.27)$$

Also:

$$\frac{R_{11}}{M_1} = u_1 = \frac{F_1 A_1}{Z_1} \qquad (5.28)$$

$$\frac{R_{12}}{M_1} = u_2 S_1 = \frac{F_2 A_2 S_1}{Z_2} \qquad (5.29)$$

Dividing (5.28) into (5.29) gives:

$$S_1 = \frac{R_{12}}{R_{11}} \times \frac{F_1}{F_2} \times \frac{A_1 Z_2}{A_2 Z_1} \qquad (5.30)$$

In this expression R_{12}/R_{11} and F_1/F_2 are known. The correction term $A_1 Z_2/A_2 Z_1$ is the same one already encountered in (5.7) above, and it can be handled in the same manner. A first estimate of S_1 is obtained by putting $A_1 Z_2/A_2 Z_1$ equal to unity, and Z_1 is calculated. This makes it possible to estimate F_1 from (5.28) whence $M = Z_1 - F_1$; then F_2 is calculated from (5.27) and $Z_2 = F_2 + M$. We are now in a position to evaluate $A_1 Z_2/A_2 Z_1$, and get an improved estimate of S_1 from (5.30). Further iteration is usually not necessary.

EXAMPLE 5.3. SURVIVAL RATE OF BLUEGILLS AT SHOE LAKE, INDIANA, COMPUTED WITH THE AID OF INFORMATION ON FISHING EFFORT. (From Ricker 1948.)

Information on the cane-pole fishing effort and tagging experiments on Shoe Lake were reported by Ricker (1945a, p. 393, 413, 419). Because of the war, effort decreased from 163 pole-hours per acre in 1941 to 106 in 1942; i.e. $F_1/F_2 = 1.43$. A representative value for rate of exploitation of bluegills in 1941 was 0.32, while in 1942 there were retaken by fishermen 0.049 of the bluegills which had been marked prior to the fishing season in 1941. Disregarding the A and Z terms in (5.30), a first estimate of S_1 is 1.54 × 0.049/0.32 = 0.236; which gives $A_1 = 0.764$, $Z_1 = 1.444$, $F_1 = 0.32 \times 1.444/0.764 = 0.605$, M = 1.444 − 0.605 = 0.839, $F_2 = 0.605/1.54 = 0.393$, $Z_2 = 0.393 + 0.839 = 1.232$, $A_2 = 0.708$. Using the whole of formula (5.30) we get:

$$S_1 = 0.236 \left(\frac{0.764 \times 1.232}{1.444 \times 0.708} \right) = 0.217$$

This value of S_1 can now be used to obtain better estimates of A_1, Z_1, A_2, and Z_2, but when these are used in (5.30) the same value for S_1 is obtained. Consequently, 0.22 is the best estimate of survival rate in the first year. For comparison, the value computed by the method of Section 5.1 was 0.24.

EXAMPLE 5.4. SURVIVAL RATE AND RATE OF FISHING FOR HALIBUT, COMPUTED WITH THE AID OF FISHING EFFORT. (Modified from Ricker 1948.)

Thompson and Herrington (1930) reported in detail the results of widespread halibut tagging in the area south of Cape Spencer, Alaska. Tagging was done during 1925 and 1926, though not on exactly the same grounds in the two years. Data for the 1925 season are described in Example 5.5 below. Of 762 fish of approximately age 8 or older tagged throughout 1926, reported recaptures were: 106 in 1926, 147 in 1927, and 52 in 1928.

Neglecting for the moment any difference in fishing effort between 1927 and 1928, a first estimate of apparent survival rate (the complement of the annual rate of disappearance, A', of tagged fish from the fishing grounds) is:

$$S_1' = \frac{52}{147} = 0.354$$

However, from data on gear used south of Cape Spencer, cited in Example 2.8, we know there was an increase in fishing from 494,100 "skates" of line set in 1926 and 498,600 in 1927, to 569,200 skates in 1928. This gives some indication of the relative magnitude of the rates of fishing in these years, so we may estimate the 1927:1928 ratio as $F_1/F_2 = 0.876$. Using (5.30) without the A and Z terms:

$$S_1' = 0.354 \times 0.876 = 0.310$$

which will be a second estimate of apparent survival rate. The corresponding instantaneous rate of disappearance is $Z' = -\log_e 0.310 = 1.17$.

A slight improvement can be made by using the whole of (5.30). A trial value of F_1 is required, but it need be only approximate; we will take trial $F_1 = 0.72^2$. Hence the trial value of $F_2 = 0.72/0.876 = 0.82$, or 0.10 more than F_1. Considering natural mortality rate constant, a trial Z'_2 is therefore equal to $Z'_1 + 0.10 = 1.17 + 0.10 = 1.27$. Consequently, using all of (5.30), and taking A/Z values from Appendix I:

$$S'_1 = 0.310 \times 0.5894/0.5663 = 0.323$$

Turning now to a serious estimate of rate of fishing, we notice first that tagging was done throughout the fishing season, which is the situation discussed in Section 4.2. The possible existence of Type C error is tested by the method of expressions (4.7) and (4.8); fortunately the test is not complicated by any significant difference in fishing effort between 1926 and 1927. As we have already called 1928 year 2 and 1927 year 1 when applying (5.30), for consistency the years 1 and 2 of the formulae of Chapter 4 must here be designated 0 and 1, respectively. The two quantities to be compared are:

$$R_0(A')^2/S_1(Z'_1 - A'_1) = 106 \times 1.012/0.323 = 332$$
$$R_1/S'_1 = 147/0.323 = 455$$

Since 332 is considerably less than 455, there is a deficiency of recaptures in the first year compared with the two later years, which means Type C error is present. This means using repeated trials with (4.18) and (4.19). Because of the change in rate of fishing after 1927, only the first term of numerator and denominator should be used in (4.18). It is convenient to rewrite these expressions with this modification, and with the subscripts reduced by 1 to conform to the present numerical designation:

$$u'_1 = \frac{Z'_0 R_1}{A'_0 M} \tag{5.31}$$

$$u'_1 = \frac{A'_1}{Z'_1}\left(Z'_1 - Z'_0 + \frac{R_0(Z'_0)^2}{M(Z'_0 - A'_0)} \right) \tag{5.32}$$

In selecting a trial Z'_0, notice that recaptures in 1926, when adjusted to a full year basis above, are 123 less than expected; thus the full-year u'_1 would be approximately $332/762 = 0.436$ instead of approximately $455/762 = 0.597$. The difference between these, 0.16, will be a useful trial difference between Z'_1 and Z'_0 to use the expressions

---

[2] This was obtained by averaging two extreme limits for F_1. Rate of exploitation in 1927 obviously cannot be less than $u_1 = 147/(762 - 106) = 0.24$, so a minimum F_1 is 0.27; a maximum value for F_1 is the instantaneous rate of disappearance, 1.17. The average of 0.27 and 1.17 is 0.72, the figure chosen; actually their geometric mean, 0.56, would be a better choice.

above. Given $S'_1 = 0.323$ and $Z'_1 = 1.13$, a trial $Z'_0 = 1.13 - 0.16 = 0.97$. Application of this and two other trial values is shown below:

Trial Z'_0	First u'_1 value	Second u'_1 value	Difference
0.97	0.302	0.321	−0.019
1.00	0.305	0.304	+0.001
1.01	0.307	0.299	+0.008

The best Z'_0 is evidently close to 1.00. From it is calculated $u'_1 = 147/762 \times 0.632 = 0.305$. Using (1.13), $F_1 = 0.305/0.599 = 0.51$, which is the instantaneous rate of fishing for 1927. It is a rather smaller figure than the $F = 0.57$ for 1926, computed by the same method from the 1925 tagging experiment (Example 5.6), though the quantity of fishing gear used was practically the same in 1926 and 1927. However, tagging was not done on exactly the same grounds, and the agreement of the two experiments can be considered very satisfactory.

Although the estimate $S'_1 = 0.323$ above involved using a trial F_1 that proves to be considerably too large, S'_1 is changed very little by using the more accurate figure 0.51: it is reduced to $S'_1 = 0.320$, which means Z'_1 would go up to 1.14. This change, however, makes no difference to the estimate of F_1 which follows.

5.6. Marking Done Throughout the Year. Natural Mortality Varies with Age

If the instantaneous natural mortality rate, M, of the fish changes with age, with or without change in F, the computation becomes even more complicated, and I have not succeeded in setting up equations in which all the unknowns would be determinable — by successive approximations or otherwise. Even so, some progress might be made using an estimate of one of the unknowns derived by analogy with another species of fish, or with the same species in another body of water. If M varies with age, it is essential to divide the marked fish into two or more groups. If age-groups are easily recognizable, we could, for example, mark ages 3 and 4 differently in year 1 of the experiment. In year 2, the same thing would be done, but of course the fish that were age 3 in year 1 are now age 4. (If age-groups are not convenient units, any other moveable dividing lines can be used provided that they are made to move at the rate at which fish of corresponding size are growing.)

Using the same symbols as formerly, the recaptures during year 1, of a unit number of age 3 fish marked during year 1, are:

$$\frac{F_3(Z_3 - A_3)}{Z_3^2} \tag{5.33}$$

Recaptures during year 2 (i.e. at age 4) of the unit number of fish marked at age 3 during year 1 are:

$$\frac{F_4 A_4 A_3}{Z_4 Z_3} \tag{5.34}$$

141

Recaptures during year 2 of a unit number of age 4 fish marked during year 2 are:

$$\frac{F_4(Z_4 - A_4)}{Z_4^2} \tag{5.35}$$

Without introducing a complicated set of symbols for the number marked and number recaptured in each category, it is obvious that the expressions above can all be evaluated from the data. The rates of fishing F_3 and F_4 (which might be equal) can be evaluated, as described in Section 5.2, from (5.33) and (5.35). However, it seems essential to put in trial values of both M_3 and M_4 simultaneously, to check survival rate against expression (5.34). This means that no definite decision can be reached concerning the size of either, though a series of corresponding values can be set up; i.e., if M_3 is so and so, M_4 must be such and such. This might have value, as showing, for example, whether or not M increases with age.

5.7. RATE OF FISHING DURING YEARS OF RECRUITMENT

5.7.1. GENERAL. Most fisheries include in their catch representatives of one or more young age-groups which are not yet fully vulnerable to fishing. That is, even when rate of fishing and natural mortality are unchanging among several older ages, the youngest fish cannot be expected to fit into the same picture. The fact they are incompletely recruited is another way of saying that their fishing mortality rate is less than the maximum or definitive rate, because some of their members are too small to be consistently taken by the kind of fishing gear in use. This being true, their total mortality should also be less than the definitive rate, other things being equal.

When recruitment occurs abruptly, there is little need to worry about the incompletely-recruited groups, because they form only a small part of the catch. In that event, marking or tagging experiments should avoid such fish, or mark them distinctively so that they will not be confounded with fully-vulnerable fish in the analysis, or make an adjustment for their smaller size such as was used, for instance, in Example 3.5. However, if recruitment extends over a period of several years, it may happen that the incompletely-recruited groups are not merely important, but actually comprise the greater part of the catch. In that event, it seems essential to mark these young fish and try to obtain some kind of information concerning them.

Fish incompletely recruited are subjected to a smaller total mortality rate than are older fish, if natural mortality does not vary with age. However, because rate of fishing is increasing with age, the ratio of one year's recaptures to the previous year's does not represent the survival rate which actually exists between the two years, but will be somewhat too great. No complete evaluation of fishing and natural mortality is possible under these circumstances, but an analysis can be made on the assumption that the *natural* mortality rate of fish of the incompletely-recruited age-groups is the same as that of those completely vulnerable. This is done by evaluating survival rate and rate of fishing first from the wholly-recruited age-groups, then working back, year by year, into those incompletely recruited.

5.7.2. SURVIVAL RATE DURING RECRUITMENT COMPUTED FROM TAG RECAPTURES. When tagging has been done on fish smaller than those fully vulnerable, it may be possible to calculate their survival rates from subsequent recaptures, provided lengths and ages can be associated, at least approximately. The method is described in Example 5.5.

EXAMPLE 5.5. SURVIVAL RATE AND RATE OF FISHING FOR INCOMPLETELY-RECRUITED AGE-GROUPS OF HALIBUT, FROM RECOVERIES OF TAGS. (From Ricker 1948.)

Data pertaining to the 1925 halibut tagging experiment off northern British Columbia and southern Alaska are shown in Table 5.5, taken from table 12 and the appendices of Thompson and Herrington (1930). The approximate age distribution indicated is from Dunlop's data in Thompson and Bell (1934). It is obvious, from the age distribution of the fish tagged (cf. Fig. 2.12), that recruitment to the fishery is not complete until about age 9. This is reflected also in the distribution of recaptured fish, for there is a relative scarcity of recaptures among the smaller fish during the year of tagging, also during the first year after tagging, and to some extent even during the second year after tagging.

TABLE 5.5 Number of halibut tagged in 1925 (excluding Cape Chacon), and the number recaptured, arranged by 5-cm length intervals. (Data from Thompson and Herrington 1930.)

| Approximate age when marked | Size group[a] | No. tagged | Number of recaptures | | | | |
			1925	1926	1927	1928	Total
....	375	1	....	....	....	....	0
....	425	6	....	....	....	....	0
....	475	28	....	....	....	3	3
4	525	66	....	7	5	1	13
5	575	188	2	17	11	10	40
6	625	293	8	55	26	10	99
7	675	330	30	61	27	4	122
8	725	212	18	55	25	3	101
9	775	142	10	35	5	2	52
10	825	63	9	11	8	1	29
11	875	37	5	13	1	1	20
12	925	25	1	4	2	3	10
13	975	21	1	7	1	1	10
14	1,025	15	1	3	0	1	5
15	1,075	12	1	1	1	1	4
16	1,125	9	1	1	0	1	3
Older	1,175 up	14	0	1	1	0	2
	Total	1,462	87	271	113	42	513
8–16	725–1,125	536	47	131	44	14	236

[a]The 375-millimeter group includes fish from 350 to 399 millimeters, etc. The "1175-up" group includes fish from the 1175 group through the 1625 group.

143

Recaptures of all sizes of fish in 1925 are scarce, indicating "Type C" error. A first step is to estimate the apparent survival rate for completely vulnerable fish. Fish tagged at age 8 will be age 9 in 1926, so we can use recaptures from them and from all older fish to estimate apparent survival rate, S'.

	1926	1927	1928
Recaptures........................	131	44	14
Ratio (S')........................		0.336	0.318

There is good agreement between the two ratios. Taking $58/175 = 0.331$ as the best representative value, the total rate of disappearance of tagged fish is $A' = 0.669$, and the instantaneous rate of disappearance is $Z' = 1.106$. Whether or not this represents the true mortality rate of the population (that is, whether or not Type B error is present), these figures must be used to obtain the estimate of rate of fishing.

Because the fish were marked throughout the 1925 fishing season, the 47 recaptures in that year would not be expected to be a member of the geometric series of later years; instead $(A')^2/S'(Z' - A')$ times 47, or 145, should be. However, this is much less than the $131/0.331 = 396$ which would be expected on the basis of later recoveries, so that "Type C" error is even more important than in the 1926 experiment discussed in Example 5.4.[3]

Using formulae (5.31) and (5.32) to obtain apparent rate of exploitation in 1926, the value $u' = 0.345$ is obtained, from which $F = u'Z'/A' = 0.345 \times 1.106/0.669 = 0.570$ — an estimate of the *true* rate of fishing.

Estimates of rate of fishing for the incompletely-recruited fish can now be found, approximately, by assuming that the instantaneous rate of disappearance of tagged fish from causes other than fishing (the other-loss rate, $M + U$) is the same prior to age 9 as it is at older ages. This value is $M + U = 1.106 - 0.570 = 0.536$, or about 0.54. The ratios of 1927 to 1926 recoveries, for successive age-intervals during the recruitment period, are as follows:

Approximate age during survival period	Recaptures in 1926	Recaptures in 1927	Ratio
5–6	7	5	0.71
6–7	17	11	0.65
7–8	55	26	0.47
8–9	61	27	0.44

[3] Part of the apparent "Type C" error results from more tagging having been done in the second half of the fishing season than in the first half. The mean date of tagging in 1925 was July 14, whereas the middle point of the fishery appears to have been about June 15 (Thompson and Herrington, p. 62). Another part of the "Type C" error might result from non-random local intraseasonal distribution of fishing effort, as described by the authors on pages 64–65 of their paper. Possibly the halibut are "off their feed" and incapable of taking baited hooks at a normal rate for a time after tagging because of harmful effects of catching and tagging. The only bias which these effects will introduce into the rate of fishing, as estimated by the procedure below, will result from the mean date of tagging being different from the mean date of apparent natural mortality; a rough computation shows that the estimated rate of fishing, 0.57, would be reduced by no more than 0.01.

If F_5, F_6, etc. represent the rate of fishing in successive years and N_5, N_6, etc. the average populations, then:

$$\frac{\overline{N}_6F_6}{\overline{N}_5F_5} = 0.71; \qquad \frac{\overline{N}_7F_7}{\overline{N}_6F_6} = 0.65; \qquad \text{etc.}$$

The use of this information to estimate successive values of F is shown in Table 5.6, which can readily be understood if the following things are kept in mind: (1) N fish at the start of any year decrease to NS' at its close, and their average abundance during the year is NA'/Z' (expression 1.15); and (2) N fish at the close of a year represent survivors of N/S' fish at its start; during that year their average abundance was NA'/Z'S', and the catch was FNA'/Z'S'. We start with the arbitrary number of 100 fish at the end of the year in which they are age 9. During that year they are subject to the definitive disappearance rate 0.67; hence at its start they numbered $N_9 = 100/$ $0.33 = 303$. Their average abundance was $\overline{N}_9 = 100 \times 0.67/1.11 \times 0.33 = 183$; the rate of fishing was the definitive rate 0.57, and the other-loss rate was 0.54. During the preceding year, the instantaneous rate of disappearance was $F_8 + 0.54$. Putting $F_8 = 0.4$ as a trial value, $Z'_8 = 0.94$, $A'_8 = 0.609$, $S'_8 = 0.391$, and hence $\overline{N}_8 = 303 \times 0.609/$ $0.94 \times 0.391 = 502$. From this $\overline{N}_9F_9/\overline{N}_8F_8 = 183 \times 0.57/502 \times 0.4 = 0.519$. But the observed value of this ratio is 0.44 (last column of Table 5.6) and thus the trial value $F_8 = 0.4$ is too small. One or two additional trials gives $F_8 = 0.46$ as the correct answer. This determines $Z'_8 = 0.46 + 0.54 = 1.00$, hence $A'_8 = 0.632$ and $S'_8 = 0.368$; and the population at the start of age 8 is $303/0.368 = 823$. All necessary data are now available to repeat the computation for age 7, and so on as far as desired. Calculations are made easier by using Appendix I, where all products involving Z, A, and S can be found. Of the series of calculated F values shown in Table 5.6, those closest to age 9 naturally have the greatest reliability. Three or four years away from age 9, both systematic and sampling error might well be excessive.

TABLE 5.6. Approximate computation of rate of fishing for years of recruitment, on the assumption that the instantaneous other-loss rate (natural mortality plus emigration plus loss of tags) remains constant at 0.54.

Age	Apparent survival rate, S'	Population	Mean population	Rate of fishing	$\dfrac{N_t F_t}{N_{t-1} F_{t-1}}$
5	0.492		$N_5 = 6640$	$F_5 = 0.17$	
		4560			0.71
6	0.454		$N_6 = 3150$	$F_6 = 0.25$	
		2070			0.65
7	0.398		$N_7 = 1351$	$F_7 = 0.38$	
		823			0.47
8	0.368		$N_8 = 520$	$F_8 = 0.46$	
		303			0.44
9	0.330		$N_9 = 183$	$F_9 = 0.57$	
		100			

5.7.3. SURVIVAL RATE DURING RECRUITMENT, COMPUTED FROM AGE COMPOSITION. A method of more general applicability is to calculate survival rates during recruitment years from a representative sample of ages in the commercial catch. For this the true natural mortality rate is required, or some assumption concerning it. It is also necessary to use age composition of the catch brought on board ship, rather than merely those landed, on the assumption that fish discarded at sea are mostly dead or will quickly succumb from injuries.

Baranov's catch equation (expression 1.17) is $C = NFA/Z$. Assume that F, A, and Z have been estimated for the youngest fully recruited age, t, so that:

$$N_t = C_t Z_t / F_t A_t \tag{5.36}$$

Now the population at the beginning of age t (N_t) is equal to S_{t-1} times that at age $t-1$; thus $N_{t-1} = N_t / S_{t-1}$. Inserting this in the catch equation we have:

$$C_{t-1} = \frac{N_t}{S_{t-1}} \times \frac{F_{t-1} A_{t-1}}{Z_{t-1}} \tag{5.37}$$

This can be solved for F_{t-1} by successive trials, knowing C_{t-1}, N_t, and M. A trial value of F yields trial $F + M = Z$, trial $S = e^{-Z}$ and trial $A = 1 - S$. Thus the right side of (5.37) is evaluated and matched with the left, until agreement is obtained. Then N_{t-1} can be calculated from N_t / S_{t-1}, and the whole routine repeated for the next younger age.

In applying this procedure to an age distribution obtained in a single year, the implied assumptions are that year-class strength and rate of fishing have both been more or less constant over the period of years during which these fish were recruited.

EXAMPLE 5.6. SURVIVAL RATE AND RATE OF FISHING FOR INCOMPLETELY-RECRUITED AGE-GROUPS OF HALIBUT, FROM THE AGE COMPOSITION OF THE CATCH. (Modified from Ricker 1948.)

The length frequency and approximate age frequency of halibut caught for tagging in 1925 and 1926 were plotted in Fig. 2.12; of these, fish less than approximately age 10 are shown in Table 5.7. For the purpose of this illustration, these fish are considered representative of the ordinary commercial *catch* of that time (including those caught but not marketed, since the latter are said to probably die). Using the estimate of Z from age distribution (Example 2.8) and the estimate of F from the 1925 tagging (Example 5.5), the true instantaneous rate of natural mortality is estimated as $M = Z - F = 0.76 - 0.57 = 0.19$. This value of M is assumed to apply also to the ages of recruitment.

In Table 5.7 the number and approximate age of fish caught are shown in column 3. Taking the ages as accurate, the definitive rate of fishing, $F = 0.57$, and the definitive rate of survival, $S = 0.47$, are entered opposite age 9. From (5.36), the initial population at age 9 is $N_9 = 359 \times 0.76 / 0.57 \times 0.532 = 900$. Consider a trial value $F_8 = 0.5$ so that trial $Z_8 = 0.5 + 0.19 = 0.69$; from Appendix I trial $A_8 = 0.498$ and trial

146

TABLE 5.7. Computation of rate of fishing for halibut during the years of recruitment, from the approximate distribution of ages in the catch taken for tagging in 1925 and 1926, on the assumption that instantaneous natural mortality rate (M) remains constant at 0.19. (Data from Thompson and Herrington 1930.)

1	2	3	4	5	6
Approx. age	Length-groups	Observed catch C	Rate of fishing F	Survival rate S	Initial population N
					11,433
3	425+475	96	0.009	0.8195	
					9,369
4	525	270	0.032	0.8009	
					7,504
5	575	740	0.115	0.7371	
					5,531
6	625	1201	0.270	0.6313	
					3,492
7	675	1175	0.456	0.5241	
					1,831
8	725	681	0.520	0.4916	
					900
9	775	359	0.57	0.47	

$S_8 = 0.502$. Thus the RHS of (5.37) is $900 \times 0.5 \times 0.498 /0.502 \times 0.69 = 647$, whereas the actual C_8 is 681. Further trials show that 0.520 is the value of F_8 which makes the RHS of (5.37) equal to 681, and the corresponding $S_8 = 0.4916$. N_8 is now calculated as $900/0.4916 = 1831$, and the computation is repeated for age 7. Table 5.7 shows the complete series.

The above is an example of a *sequential* computation by age; such computations are considered further in Section 8.6.

5.8. ESTIMATION OF PERCENTAGE OF THE YOUNGER AGE-GROUPS PRESENT ON THE FISHING GROUNDS

Comparisons of rates of fishing calculated by the methods of Examples 5.5 and 5.6 might be used to decide to what extent the reduced vulnerability of the various younger age-groups is due to their reaction to the fishing gear, and to what extent it results from their absence from the fishing grounds. If the two estimates of F agree at a given age during the recruitment period, it indicates that the fish are present on the grounds but are less vulnerable to the gear than the fully-recruited stock. If the estimate of F from tagging is greater than that from age composition, it indicates that the age-group in question was not yet completely present on the fishing grounds at the time tagging was done. The limiting situation, where all of the reduced vulnerability of recruitment years is due to absence from the fishing grounds, would be indicated by F-values from tagging which are the same for recruitment ages as for fully-vulnerable ages. For such comparisons it would of course be necessary to be sure that there was no extra tagging mortality among the younger fish.

147

In the actual example of Tables 5.6 and 5.7, since the F-values from tag recoveries tend to be even somewhat *less* than those from age composition for ages 6 to 8, it would be concluded that a lesser susceptibility to capture by longlining, rather than absence from the fishing grounds, accounts for the incomplete recruitment of those ages. Very young halibut are of course likely to be at least partly absent from grounds frequented by old fish, and the age 5 comparison is in that direction.

CHAPTER 6. — ESTIMATION OF SURVIVAL AND RATE OF FISHING FROM THE RELATION OF FISHING SUCCESS TO CATCH OR EFFORT

6.1. Principles of Fishing-success Methods

6.1.1. General and historical. The methods of this chapter are applicable when a population is fished until enough fish are removed to reduce significantly the catch per unit effort, the latter being considered proportional to stock present. For example, if removal of 10 tons of fish reduces C/f by a quarter, the original stock is estimated as $10/0.25$ or 40 tons. Instead of estimating C/f only at the start and finish of the experiment, a series of estimates is usually made. That is, a number of points are used to determine the rate of decrease of C/f, and hence of the stock. The reason is that variables such as weather, which affect vulnerability, tend to make single estimates of C/f unreliable for this purpose.

The origin of such methods can be traced to Helland (1913–14). From hunting statistics he made estimates of the population of Norwegian bears, assuming that the number killed per year was proportional to the number in the population, and that births were balanced by natural mortality. Hjort et al. (1933) applied a similar principle, using annual catch per unit of effort, to whale stocks off Iceland and elsewhere; but they made adjustments for recruitment based on known fecundities and various estimates of pre-adult mortality. In either case the instantaneous rate of kill (F) was equal to the catch divided by the mean population. Both the bears and the whales were decreasing fairly rapidly, which made the assumptions used reasonably realistic.

Applications of a similar principle to smaller animals apparently began with Leslie and Davis's (1939) computations for a rat population and Smith's (1940) paper based on a starfish reduction program. Fish populations were apparently treated in this manner by Shibata in 1941 (Kawasaki and Hatanaka 1951), but the first fishery applications to attract wide attention in the west were DeLury's two comprehensive papers (1947, 1951) and a shorter study by Mottley (1949).

6.1.2. Types of computation and symbols. The procedures and computations in common use are of two main types. The first, introduced by Leslie and Davis (1939), involves plotting catch per unit effort against cumulative catch over a period of time; from the resulting straight line, initial population and catchability can be estimated. In the second method, first described by DeLury in 1947, the logarithm of catch per unit effort is plotted against cumulative effort, and the fitted straight line yields the same statistics. Both methods can be improved by a minor change suggested by Braaten (1969), and are described here in that form.

The concepts and symbols to be employed are as follows:

N_0 original population size
N_t mean population surviving during time interval t
C_t catch taken during time interval t
K_t cumulative catch to the start of interval t plus half of that taken during the interval

C   total catch (ΣC_t)
q catchability — the fraction of the population taken by 1 unit of fishing effort (k of DeLury)
p $1-q$; the complement of catchability
f_t fishing effort during time interval t
E_t cumulative fishing effort up to the start of interval t, plus half of that during the interval
f total fishing effort for the whole period of the experiment (E of DeLury)
C_t/f_t catch per unit effort during the interval t (C_t of DeLury)

6.2. POPULATION ESTIMATES FROM THE RELATION OF FISHING SUCCESS TO CATCH ALREADY TAKEN — LESLIE'S METHOD[1]

6.2.1. GENERAL CASE. By definition, catch per unit of effort during time interval t is equal to catchability multiplied by mean population present during the interval[2]; that is:

$$\frac{C_t}{f_t} = qN_t \tag{6.1}$$

The population at the time K_t fish have been caught is equal to the original population less K_t:

$$N_t = N_0 - K_t \tag{6.2}$$

From (6.1) and (6.2):

$$\frac{C_t}{f_t} = qN_0 - qK_t \tag{6.3}$$

Equation (6.3) indicates that catch per unit effort during interval t, plotted aginst the cumulative catch K_t, should give a straight line whose slope is the catchability, q. Also, the X-axis intercept is an estimate of the original population, N_0, since it represents the cumulative catch if C_t/f_t, and thus the population also, were to be reduced to zero by fishing. The Y-axis intercept is the product of the original population, N_0, and the catchability, q.

Confidence limits for the estimate of N_0 can be calculated using formulae 2.6 and 2.7 of DeLury (1951). Upper and lower limits of confidence for any level of probability (P) are the roots of the equation:

$$N^2(q^2 - t_P^2 s_{yx}^2 c_{22}) - 2(q^2 N_0 - t_P^2 s_{yx}^2 c_{12})N + (q^2 N_0^2 - t_P^2 s_{yx}^2 c_{11}) = 0 \tag{6.4}$$

---

[1] Leslie and Davis also had to deal with a complication not considered here: their unit of effort, a break-back trap, could catch only one rat at a time. For any given number of traps in use, this means that C/f increases less rapidly than population, because at higher densities encounters of rats with sprung traps are relatively more frequent than at lower densities.

[2] This is Braaten's (1969) modification of Leslie's method. Leslie and DeLury both put N_t and K_t equal to population and catch at the *beginning* of each time interval. This approximation is close to Braaten's procedure when time intervals are short and q is small.

where:

$$c_{11} = \Sigma X^2/n\Sigma x^2$$
$$c_{12} = \Sigma X/n\Sigma x^2$$
$$c_{22} = 1/\Sigma x^2$$

t_P = the t value corresponding to a given probability P for $n-2$ degrees of freedom, found from a t-table such as Snedecor's table 3.8

n = the number of days of fishing

6.2.2. SPECIAL CASE. A special case of the Leslie method occurs when equal units of effort are used to make the successive catches, so the latter can be plotted directly against cumulative catch:

$$C_t = qN_0 - qK_t \qquad (6.5)$$

This situation has been studied by Hayne (1949), Moran (1951), and Zippin (1956).

In fitting a line to (6.5), Zippin shows that the statistical weighting for catches should be:

$$\frac{1}{N_0 - K_t} \qquad (6.6)$$

where N_0 is a preliminary estimate obtained by eye.

A comparable weighting formula for the general situation (6.3) would be:

$$\frac{f_t}{N_0 - K_t} \qquad (6.7)$$

However, factors other than size of sample and number of marked fish at large usually play a big part in determining the scatter of the points about the regression line — for example, day to day variation in vulnerability of the fish. Thus it may often be more accurate, and it is always less trouble, to fit a line without weighting. The same considerations apply to Moran's (1951) maximum likelihood estimate of N_0, for which Zippin (1956, p. 168-169) prepared charts to simplify the calculation when the number of successive catches is from 3 to 7.

6.2.3. EFFECT OF VARIABILITY. From the discussion in Appendix IV it appears that an ordinary predictive regression line fitted to expression (6.3) or (6.5) will provide unbiased estimates of q and N_0 only if there is no error in K_t. That is, the catch statistics must be completely reliable, for practical purposes. When this is so, all the variability lies in C_t/f_t and the predictive regression is also the functional one. In many situations this is the actual state of affairs. If not, however, an estimate of catchability will tend to be too small and the initial population too large.

EXAMPLE 6.1. SMALLMOUTH BASS POPULATION OF LITTLE SILVER LAKE ESTIMATED BY THE LESLIE METHOD. (Modified from Ricker 1958a, after Omand 1951.)

Little Silver Lake in Lanark County, Ontario is of 100–125 acres extent. It was trapped intensively for 10 days in September, 1949, and Leslie estimates of the fish populations were made. Since the same number of traps (7) were used on all 10 days of fishing, they can be considered collectively as a single unit of effort, so that the

151

daily catch is also the catch per unit of effort — thus avoiding division of each catch by 7. The data for smallmouth bass (*Micropterus dolomieui*) are shown in Table 6.1.

TABLE 6.1. Catches and fishing efforts for the experiment of Examples 6.1 and 6.2; explanation in the text. (From data of Omand 1951.)

1	2	3	4	5	6	7	8
Day	C_t	$C_t/2$	K_t	f_t	E_t	C_t/f_t	$\log_e(C_t/f_t)$
1	131	65.5	65.5	7	3.5	18.71	2.929
2	69	34.5	165.5	7	10.5	9.86	2.288
3	99	49.5	249.5	7	17.5	14.14	2.649
4	78	39.0	338.0	7	24.5	11.14	2.410
5	56	28.0	405.0	7	31.5	8.00	2.079
6	76	38.0	471.0	7	38.5	10.86	2.385
7	49	24.5	533.5	7	45.5	7.00	1.946
8	42	21.0	579.0	7	52.5	6.00	1.792
9	63	31.5	631.5	7	59.5	9.00	2.197
10	47	23.5	686.5	7	66.5	6.71	1.904
Totals	710	355.0	4,125.0	70	. . . .	. . . .	. . . .

Representing the K_t values of expression (6.5) by X, and C_t values by Y, and representing the same quantities *measured from their means* by x and y, formulae for the squares, products, and primary regression statistics are as below, using the symbols of Snedecor (1946, Sections 6.5–6.9; n = number of observations):

$$\Sigma xy = \Sigma(XY) - (\Sigma X)(\Sigma Y)/n$$
$$\Sigma y^2 = \Sigma(Y^2) - (\Sigma Y)^2/n \qquad (6.8)$$
$$\Sigma x^2 = \Sigma(X^2) - (\Sigma X)^2/n$$

$$\text{Slope} = b = \frac{\Sigma xy}{\Sigma x^2} \qquad (6.9)$$

$$\text{Y-axis intercept} = a = \frac{\Sigma Y - b\Sigma X}{n} \qquad (6.10)$$

$$\text{Variance of the points from the regression line} = s_{yx}^2 = \frac{\Sigma y^2 - b\Sigma(xy)}{n-2} \qquad (6.11)$$

$$\text{Variance of the regression coefficient } b = s_b^2 = \frac{s_{yx}^2}{\Sigma(x^2)} \qquad (6.12)$$

152

A desk-size computer will print out all this information in one operation. From the data of columns 2 and 4 of Table 6.1 we obtain·

$\Sigma x^2 = 382416$
$\Sigma y^2 = 6652$
$\Sigma xy = -40825$
$b = -0.10676$
$a = 115.04$
$s_{yx}^2 = 286.71 \quad s_{yx} = 16.93$
$s_b^2 = 0.0007497 \quad s_b = 0.02738$

The regression equation, in the original symbols, becomes:

$$C_t = 115.04 - 0.10676K_t$$

Comparing with (6.5), $q = 0.10676$ and $N_0 = 115.04/0.10676 = 1078$, which is the estimated initial population. The instantaneous rate of removal of the bass is estimated as 0.10676 per day's fishing, corresponding to an actual removal of 10.13% of the population (from columns 1 and 3 of Appendix I).

The standard error of b, and thus of q, is 0.02738. To obtain 95% confidence limits for the estimate of N_0, the t value is looked up for $P = 0.05$ and 8 degrees of freedom: $t_P = 2.306$. The multipliers for expression (6.4) are:

$c_{11} = 2083978/10 \times 382416 = 0.54495$
$c_{12} = 4125/10 \times 382416 = 1.0787 \times 10^{-3}$
$c_{22} = 1/382416 = 2.6150 \times 10^{-6}$

Inserting these values in expression (6.4) and solving for N, the confidence limits for the estimate $N_0 = 1078$ are 814 and 2507. These are not at all symmetrical with respect to the best estimate.

6.3. POPULATION ESTIMATES FROM THE RELATION OF FISHING SUCCESS TO CUMULATIVE FISHING EFFORT — DELURY'S METHOD

6.3.1. GENERAL CASE. Equation (6.1) can be written in the form:

$$\frac{C_t}{f_t} = qN_0\left(\frac{N_t}{N_0}\right) \tag{6.14}$$

or, $$\log_e(C_t/f_t) = \log_e(qN_0) + \log_e(N_t/N_0) \tag{6.15}$$

When the fraction of the stock taken by a unit of effort is small — for example, 0.02 or less— it can be used as an exponential index to show the fraction of stock remaining after E_t units have been expended:

$$\frac{N_t}{N_0} = e^{-qE_t} \tag{6.16}$$

Substituting (6.16) in (6.15):

$$\log_e(C_t/f_t) = \log_e(qN_0) - qE_t \tag{6.17}$$

In terms of base-10 logarithms (6.17) is:

$$\log_{10}(C_t/f_t) = \log_{10}(qN_0) - 0.4343qE_t \tag{6.18}$$

Thus a plot of the (base-10) logarithm of catch per unit effort during each interval, against cumulative effort up to the middle of the interval, yields a line whose slope is $0.4343q$ and whose Y-axis intercept is the logarithm of qN_0. From these two both q and N_0 can be estimated.

6.3.2. SPECIAL CASE. When effort is measured in larger units, so that each unit takes some appreciable fraction of the stock, q cannot be used in the exponential formula (6.16). In that event the slope of the regression line of $\log(C_t/f_t)$ against E_t can be antilogged to give the fractional survival of the stock, S, after the action of one unit of effort. Since f units of effort are used altogether, the estimate of survival to the end of the experiment is S^f, and the fraction of the stock removed is $1 - S^f$. This can be divided into the total removals, C, to give an estimate of initial population:

$$N_0 = \frac{C}{1 - S^f} \tag{6.19}$$

Expression (6.19) is applicable with *all* values of S and q, but if q is really small the procedure of Section 6.3.1 is more convenient. An example has been given by Ricker (1949a, 1958a).

6.3.3. EFFECTS OF VARIABILITY. In the above expressions the regression is of fishing success on fishing effort, so that the latter is on the right-hand side of the equation. Measures of effective fishing effort tend to be less accurate than catch statistics, so in most cases there will be error in E_t and a predictive regression line will underestimate q and overestimate N_0 (Appendix IV). Nor is it easy to obtain an unbiased estimate of the functional regression, since the relative errors of $\log(C_t/f_t)$ and E_t will usually be unknown. *For this reason the Leslie procedure of Section 6.2 is usually the preferable one.* In the example below the physical effort applied is known precisely, but variations in its effectiveness due to changes in weather or other factors are of course unknown.

EXAMPLE 6.2. SMALLMOUTH BASS POPULATION OF LITTLE SILVER LAKE ESTIMATED BY THE DELURY METHOD.

Table 6.1 shows fishing effort in trap-days (f_t), cumulative effort to the half-way point of each day (E_t), catch per unit effort each day (C_t/f_t), and the natural logarithm of the latter. Referring to expression (6.17), $\log_e(C_t/f_t)$ is regressed against E_t to obtain the following equation:

$$\log_e(C_t/f_t) = 2.7195 - 0.013189E_t$$

154

Thus the catchability of the fish is estimated as 1.32% per trap-day, and the initial population is $N_0 = (\text{antilog}_e 2.7195)/0.013189 = 15.173/0.013189 = 1150$.

The ordinary regression used above tends to underestimate q and overestimate N_0, and the two estimates above differ from those of Example 6.1 in the expected direction. The comparison is as follows:

	DeLury method (Example 6.2)	Leslie method (Example 6.1)
Catchability, q	0.0132	$0.10676/7 = 0.0153$
Initial population, N_0	1150	1078

6.4. Systematic Errors in Fishing-success Methods

Inconstant catchability is perhaps the greatest potential source of error in applying methods of estimation based on secular change in catch per unit of effort. Many populations have been found not to be amenable to this treatment, either because catchability varies with seasonal changes in environmental conditions or the fish's reactions, or because individual fish differ in vulnerability and those more vulnerable are more quickly removed. Either effect may produce changes in catch per unit effort which cannot be distinguished from those produced by changed abundance.

Less serious, but of widespread occurrence, is day-to-day or other short-term variation in catchability. Usually this merely increases the scatter of points along the line of graphs such as those in Fig. 6.2, below. Occasionally it may be possible to relate it to other measurable factors and make appropriate adjustments. For example, in a sport fishery catchability may decrease on holidays when total effort is high, because of interference between fishermen or temporary fishing-out of the more accessible pools. On the other hand, effort may become greater whenever, and because, success is good. To adjust for the latter effect Mottley (1949) in one example used the square root of the catch, divided by effort, as the variable in the left-hand side of expression (6.3); however, an adjustment of fishing effort to some standard base would be more consistent with the theory of the method.

Obviously recruitment and natural mortality, or immigration and emigration, can introduce serious error into Leslie or DeLury calculations, unless opposed tendencies happen to be in balance. It is of course unlikely that the incidence of either recruitment or mortality would exactly coincide in time with the application of fishing effort; hence we should usually expect them to make lines such as those of Figure 6.2 curved or irregular in shape. Experience shows, however, that points used to determine such lines seldom lie close to them, so that it is usually impossible to detect recruitment or natural mortality by any curvature which they may introduce. Evidently it is advantageous to concentrate the fishing effort into a rather short period of time, so that these disturbing effects will be minimized.

6.5. Use of Fishing-success Methods with Marked Populations

Usually there is sufficient likelihood of significant departure from the conditions required for fishing-success estimates that it is essential to check them. DeLury (1951) points out that such a check is provided by a concurrent analysis of a group of *marked* fish similar in other respects to the population being estimated. The estimated population of marked fish is then compared with the actual number marked. Quite a variety of causes may produce a discrepancy between the actual and the estimated number. Among these are:

1. Change in catchability, q, during the experiment, either (a) among the population as a whole because of seasonal change in habits or habitat; or (b) because of selective removal of the temperamentally more vulnerable individuals; or (c) because catchability is itself a function of stock density, and decreases as the stock is thinned out.

2. Natural mortality during the experiment.

3. Mortality caused by fishing gear during the experiment (e.g. fish held in a net may be removed by predators).

4. Mortality caused by the marking procedure or the tag or mark itself.

5. Emigration of fish from the population in the area of study.

All these causes tend to produce a deviation in the same direction — toward too small an estimate of population and too large an estimate of catchability — except that 1(a) *may* operate in the reverse manner. Therefore, unless there is reason to suspect a progressive increase in vulnerability, agreement of estimated and actual numbers of fish marked can be taken as fairly convincing evidence that errors (1b)–(5) above are inconsequential. On the other hand, if the estimated figure is too low, there are a number of possible reasons for it. However, usually one or more may be eliminated as improbable, and quite often a single one stands out as the only likely cause of the observed discrepancy. In that event the difference between calculated and observed population provides a means of obtaining a numerical estimate of the effect in question.

Computations applicable to situations of this sort were developed by Ketchen (1953) for a population in which both immigration and emigration were possible. For a simple treatment it is necessary to postulate that immigration and emigration occur at constant instantaneous rates, proportional to the number of fish present in the fishing area. Let F, y, and z be the instantaneous rates of fishing, emigration, and immigration, respectively, based on the whole fishing season as a unit of time. (Note that immigration *adds to* the population and is given the opposite sign to F and y). Based on a unit of fishing effort, these instantaneous rates become F/f, y/f, and z/f; F/f is the catchability, q, of the fish, while the other two are analogous quantities not easy to name.

In Fig. 6.1, the marked population, originally M_1 in number, is affected by emigration and fishing; thus the slope of the line BC^3 is equal to:

$$F/f + y/f = (F + y)/f \qquad (6.20)$$

---

[3] Ketchen uses symbols for the slopes of the two observed regressions, as follows: for untagged fish (line AE), slope $= \hat{k} =$ our $(F+y-z)/f$; for tagged fish (line BC), slope $= k' =$ our $(F+y)/f$; while k is used for catchability (our F/f).

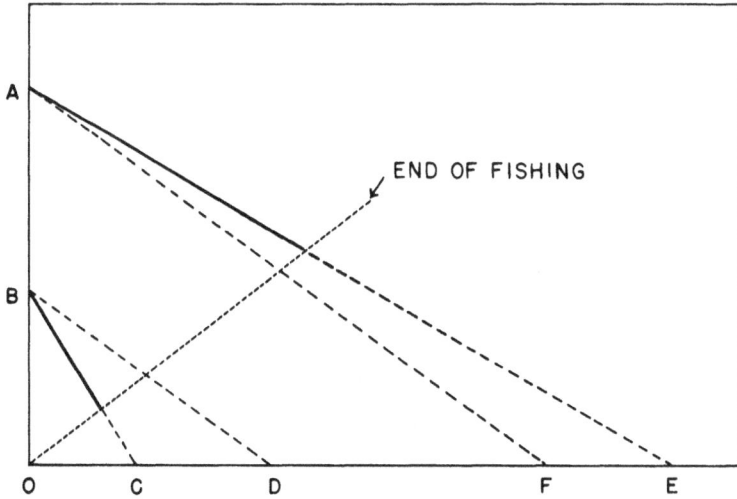

FIG. 6.1. Diagram showing relationship between Leslie estimates of the whole population (above) and the marked population (below). Ordinate — catch per unit of effort; abscissa — cumulative catch. (Modified from Ketchen 1953.)

from which $F + y$ can be calculated. The estimate of the *apparent* original number, M_{1a}, is the X-axis intercept, OC. Had there been no emigration, C/f for the marked population would have decreased along line BD, having slope F/f, and the intercept OD would have been an unbiased estimate of the number marked, M_1. We note that:

$$\frac{F/f}{(F + y)/f} = \frac{\text{slope of BD}}{\text{slope of BC}} = \frac{OB/OD}{OB/OC} = \frac{OC}{OD} = \frac{M_{1a}}{M_1} \qquad (6.21)$$

Thus F can be estimated from:

$$F = \frac{M_{1a}}{M_1}(F + y) \qquad (6.22)$$

The untagged population can now be treated in a similar manner. It is affected by fishing, emigration, and immigration, so that the slope of the line AE is equal to:

$$F/f + y/f - z/f \qquad (6.23)$$

from which z can be calculated since f is known and F and y were found above. If there had been no immigration or emigration, fishing success should have decreased along line AF, which is parallel to BD. The estimated apparent initial population, N_a ($= OE$), is to the true initial population, $N(= OF)$, as the slope of AF is to the slope of AE, or as F/f is to $(F + y - z)/f$; thus an estimate of N is:

$$N = N_a(F + y - z)/F \qquad (6.24)$$

157

Using $\overline{N}$ for the actual average population and $\overline{B}$ for its average biomass, we may write, as in expressions (1.17) and (1.40) of Chapter 1:

$$F\overline{N} = C = \text{catch}; \quad F\overline{B} = Y = \text{weight of catch (yield)} \tag{6.25}$$

$$y\overline{N} = \text{number of emigrants}; \quad y\overline{B} = \text{weight of emigrants} \tag{6.26}$$

$$z\overline{N} = \text{number of immigrants}; \quad z\overline{B} = \text{weight of immigrants} \tag{6.27}$$

From (6.25) $\overline{N}$ or $\overline{B}$ can be evaluated, and the number or biomass of emigrants and immigrants is then obtained from (6.26) and (6.27).

EXAMPLE 6.3. RATE OF FISHING, IMMIGRATION, AND EMIGRATION IN A MIGRATORY POPULATION OF LEMON SOLES. (From Ricker 1958a, after Ketchen 1953.)

Ketchen worked with a population of lemon soles (*Parophyrs vetulus*) which was in process of migration, so that individuals were entering and leaving the fishing area during the course of the fishery. Described in a somewhat simplified form, the experiment consisted of marking 2190 fish ($= M_1$) immediately prior to April 29, near the beginning of the fishery. Daily record was kept of fishing effort, number of fish caught, and number of tags caught. The plot of catch per unit effort against cumulative catch is shown for the whole stock in the top panel of Fig. 6.2, while that for the tagged fish is in the lower panel, and ordinary least-squares lines give the statistics below:

Slope of BC $= (F + y)/f = 0.000695$
$M_{1a} = 958$ pieces
Slope of AE $= (F + y - z)/f = 0.000246$
$N_a = 5.83$ million lb

We know also:

$f = 2285$ boat-hours
$M_1 = 2190$ pieces
$Y = 2.54$ million lb

The first entry above gives:

$$F + y = 0.000695 \times 2285 = 1.588$$

By (6.22), $F = 958 \times 1.588/2190 = 0.695$

Hence, $y = 1.588 - 0.695 = 0.893$

From (6.23), $F + y - z = 2285 \times 0.000246 = 0.562$

and $z = 1.588 - 0.562 = 1.026$

From (6.24), $B = 5.83 \times 0.562/0.695 = 4.72$ million lb

From (6.25), $\overline{B} = 2.54/0.695 = 3.65$ million lb

From (6.26), quantity of emigrants $= 0.893 \times 3.65 = 3.26$ million lb

From (6.27), quantity of immigrants $= 1.026 \times 3.65 = 3.74$ million lb

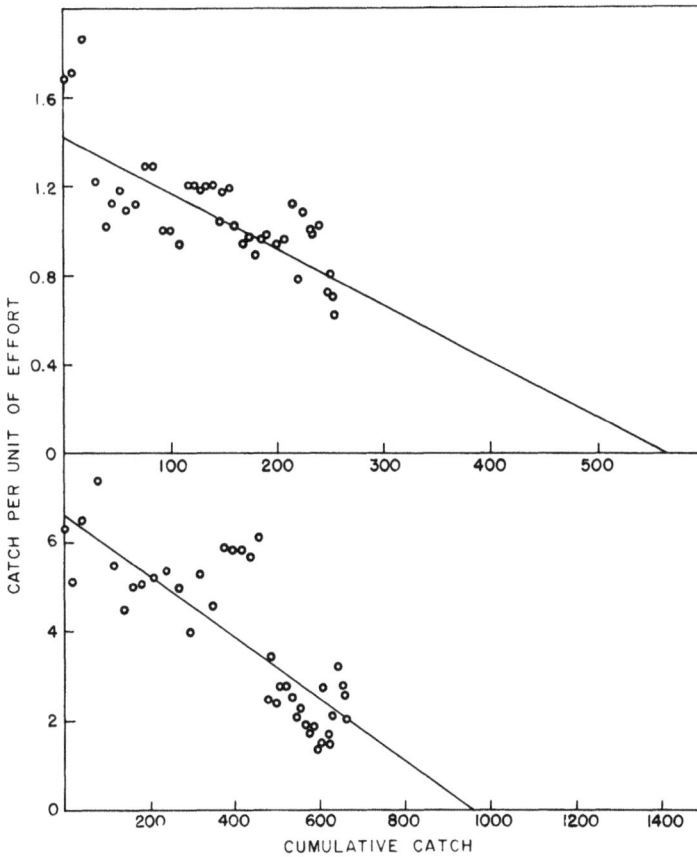

FIG. 6.2. Leslie graphs of catches of unmarked lemon soles, in 10,000's of pounds (above) and of tagged soles, in 10,000's of fish (below) in Hecate Strait, B.C. (After Ketchen 1953.)

The total quantity of fish involved during the experimental period is the initial number plus the immigrants, or $4.72 + 3.74 = 8.46$ million lb.

Dr Ketchen's original account should be consulted for estimates of the total stock for the season, and for some of the consequences of possible variation in rate of immigration or emigration. In Example 3.5 an estimate of B was obtained from tag recaptures, as 4.4 million lb; this is an independent estimate of the same quantity as the 4.72 million lb obtained above.

6.6.   FISHING-SUCCESS METHODS WHEN MORE THAN ONE KIND OF FISHING EFFORT IS USED, OR WHEN THERE ARE DATA FOR ONLY A PART OF THE EFFORT

It often happens that catch/effort statistics are available for only part of the fishing effort used on a population, or two different kinds of effort may be used which cannot be summed directly. The general formula for handling such data is given by DeLury (1951), and it was applied to this situation by Dickie (1955). Suppose:

$f_t, f'_t, f''_t,$ etc. — quantities of different kinds of effort applied each day (or other short interval)

$q, q', q'',$ etc. — catchabilities of the stock by the above kinds of effort

159

Then for any selected kind of effort, f, the catch per unit effort, C_t/f_t, declines as:

$$C_t/f_t = qN_0 - \left(q + \frac{q'f_t'}{f_t} + \frac{q''f_t''}{f_t} + \ldots\right)K_t \qquad (6.28)$$

$$= qN_0 - \bar{q}K_t \qquad (6.29)$$

where $\bar{q}$ is the slope of the Leslie graph ($= \hat{k}$ of Dickie).

Usually we will be most interested in one particular type of gear, or will have catch per effort data concerning only one: let it be f, and let all others be f'. From the definition of $\bar{q}$ in (6.28) and (6.29) we have:

$$q = \frac{\bar{q}}{1 + \dfrac{q'f_t'}{qf_t}} \qquad (6.30)$$

Further, as $C_t/f_t = qN_t$, then $qf_t = C_t/N_t$; similarly, $q'/f_t' = C_t'/N_t$. Thus the denominator term $q'f_t'/qf_t$ is equal simply to C_t'/C_t, the ratio of the catches taken by the two kinds of effort in successive fishing intervals. It follows that (6.30) becomes:

$$q = \frac{\bar{q}}{1 + C_t'/C_t} \qquad (6.31)$$

If the ratio C_t'/C_t remains reasonably constant throughout the fishing season, it is also true that:

$$q = \frac{\bar{q}}{1 + C'/C} \qquad (6.32)$$

Thus from the total catch of the two kinds of gear, plus the slope of the Leslie graph, the true catchability, q, can be obtained.

The condition that the two kinds of gear operate in at least approximately proportional quantities throughout the season can be checked by examining the seasonal distribution of the catch of each. In addition, if there is any serious deviation from this requirement, the Leslie line will not be straight, especially if C' is large relative to C.

If there is natural mortality in the population during the time of the experiment, it too will contribute to the value of q. An adjustment is possible if this can be estimated independently. Still following Dickie, let the instantaneous rate of natural mortality for the duration of the experiment be M; so that, in terms of a unit of effective fishing effort, it is M/f. Then:

$$q = \frac{\bar{q} - M/f}{1 + C'/C} \qquad (6.33)$$

160

EXAMPLE 6.4. ABUNDANCE AND MORTALITY OF BAY OF FUNDY SCALLOPS (*Placopecten magellanicus*) BY THE LESLIE METHOD, USING CATCH AND EFFORT DATA FOR PART OF THE FLEET. (From Ricker 1958a, after Dickie 1955.)

To reduce variability in catch per effort data, Dickie used the catch and fishing statistics of a part of the scallop dragger fleet which kept good records, and used only those pertaining to calm days when dragging could be done with something approaching a standard or maximum efficiency. Of his eleven years' data (Dickie's figure 6 and tables III and IV), we will illustrate those for 1944–45. Since catches are in weight units, Y and Y' are substituted for C and C' in the formulae above.

Catch of sampled fleet	Y	= 130,447 lb
Catch of remainder of fleet	Y'	= 563,783 lb
Fishing effort of sampled fleet	f	= 320 boat-days
Slope of Leslie graph	$\bar{q}$	= 0.001399
Y-axis intercept (initial fishing success)	qB_0	= 589.6 lb/boat-day
Instantaneous rate of natural mortality for the season	M	= 0.06
Instantaneous natural mortality per unit of sampled fishing effort	M/f	= 0.0001875

From (6.33):

$$q = \frac{0.001399 - 0.0001875}{1 + 563783/130447}$$

$$= 0.0002276$$

The initial population is estimated as:

$$B_0 = \frac{589.6}{0.0002276} = 2{,}591{,}000 \text{ lb}$$

CHAPTER 7. — ESTIMATION OF SURVIVAL AND RATE OF FISHING FROM CATCH AND FISHING EFFORT IN SUCCESSIVE YEARS

7.1. DIRECT COMPARISON OF CATCH AND FISHING EFFORT

If fishing effort is sufficiently great to remove at least a moderately large fraction of a stock in a year, and if it varies considerably between years, accompanying changes in mortality and survival of the stock can provide a basis for estimating rate of fishing.

In almost any situation a first step will be to plot catch, C, against effort, f, for successive years, and see what indication there is of regression or correlation between the two. Catch can increase with increasing effort only as long as there are reserves of stock to draw from. Therefore if a fairly large significant correlation is found between C and f, it suggests that rate of exploitation has not been really severe — has been less than 70–75% over most or all of the range of efforts represented.

If a correlation is indicated, favorable circumstances may permit an estimate of rate of fishing from the curvature of the line relating catch to effort (Ricker 1940). This method is applicable primarily to Type 1 fisheries — those in which the combined action of recruitment and natural mortality has a negligible effect on the stock while fishing is in progress, so that the whole population change is due to fishing. It can be used in two somewhat different situations:

1. Catch, C, and effort, f, for the whole fishing season of at least two years are available; as well as an index of relative initial abundance of the stock, N, in the same years, such as might sometimes be available from a measurement of C/f made early in each fishing season.

2. Catch, C, and effort, f, are available for a moderately long series of years during which there have been no trends in abundance having a duration comparable to the length of the available series.

In either situation the effort data are assumed to measure *effective* effort: that is, instantaneous rate of fishing mortality, F, is taken as proportional to fishing effort, f.

Since all mortality is from fishing while fishing is in progress, catch is equal to population times the rate of exploitation (C = Nu). In any two years, not necessarily consecutive, we have:

$$\frac{u_2}{u_1} = \frac{C_2/N_2}{C_1/N_1} = \frac{C_2 N_1}{C_1 N_2} \tag{7.1}$$

Since F varies as f, and $F = -\log_e(1-u)$ in the absence of natural mortality:

$$\frac{\log(1-u_2)}{\log(1-u_1)} = \frac{F_2}{F_1} = \frac{f_2}{f_1} \tag{7.2}$$

In situation A above, the ratio of N_1 to N_2 is known, as is C_1, C_2, f_1, and f_2; thus the right hand sides of (7.1) and (7.2) are both known, and the two equations can be solved simultaneously for u_1 and u_2, by trial. A graph from which a two-place solution can usually be obtained is given in Fig. 7.1.

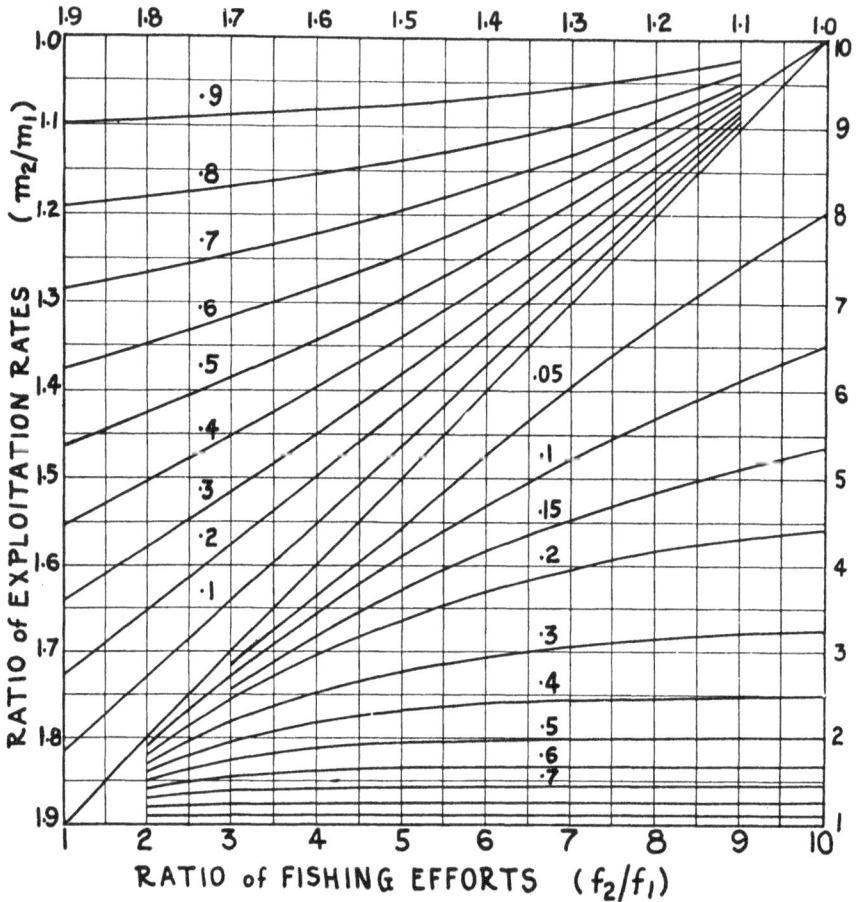

FIG. 7.1. Relation between the ratio of fishing efforts (f_2/f_1) and the ratio of rates of exploitation (m_2/m_1), for Type I populations (in which $m = u$). Curved lines indicate even values of m_1, the rate of exploitation in the year having the smaller effort. (From Ricker 1940.)

In situation B, the best procedure is to fit a line to a graph of catch against effort. The fact that the line must pass through the origin serves as a guide to the amount of curvature to be expected (Fig. 7.2 below). Using the adjusted catches, C', corresponding to the maximum and minimum efforts, f, in the series, values are obtained appropriate to equations (7.1) and (7.2):

$$\frac{u_2}{u_1} = \frac{C'_2}{C'_1}; \text{ and } \frac{f_2}{f_1}$$

The equations can then be solved by trial, or by using Fig. 7.1.

EXAMPLE 7.1 RATE OF EXPLOITATION USING FIGURE 7.1. (From Ricker 1958a, after Ricker 1940, p. 56.)

Figure 7.2 shows catches and efforts modelled after data for a chinook salmon (*Oncorhynchus tshawytscha*) troll fishery, described to the writer by Dr A. L. Tester. Catch tends to increase with gear, but not proportionally: that is, catch per unit effort is less at greater efforts. Comparison of years in which effort was approximately the same provides no indication of progressive change in C/f with time; thus the stock too cannot have had any sustained trend in abundance, though there is evidently year-to-year variation. A line was fitted freehand to the points on Fig. 7.2 and mean catches for the maximum and minimum effort were read as 4000 and 2830 fish, respectively. Their ratio is 1.41, compared with an effort ratio of 660 to 300, or

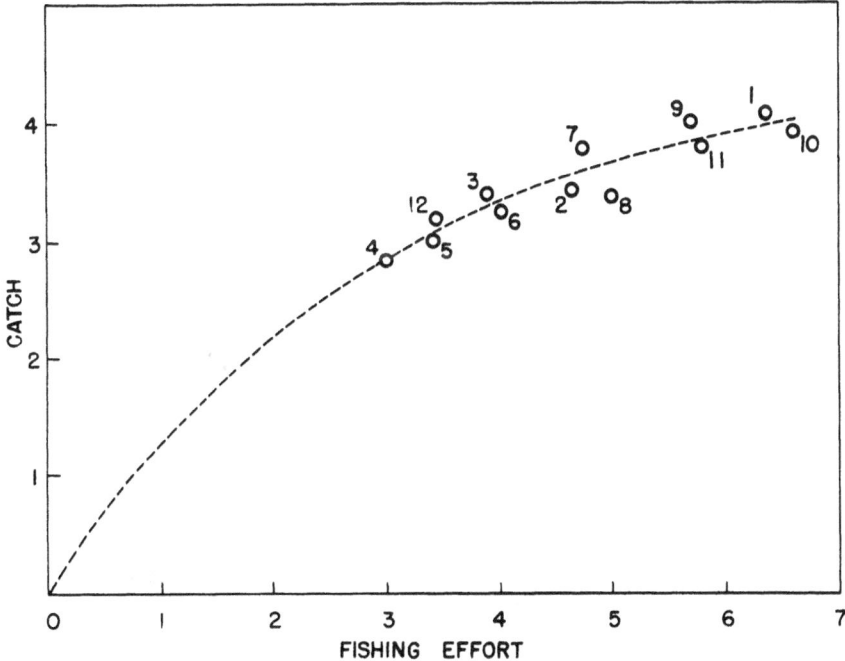

FIG. 7.2. Graph of catch (in 1000's of fish) against fishing effort (in 100's of lines) for the salmon fishery of Example 7.1.

165

2.20. From Fig. 7.1 a preliminary value of $u_1 = 0.6$ can be read, and this can be improved to 0.63 by successive approximations in expressions (7.1) and (7.2). The other u-values are then calculated from (7.2), as shown in column 4 of Table 7.1.

TABLE 7.1 Effort and catch in a troll fishery, and computed rates of exploitation and initial populations (Example 7.1).

Year	Effort (lines)	Catch (pieces)	Exploitation (u)	Population (pieces)
1	636	4080	0.88	4600
2	465	3430	0.79	4300
3	390	3390	0.72	4700
4	300	2830	0.63	4500
5	342	3000	0.68	4400
6	402	3240	0.74	4400
7	474	3780	0.79	4800
8	501	3370	0.81	4200
9	570	4000	0.85	4700
10	660	3919	0.89	4400
11	579	3780	0.85	4400
12	345	3190	0.68	4700

7.2 RATE OF EXPLOITATION ESTIMATED BY EQUALIZATION OF FISHING EFFORT — SETTE'S METHOD

Still considering Type 1 fisheries (Section 1.5), when a breakdown of catch and fishing by days or weeks is available, it is possible to total up, for each of two or more years, the catches taken during the time that some standard amount of effort was used (O. E. Sette, in Ricker 1940, p. 53). The most efficient amount to use is that which was actually expended in the year of least fishing. Assuming this constant effort is proportional to rate of fishing while it operates, the rates of fishing *for the indicated periods of time* must be the same. In the absence of natural mortality, equal rates of fishing mean equal rates of exploitation; and since $C = uN$, the catches of those periods are proportional to the initial populations. This gives the ratio $N_1:N_2$ needed in (7.1), and the actual population size can be obtained as in Section 7.1.

EXAMPLE 7.2. RATE OF FISHING AND SIZE OF STOCK COMPUTED BY SETTE'S METHOD. (From Ricker 1958a.)

In three years, overall statistics of a fishery attacking a circumscribed population (recruitment being absent during the fishing season) were as follows:

Year	Effort	Catch (tons)	Reduced catch
1923	2268	248	186
1924	1549	200	200
1925	1684	283	265

166

The column "reduced catch" is the catch taken up to the time that 1549 units of effort were expended in each year, some minor interpolation being made within a statistical catch period in order to obtain a catch corresponding to exàctly 1549 effort units. The initial populations in the three years were in the ratio of the reduced catches, 186:200:265.

Applying (7.1) to the first two years above, with 1923 = year 2 because it had the greater fishing effort:

$$\frac{u_2}{u_1} = \frac{248}{200} \times \frac{200}{186} = 1.333$$

$$\frac{f_2}{f_1} = \frac{2268}{1549} = 1.464$$

Entering the upper half of Fig. 7.1 with 1.464 on the abscissa and 1.333 on the ordinate, the value $m_1 = u_1 = 0.36$ is obtained; from which $F_1 = 0.446$ (Appendix I), $F_2 = 1.464 \times 0.446 = 0.653$, and $u_2 = 0.48$. Also the 1924 original population is estimated as $200/0.36 = 560$ tons, and that for 1923 as $248/0.48 = 520$ tons.

The years 1924 and 1925 differ so little in effort that a similar calculation is not likely to be useful; however 1925 can be compared with 1923. Better, all three years could be included in one analysis, as described for "situation B" in Section 7.1.

7.3. RATE OF FISHING AND NATURAL MORTALITY FROM COMPARISON OF SURVIVAL RATES AT TWO LEVELS OF FISHING EFFORT — SILLIMAN'S METHOD

7.3.1. BASIC FORMULAE. A method proposed by Silliman (1943) is applicable to fisheries of either Type 1 or Type 2 — that is, natural mortality and recruitment may occur either during or outside of the fishing season. What is needed is that in the history of the fishery there shall have been two different levels of fairly uniform fishing effort, each persisting long enough to give a reliable estimate of the prevailing survival rate, S. The assumptions required are that instantaneous rate of natural mortality, M, be the same under both regimes; and, as usual, that rate of fishing, F, be proportional to the available physical measure of fishing effort, f. The computations can be simplified (Ricker 1945c) by using instantaneous rates, as follows:

$$F_1 + M = Z_1 = -\log_e S_1 \qquad (7.3)$$

$$F_2 + M = Z_2 = -\log_e S_2 \qquad (7.4)$$

$$F_1/F_2 = f_1/f_2 \qquad (7.5)$$

The right-hand sides are known, so the equations can be solved directly for F_1, F_2, and M; also $q = F_1/f_1 = F_2/f_2$.

7.3.2. COMPARISON OF ADJACENT YEARS. In Silliman's use of this method, fishing effort had been stabilized for long enough at each of two levels that the survival rate for each could be read from a catch curve. However, the method can also be applied under less restrictive conditions. If there are two adjacent years at each of two stable levels of effort, the catches *of the same year-class* or group of year-classes can be compared in consecutive years, using only ages that are fully recruited in both years. This can be represented by the expression:

$$S = \frac{(C_b + C_c + \ldots + C_j)_2}{(C_a + C_b + \ldots + C_i)_1}$$
(7.6)

The subscripts a, b, etc., refer to age, while 1 and 2 refer to the two successive years being compared.

Conceivably, more than two pairs (or short sequences) of years might be available for estimates of survival rate — each pair being characterized by constant fishing effort. The best procedure then is to estimate S from each pair of adjacent years, convert each S to the corresponding instantaneous mortality rate Z, and compute F_1, F_2, M and q for each combination of two pairs of years by the method of Section 7.3.1. The values of M and of q can then be averaged to obtain a common value, if this seems justified. Widrig (1954b, p. 143) suggested obtaining a single estimate by regressing Z on f for all pairs of years available, but as there is sure to be sampling error in f, an ordinary regression will give too small an estimate of q and too large an estimate of M (compare Section 7.4.2).

EXAMPLE 7.3. RATE OF FISHING AND NATURAL MORTALITY RATE FOR PACIFIC SARDINES (*Sardinops sagax*), FROM COMPARISON OF TWO LEVELS OF FISHING EFFORT AND THE CORRESPONDING SURVIVAL RATES. (From Ricker 1958a, modified from Silliman 1943.)

Survival rates were calculated from catch curves for two periods of the sardine fishery, as follows:

Period	Relative fishing effort	Survival rate	Instantaneous mortality rate $(= -\log_e S)$
1927–33	$f_1 = 1$	$S_1 = 0.90$	$Z_1 = 0.511$
1937–42	$f_2 = 4$	$S_2 = 0.20$	$Z_2 = 1.609$

The equations (7.3)–(7.5) are:

$$F_1 + M = 0.511$$
$$F_2 + M = 1.609$$
$$F_1/F_2 = 1/4$$

Solving these, $F_1 = 0.366$, $F_2 = 1.464$, and $M = 0.145$.

Later work suggested that these preliminary results gave figures too low for natural mortality and too high for fishing mortality (Clark and Marr 1956). There are various possible reasons for the discrepancy, among them (1) a temporary progressive increase in recruitment among the year-classes from which the 1937–42 survival rate was estimated, making S_2 too low; and (2) the possibility that the unit of gear used became more efficient over the time compared — perhaps because of better cooperation in locating the sardine schools.

7.4. Rate of Fishing and Natural Mortality from Catch and Effort Statistics, When Effort Varies Continuously — Beverton and Holt's Method

7.4.1. Procedure. Beverton and Holt (1956, 1957) proposed a method that can be regarded as an extension of Silliman's (Section 7.3) to the situation where fishing effort varies from year to year. Necessary conditions are: (1) that catch per unit effort (C/f) is proportional to mean population present (N) during the fishing season; (2) that neither natural mortality rate (M) nor catchability (q) vary with age after the age of full recruitment (complete vulnerability) is reached; and (3) that natural mortality rate does not change from year to year. A less rigid alternative to (3) is that M should not have unidirectional trends lasting for a length of time greater than about one-third of the number of years' data available.

The basic information required is fishing effort (f), catch (C), and age composition of the catch, for a series of at least 3 years. Age composition is required in sufficient detail to: (1) identify and separate off all incompletely-recruited ages, and (2) identify the number of fish of a given year's youngest completely-recruited year-class in the following year. It is not necessary that older ages be distinguished from each other. A Type 2 fishery is postulated; that is, fishing and natural mortality are assumed to act concurrently throughout the year[1].

For successive pairs of years a series of expressions are computed as follows:

$$\left.\begin{array}{l} \dfrac{(C_b + C_c + ... + C_j)_2/f_2}{(C_a + C_b + ... + C_i)_1/f_1} = \dfrac{(C/f)_2}{(C/f)_1} \\[3mm] \dfrac{(C_b + C_c + ... + C_j)_3/f_3}{(C_a + C_b + ... + C_i)_2/f_2} = \dfrac{(C/f)_3}{(C/f)_2} \end{array}\right\} \tag{7.7}$$

etc.

The subscripts a, b, etc., refer to age, while 1, 2, etc., refer to successive years of the fishery. Age a is the first fully-recruited age present in the first year of each pair, age b is the next older age, and so on. Age i is the oldest age present in the first year of each pair, and age j is the next older age in the following year, for which the catch will

---

[1] If fishing mortality precedes natural mortality (Type 1 fishery) the procedure is the same, but $m = 1 - e^{-F}$ is substituted for A, and F for Z, in expressions 7.8–7.10 below.

usually be zero. Thus the expressions in (7.7) represent the ratio of catch per unit effort *of the same year-classes in two successive years.*

Each line of expression (7.7) would be an estimate of survival rate, S, if fishing effort were the same in the two years involved; but since effort changes, (7.7) is not any simple function of survival in either the earlier or the later year. Since the average population of any fully-recruited age-group in a season is equal to initial population multiplied by A/Z (expression 1.15), we can write:

$$\frac{(C/f)_2}{(C/f)_1} = \frac{\overline{N}_2}{\overline{N}_1} = \frac{N_2 A_2 / Z_2}{N_1 A_1 / Z_1}$$

$$S_1 = \frac{N_2}{N_1} = \frac{(C/f)_2}{(C/f)_1} \times \frac{A_1 Z_2}{A_2 Z_1} \tag{7.8}$$

The "correction term" $A_1 Z_2 / A_2 Z_1$ is the same as appeared in exactly comparable situations involving tag recaptures (expressions 5.7, 5.30). Taking logarithms:

$$\log_e S_1 = \log_e[(C/f)_2/(C/f)_1] + \log_e(A_1 Z_2 / A_2 Z_1) \tag{7.9}$$

Since $-\log_e S_1 = Z_1 = F_1 + M$, and $F_1 = qf_1$, this becomes, with some transposition:

$$Z_1 = -\log_e[(C/f)_2/(C/f)_1] - \log_e(A_1 Z_2 / A_2 Z_1) = M + qf_1$$

Similar equations can be developed for each pair of years represented in the data, resulting in the series:

$$\left. \begin{array}{l} -\log_e[(C/f)_2/(C/f)_1] - \log_e(A_1 Z_2 / A_2 Z_1) = M + qf_1 \\[2mm] -\log_e[(C/f)_3/(C/f)_2] - \log_e(A_2 Z_3 / A_3 Z_2) = M + qf_2 \end{array} \right\} \tag{7.10}$$
$$\text{etc.}$$

Thus we obtain a series of linear equations in f, whose slopes are an estimate of catchability, q, and whose Y-intercepts estimate rate of natural mortality, M.

To estimate both q and M, a minimum of 3 years' data, that is, two equations, are required. For a first estimate the correction terms in (7.10) are ignored, that is, $-\log_e[(C/f)_2/(C/f)_1]$ is taken as a trial estimate of Z in year 1, $-\log_e[(C/f)_3/(C/f)_2]$ in year 2, and so on. These Z values are plotted against f_1, f_2, etc., and a regression line is fitted. The slope of this line is used as a first estimate of q, and its Y-axis intercept as a first estimate of M. From these a value of Z for each year is estimated from $Z = M + qf$, and from this series trial values for the log correction terms are computed, hence adjusted values for the left hand sides of (7.10). The latter are plotted against successive f's, a new line is fitted, and from the adjusted values of M and q a still better relationship can be obtained. The third or fourth fitting is likely to be the last that will be at all rewarding.[2,3]

7.4.2. SELECTION OF A REGRESSION LINE. A difficulty with this method lies in deciding what regression line to use. To date the predictive regression of Z on f has always been employed, which is correct if f is measured without error. By contrast, if f were subject to error and Z were not, the regression of f on Z would be appropriate. Although the physical measurement of fishing effort may be quite accurate, its effectiveness can vary greatly due to variable weather or changes in distribution patterns of the fish; thus it must usually be regarded as subject to considerable random variation from year to year. Estimates of Z, also, are apt to be quite uncertain. Thus with most actual data both Z and f will be subject to error but the relative magnitude of these errors will be unknown. One of the "AM regressions" (Appendix IV) is theoretically correct in this situation, if calculated symmetrically, but their confidence limits tend to be wide.

7.4.3. EXAMPLES. Applications of the Beverton–Holt procedure (using the ordinary regression of Z on f) can be found in their papers, or in Example 7D of Ricker (1958a). Insofar as the fishing effort exerted may be subject to error, these estimates of q will be less than the best estimate, while the estimates of M will be too large.

7.5. RATES OF FISHING AND NATURAL MORTALITY FROM CATCH AND EFFORT DATA — PALOHEIMO'S METHOD.

Paloheimo (1961) showed that the procedure of Section 7.4 can be simplified by referring the computation to time intervals that include half of one year and half of the next. He used:

$$\bar{Z} = -\log_e[C/f)_2/(C/f)_1] \tag{7.11}$$

---

[2] Expressions describing the further complication of a rate of fishing that varies with size of the fish have been developed by Beverton and Holt (1956, p. 72).

[3] Schuck (1949) made an interesting analysis that superficially resembles the Beverton–Holt procedure and is also related to the fishing-success methods of Chapter 6. He plotted the absolute decrease in catch per unit of effort from the beginning to the end of a year (of year-classes that were fully vulnerable throughout the year) against the absolute catch, over a series of 15 years of the New England haddock fishery. When a straight line was fitted it ran almost exactly through the origin, whereas Schuck evidently expected it to have a positive intercept that would reflect the existence of natural mortality; however, the random error in the data would have accommodated a small natural mortality rate. The defect of this method is that both size of initial population and intensity of fishing contribute to the absolute decrease in C/f in any year. In the haddock fishery, initial populations varied by a factor of 3 or 4, whereas variation in *percentage* decrease in C/f (due to different rates of fishing) was only 2:1; thus most of the slope of the line reflected different population sizes and the effects of fishing were swamped. The same data can readily be treated by the methods of Sections 7.4 or 7.5.

where $(C/f)_1$ is the catch per unit effort for year 1 of each pair, and $(C/f)_2$ is the catch per unit effort *of the same year-classes* in year 2. $\overline{Z}$ thus approximates the mean of the instantaneous mortality rates of years 1 and 2. He also defined:

$$\bar{f} = \frac{f_1 + f_2}{2} \tag{7.12}$$

Then the predictive regression of $\overline{Z}$ on $\bar{f}$ was computed (Paloheimo's expression 16), to give estimates of q and M without iteration. In trials with numerical models this procedure gave estimates having much less variance from the true value than did the Beverton–Holt procedure. However, the problem remains of choosing the best regression, as discussed in Section 7.4.2 (see also Example 7.4).

7.6. Rate of Fishing from Catch and Effort Data When Natural Mortality Rate Is Known or Postulated

7.6.1. PROCEDURE. Experience with the methods of Sections 7.4 and 7.5 shows that frequently the natural mortality estimate obtained (M) lies outside the bounds of reasonable expectation; sometimes it is negative. Even when it appears reasonable, its limits of error may be so broad that it adds nothing to available information on natural mortality rate in the population. In either of these situations it is better to use the best available estimate of M to help in obtaining the best possible value for q. Such an estimate of M might come, for example, from the age structure of the population at an early stage of exploitation, or we might use the value of M obtained from a stock of the same species in a different (but similar) region. If there is no such guidance, several M values that span the likely range can be tried. With M given, only two years' data are needed, but ordinarily a series of years is available. Using Paloheimo's method (Section 7.5), $\overline{Z}$ is plotted against $\bar{f}$ and a line fitted that will pass through the point $(\bar{f} = 0, \overline{Z} = M)$. The functional estimate of the slope of this line is an estimate of catchability:

$$q = \frac{\Sigma(\overline{Z} - M)}{\Sigma \bar{f}} \tag{7.13}$$

7.6.2. ILLUSTRATION. Table 7.3 below gives data for three years of a constructed fishery in which M $= 0.3$, $q = 0.04$, and the fishing efforts are 10, 20, and 30, respectively. In the first year 4372 fish are caught, and $0.497 \times 7304 = 3630$ of the first year's fully-recruited ages and "platoons" were taken in the second year. Similarly the 7304 fish caught in year 2 are represented by $0.397 \times 7750 = 3077$ in year 3. Using expression (7.11) we may write:

$$Z_{1-2} = -\log_e\left[\frac{3630}{20} \Big/ \frac{4372}{10}\right] = 0.879$$

$$Z_{2-3} = -\log_e\left[\frac{3077}{30} \Big/ \frac{7304}{20}\right] = 1.270$$

The mean efforts are $f_{1-2} = 15$ and $f_{2-3} = 25$. From (7.13):

$$q = \frac{(0.879 - 0.3) + (1.270 - 0.3)}{15 + 25} = 0.0387$$

This estimate of q is close to the true value 0.04. Estimated values of F are 0.387, 0.774, and 1.161 respectively. Since it is a Type 1 fishery, rates of exploitation are equal to $1 - e^{-F}$, i.e. 0.321, 0.539, and 0.687. Dividing these into the catches (Table 7.3), the initial fully-vulnerable populations were 13,620, 13,550 and 11,280, respectively.

EXAMPLE 7.4. CATCHABILITY COEFFICIENT AND NATURAL MORTALITY RATE OF ARCTIC COD.

Garrod (1967) computed fishing efforts and total mortality rates for Arcto-Norwegian cod by age, using an ingenious combination of data of three different fisheries. The effort was totalled from mid-year to mid-year, so that it corresponds closely to the Paloheimo situation of Section 7.5. Mean values for ages 6–7 to 9–10 are shown in Table 7.2 and Fig. 7.3.

TABLE 7.2. Mean values of fishing effort (f, in 10^8 vessel ton-hours) and total mortality rate (Z) for ages 6–7 through 9–10, for the Arcto-Norwegian cod fishery. (Data from Garrod 1967, table 3.)

Year	$\bar{f}$	$\bar{Z}$
1950–51	2.959	0.734
1951–52	3.551	0.773
1952–53	3.226	0.735
1953–54	3.327	0.759
1954–55	4.127	0.583
1955–56	5.306	1.125
1956–57	5.347	0.745
1957–58	4.577	0.859
1958–59	4.461	0.942
1959–60	4.939	1.028
1960–61	6.348	0.635
1961–62	5.843	1.114
1962–63	6.489	1.492
Totals	60.500	11.524
Means	4.6538	0.8865

Garrod computed the predictive regression of $\bar{Z}$ on $\bar{f}$ and obtained the equation:

$$\bar{Z} = 0.341 + 0.1172\,\bar{f}$$

Although most of the variability in position of points in Fig. 7.3 may stem from error in $\bar{Z}$, there is certain to be some error in $\bar{f}$ as well (either actual mistakes in reported amount of fishing, or variations in effectiveness of gear from year to year). Thus the estimate M = 0.341 will tend to be too large, and it is in fact unreasonably large for long-lived fish like northern cod.

173

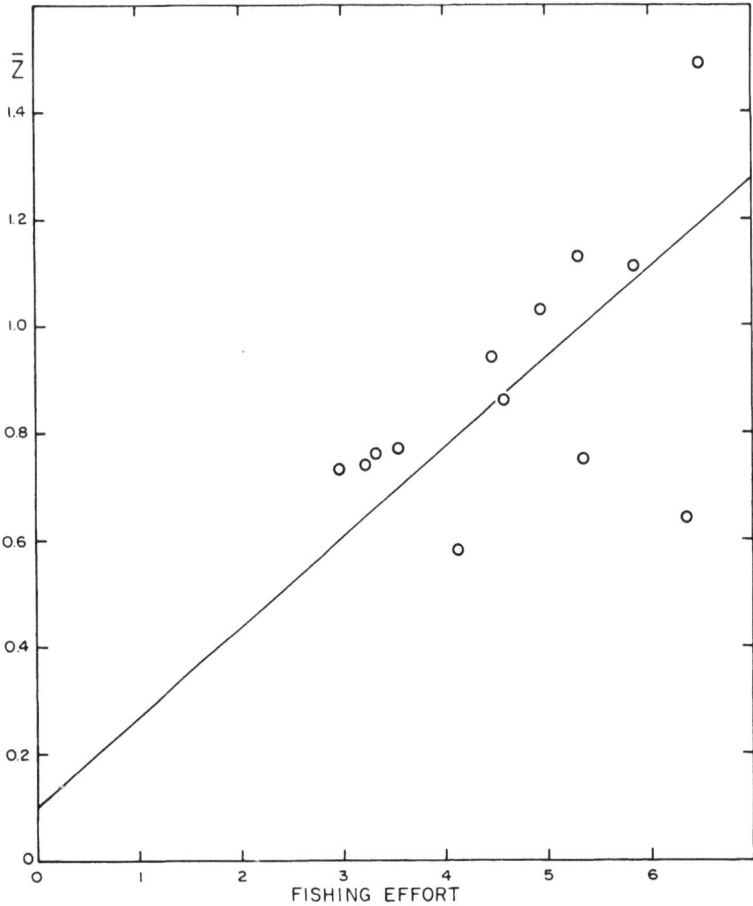

FIG. 7.3. Total mortality estimates ($\bar{Z}$) plotted against fishing efforts for Arcto-Norwegian cod, and a regression line fitted to pass through $M = 0.1$.

A GM functional regression (Appendix IV) would be expected to give too small a value for M (if Z is more variable than f), and in fact it produces a negative estimate ($q = 0.2110$, $M = -0.085$). The Wald estimate is also impossible ($q = 0.2106$, $M = -0.094$), but the Nair–Bartlett estimate is $q = 0.1811$, $M = +0.044$. This value of M is probably reasonable, but only accidentally so, since its limits of sampling error are very wide.

Evidently we will do better to adopt the procedure of Section 7.6. The value of M for a northern cod population should be between about 0.05 and 0.15, because any larger value would not permit the accumulation of large old fish (age 20 and up) that originally existed in considerable numbers. Using $M = 0.1$, a value of q can be estimated from (7.13) using the totals in Table 7.2:

$$q = \frac{11.524 - 13 \times 0.1}{60.5} = 0.169$$

174

From this and the efforts of Table 7.2 a smoothed rate of fishing can be estimated for each year. These exhibit a fairly steady increase from 0.50 in 1950–51 to 1.10 in 1962–63.

7.7. ESTIMATES OF CATCHABILITY MADE WITH INCOMPLETE INFORMATION ON CATCH AND EFFORT

Frequently there are situations in which part of the effort used in capturing a stock, and the corresponding catch, cannot be measured. For example, a stock may be attacked by both commercial and sport fisheries, but catch and effort are known in detail only for the former. A Beverton–Holt or Paloheimo analysis of commercial data will reflect the activity of both fisheries, even if they are separated in time; that is, the decrease in C/f observed in a year-class from one year to the next reflects the losses by all kinds of mortality. Thus there is a need for careful interpretation.

Consider a stock attacked by two kinds of fishing effort, X and Y; there are catch data and effort data only for X, and an attempt is made to estimate Z each year by the method of Section 7.4, 7.5, or 7.6. Figure 7.4 shows some of the results possible when there is no natural mortality (M = 0). Let the slope of the line in each case be b, and the Y-axis intercept = a; also let F_x and F_y be the rates of fishing generated by the two gears, and $F = F_x + F_y$.

1. If Y = 0 we get line A, through the origin; its slope measures the true catchability q_x of fish by gear X ($b = q_x$).

2. If Y is appreciable and there is perfect correlation ($r = 1$) between the efforts X and Y from year to year, we get line B which runs through the origin and has a steeper slope than A; the angle between lines A and B depends on the relative size of efforts X and Y. In this case the catchability coefficient estimated from b includes effects of both X and Y, and it is too large if it is referred to the X effort data alone ($b = q_x + q_y$ if X and Y are measured in the same units).

3. If there is no correlation between efforts X and Y ($r = 0$) we get line C, parallel to A. The slope b estimates q_x without systematic bias, while the intercept a is an estimate of the average instantaneous mortality rate generated by effort Y over the period concerned, i.e. F_y.

4. If r is positive and less than 1, we get a line such as D, where the effect of effort Y appears partly in the slope and partly in the intercept; thus the slope is too large to be an estimate of q_x.

5. If r is negative and larger than –1, we get a line such as E, with a slope less than A or C, which is too small for an estimate of q_x. The position of the intercept depends on the value of r and the relative size of X and Y, but it *can* be above that for line C (as shown).

6. In an extreme case, with $\overline{Y} > \overline{X}$ and r negative, there might even be a reversed slope and negative value of b (line F).

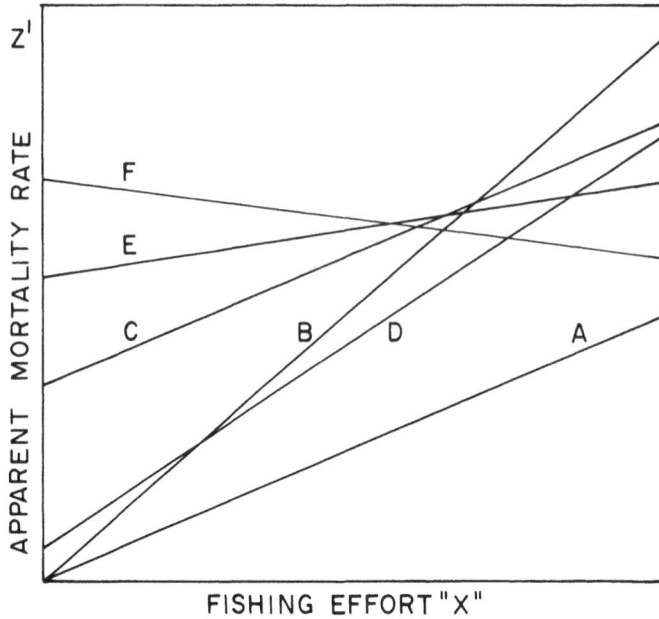

FIG. 7.4. Types of regression lines obtained when two kinds of fishing effort, X and Y, attack a stock, of which Y is unknown. Natural mortality = 0 (see the text).

In practice there will be natural mortality (M) in addition to fishing mortality. If M is invariable from year to year, or if it varies but is not correlated with either X or Y, all the lines in Fig. 7.4 will be shifted upward by the average magnitude of M, but will have the same slopes and relative positions.

Usually the effect of fishing effort Y appears in varying proportion in the slope b and in the intercept a. For all such lines we can write:

$$F = F_x + F_y = a - M + bf_x \qquad (7.14)$$

Obviously no analysis can be made if there is a body of catch and effort whose magnitude is completely unknown but may be significant, and whose effort may be correlated (positively or negatively) with the known effort X. However, if the catch taken by Y is known but the effort is not, an analysis can be made on the basis that rates of fishing generated by the X and Y efforts in each year are proportional to their respective total catches (Section 6.6). This will be true only if the seasonal distribution of the two kinds of effort is similar.

7.8. RATE OF FISHING FROM CATCH AND EFFORT DATA, RECRUITMENT, AND NATURAL MORTALITY RATE — ALLEN'S METHOD

Allen's (1966b, p. 1562) method of analysis, like that of Section 7.6, requires an independent estimate of natural mortality rate, which is assumed constant for all vulnerable fish in all years. His model implies a Type 1 population, in which three

176

events of the biological year occur separately and in the following order: recruitment of fish previously not vulnerable, fishing mortality, and natural mortality. In such circumstances $1 - e^{-F} = m = u$, and the mean population present during the fishing season is equal to u/F times the initial population; in the model this is approximated by the initial population less half the catch.

7.8.1. TWO YEARS' DATA AVAILABLE. Natural mortality rate (M) and catchability (q) of the fish are assumed to be the same in both years. Symbols are as follows:

	Year 1	Year 2
Fraction of new recruits in catch	$\ldots$	W_2
Initial population	N_1	$N_2 = \dfrac{(N_1 - C_1)e^{-M}}{1 - W_2}$
Population after fishing	$N_1 - C_1$	$N_2 - C_2$
Mean population during fishing	$N_1 - C_1/2$	$N_2 - C_2/2$
Catch	$C_1 = qf_1(N_1 - C_1/2)$	$C_2 = qf_2(N_2 - C_2/2)$
Final population	$(N_1 - C_1)e^{-M}$	$(N_2 - C_2)e^{-M}$

Of the above, data are available concerning catches C_1 and C_2, and efforts f_1 and f_2. The fraction of new recruits, W_2, is estimated by any available method, for example that of Section 11.3. For the natural mortality rate, M, an estimate or trial value is used as described in Section 7.6. This leaves N_1, N_2, and q to be obtained by solving the expressions for C_1, C_2, and N_2 above.

7.8.2. SEVERAL YEARS' DATA AVAILABLE. For this situation Allen (1966b, p. 1564) shows how to obtain an overall best estimate of q, using normal equations. The criterion used for the best estimate is that which minimizes the sum of squares of differences between actual and expected catches. From this q other population statistics can be calculated for each year. The computations are fairly arduous and are best carried out by computer. An appropriate program is THPOP by Allen, available in Fortran IV at the Pacific Biological Station of the Fisheries Research Board of Canada, Nanaimo, B.C., or from CSIRO at Cronulla, Australia.

If computer assistance is not available, it is always possible to solve the data for successive years by pairs as in Section 7.8.1, thus obtaining a series of q values. This has the advantage that it will show up any trend in q that might result from subtle improvements in fishing technique or in the skill of the fishermen—these being difficult to quantify in estimating effective fishing effort. If no trend appears, the q values obtained can be averaged. This average should be close to that obtained using Allen's normal equations and computer program, although usually not exactly the same.

7.8.3. ILLUSTRATION OF ALLEN'S METHOD. Table 7.3 shows three successive years of the constructed population of Tables 11.1 and 11.2. Using computer program THPOP, the estimates are $N_1 = 12,987$, $N_2 = 12,841$, $N_3 = 10,332$, and $q = 0.0400$.

TABLE 7.3. Data for estimating populations by Allen's method, from Tables 11.1 and 11.2. Catchability $= q = 0.04$ in all years.

	Fishery year		
	1	2	3
Fishing effort	10	20	30
Catch, C	4,372	7,304	7,750
Fraction of new recruits, W	....	0.503	0.603
Natural mortality, M	0.3	0.3	0.3
e^{-M}	0.7408	0.7408	0.7408
C/2	2,186	3,652	3,875
N	13,259	13,259	11,090

Compared with the true figures in the last line of Table 7.3 these population estimates are $2 - 7\%$ low, presumably because $N - C/2$ is used for the average population during fishing instead of Nu/F which was used in calculating them. However, the estimate of catchability is exact, and this provides a means of obtaining correct population estimates. In a Type 1 fishery the rate of exploitation u is equal to $1 - e^{-qf}$, and catch divided by u is the initial population, as shown below:

Year	$F = qf$	u	C	N
1	0.4	0.3297	4372	13,260
2	0.8	0.5507	7304	13,260
3	1.2	0.6988	7750	11,090

In any real example there would of course be sampling error. This can be estimated approximately from the number of fish available in the age samples used to estimate W.

The value of q obtained from the same data by Paloheimo's method was 0.0387 (Section 7.6.2.)

7.8.4. ILLUSTRATION OF YEAR-BY-YEAR COMPUTATION. If no computer is available, population parameters can be estimated from Table 7.3 by the method of Section 7.8.1. For years 1 and 2 the three necessary equations are:

$$4372 = 10q(N_1 - 2186)$$

$$7304 = 20q(N_2 - 3652)$$

$$N_2 = \frac{0.7408(N_1 - 4372)}{1 - 0.503}$$

Solving these gives $N_1 = 12,720$, $N_2 = 12,460$, $q = 0.0414$.

178

For years 2 and 3 the equations are:

$$7304 = 20q(N_2 - 3652)$$
$$7750 = 30q(N_3 - 3875)$$
$$N_3 = \frac{0.7408(N_2 - 7304)}{1 - 0.603}$$

Solving these, $N_2 = 12{,}880$, $N_3 = 10{,}400$, $q = 0.0396$.

The computed estimates, averaged where possible, are $N_1 = 12{,}720$, $N_2 = 12{,}670$, $N_3 = 10{,}400$, and $q = 0.405$. All these estimates except N_3 are not as good as those obtained from the computer program: the negative error in N is never less than 4%. However, the error in q is only a 1.2% overestimate, so again improved estimates of N can be had by calculating rates of exploitation and dividing them into the catches, as described in the previous section. The values obtained are $N_1 = 12{,}130$, $N_2 = 12{,}160$, and $N_3 = 11{,}020$.

CHAPTER 8. — ESTIMATION OF STOCK AND MORTALITY FROM STATISTICS OF THE CATCH AND ITS QUALITATIVE COMPOSITION

8.1. DERZHAVIN'S BIOSTATISTICAL METHOD OF POPULATION ANALYSIS

If fishing is the major cause of mortality, much information about the dynamics of a population can be obtained by making an age census of the catch and dividing the fish among their appropriate year-classes. The sum of catches from a particular year-class, during the years it contributes to the fishery, is a minimum estimate of its abundance at the time it was just entering the catchable size range. Similarly, partial sums will give the minimum number of fish in existence, of each year-class, in any given calendar year. The total of these sums in a particular year represents the minimum number of catchable fish present in that year. This figure has been called the *utilized stock* (*izpolzuemyi zapas*) by Voevodin (1938) and the *virtual population* by Fry (1949): it represents the population present at a given time, with the exception of all fish that will subsequently die of natural causes.

The above procedure is often called the *biostatistical method* of population analysis, and is associated particularly with A. N. Derzhavin. A precursor was described in a paper by Tereshchenko (1917) concerning the Volga bream (*Abramis brama*), is which Baranov's assistance is acknowledged. The assumptions used there include: (1) representative sampling of bream stock during the autumn-to-spring fishery; (2) complete recruitment of all bream of age 2+ and older (i.e. those which had completed 3 growing seasons); (3) the same rate of exploitation for all ages; (4) negligible natural mortality at all ages greater than 1; and (5) constant recruitment at age 2, from year to year. Under these conditions the catch in a year must equal the number of age 2 recruits; consequently the total commercial stock is equal to catch divided by the ratio of age 2+ individuals to the total sample. In Tereshchenko's example, catch was taken as 20 million pieces, of which 66% were age 2+. Thus the total stock was calculated as 20/0.66 = 30 million pieces. (*See also* Baranov 1918, p. 100, for this computation).

Derzhavin (1922) freed this procedure of some of the restricting conditions above and in effect developed a new approach. He did this by: (1) using an age composition based on age and length data calculated over a period of years, so that possible short-term variability in year-class strength was smoothed out; (2) using catch data for many years and calculating absolute abundance by ages for each year separately; and (3) computing a separate rate of exploitation for each age-class, calculated from the mean age composition. He perforce retained the assumption of no long-term trends in percentage age composition, while pointing out that this was not strictly in accord with the observed fact of moderate trends in the catch. Again, fish dying from natural mortality were not considered, but natural deaths are likely to have been relatively fewer, among his potentially long-lived sturgeon, than among Tereshchenko's short-lived bream. Outside of the USSR Derzhavin's method was apparently first used by Bajkov (1933) in an application to whitefish of Lake Winnipegosis, but the results were misleading because that population was far from having a steady age composition (*see* Example 2.9).

The total population considered by Derzhavin comprised all fish in the population at the start of a calendar year, including those of the previous year's hatch, since for these "the large 'infant' mortality among the newly-hatched young no longer plays a role, and the surviving fingerlings have grown . . . enough that they can become the prey of large fishes only." Let us designate fish as "age 1" from January 1 of their first year of life, etc., and define:

C_1, C_2, etc.   catches in years 1, 2, etc.

x_1, x_2, etc.   fractional representation of each age, t, in the catch of a given year
($x_0 = 0$)

r greatest age involved

Ignoring fish which die naturally from age 1 onward, the total stock at the start of a year will be the sum of: that year's catch, the next year's catch diminished by the number of age 1 fish in it (because these were not yet hatched at the beginning of the base year), the next year's catch not including the age 1 and age 2 fish, and so on.

Using average age composition for estimating the fraction of 1's, 2's, etc., the expression for population at the start of age 1 becomes (cf. Derzhavin, p. 15):

$$N = C_1 + C_2(1 - x_1) + C_3(1 - x_1 - x_2)$$
$$+ \ldots + C_r(1 - x_1 - x_2 - \ldots - x_{r-1})$$
$$= \sum_{t=1}^{t=r} C_t(1 - x_1 - x_2 - \ldots - x_{t-1}) \tag{8.1}$$

If fish do not begin to appear in the fishery until after some years have elapsed — for example, at age 5 — then x_1, x_2, x_3, and x_4 are equal to zero, and all of the first four years' catch is included in the total stock. However, if only stock *of commercial age* were needed, the first four terms of (8.1) would then be omitted from the total.

The ratio of the catch in a given year to the utilized stock present at the start of that year can be called the *biostatistical rate of exploitation* (Ricker 1970, 1971b). It is always greater than the true rate of exploitation.

EXAMPLE 8.1. UTILIZED STOCK AND EXPLOITATION OF KURA STURGEON. (From Ricker 1958a, after Derzhavin 1922.)

From data on length composition and age determinations made in 1912–19, described in Example 2.9, Derzhavin (p. 229) constructed a table of the probable absolute age structure of sevryuga (*Acipenser stellatus*) catches taken in the Kura River from 1881 to 1915. The year-classes 1854–1906 were represented, at ages 9–27. Too extensive to be reproduced here, the columns of this table, when totalled vertically, provide estimates of the complete contribution of the year-classes 1872 through 1888 to the catch; and also the contributions of substantial portions of several adjacent broods.

Derzhavin, however, was most interested in estimating total stock present in the sea in successive years. The percentage age composition was summed cumulatively from the oldest back to the youngest, a few rare age-groups being ignored at either end (Table 8.1, column 3). These sums comprise the terms $(1 - x_1)$, $(1 - x_1 - x_2)$, etc., of expression (8.1). Each is then multiplied by the catch of the corresponding year. As an

182

example, the stock for 1881 is computed in Table 8.1, column 5. The total is 9,383,000 sturgeon, 5,024,000 being age 9 and older. The 1881 catch of 427,000 is 8.5% of the latter, which is the overall biostatistical rate of exploitation. The biostatistical rate of exploitation of fully-vulnerable fish is of course greater than this: it can be computed for individual ages using Derzhavin's complete table. For example, 22,200 fish of age 20 were taken in 1881, and 50,400 of this year-class were captured in later years. The biostatistical rate of exploitation of age-20 fish in 1881 was therefore $C/V = 22,200/(22,200 + 50,400) = 31\%$.

TABLE 8.1. Computation of the 1881 stock of Kura River stellate sturgeon, after Derzhavin (1922); catches are in thousands.

1	2	3	4	5
Age	Mean age composition of the catch	Cumulative age composition	Catches 1881–1907	Contribution to population at the beginning of 1881
1	0	1.000	427(1881)	427
2	0	1.000	405	405
3	0	1.000	437	437
4	0	1.000	539	539
5	0	1.000	591	591
6	0	1.000	589	589
7	0	1.000	720	720
8	0	1.000	651	651
9	0.006	1.000	699	699
10	0.027	0.994	738	734
11	0.061	0.967	814	787
12	0.107	0.906	694	629
13	0.118	0.799	544	435
14	0.110	0.681	540	368
15	0.093	0.571	451	258
16	0.080	0.478	573	274
17	0.076	0.398	702	279
18	0.090	0.322	621	200
19	0.076	0.232	564	131
20	0.052	0.156	583	91
21	0.042	0.104	745	77
22	0.030	0.062	548	34
23	0.018	0.032	517	17
24	0.007	0.014	503	7
25	0.004	0.007	490	3
26	0.002	0.003	403	1
27	0.001	0.001	292(1907)	0
Total				9383

In Section 8.2 it is shown that, under the equilibrium conditions that Derzhavin postulated, this 31% is really an estimate of *total* annual mortality rate, A. The overall biostatistical rate of exploitation, 8.5%, is a more complex statistic that has no easy interpretation, because many incompletely-recruited ages are included in it.

8.2. UTILIZED STOCK AND BIOSTATISTICAL RATE OF EXPLOITATION WHEN AGE COMPOSITION VARIES — BOIKO'S METHOD

A considerable refinement of the Derzhavin method was made by Boiko (1934, 1964), Monastyrsky (1935), Chugunov (1935), and Fry (1949), all working independently. They took representative age samples of the catch *each year*, and thus were able to total up a more accurate series of utilized populations.

Analyses of this sort provide a minimum estimate of population size, V, and an estimate of the biostatistical rate of exploitation, C/V or $u(max)$, which is greater than the true rate of exploitation. C/V can be calculated for the whole stock or for individual ages separately. The catch of fish of a given age, t, in a given year, divided by the utilized stock of that age at the start of the year, is:

$$u(\text{max})_t = \frac{C_t}{V} \qquad (8.2)$$

From this a *biostatistical rate of fishing* can be estimated as:

$$F(\text{max})_t = -\log_e[1 - u(\text{max})_t] \qquad (8.3)$$

An estimate of *biostatistical catchability* of fish of age t is:

$$q(\text{max})_t = \frac{F(\text{max})_t}{f_t} \qquad (8.4)$$

Models can be used to compare the magnitude of the biostatistical with the true rate of exploitation, and with other population statistics. Three are shown in Tables 8.2–8.4, and others have been published by Ricker (1970, 1971b). From these and other models the conclusions below were obtained. They apply only to situations where neither F nor M for a given age change with time (though they may be different at different ages):

1. The biostatistical rate of exploitation, C/V, for the oldest age represented will always be 100%, by definition; thus it will be larger than the actual rate of exploitation, u. The next younger age will also have a considerable bias from the same source.

2. When neither natural nor fishing mortality rates vary with age, C/V exceeds u by a constant factor, apart from sampling fluctuation. The ratio of C/V to u is equal to Z:F or A:u; in other words, biostatistical rate of exploitation is equal to the true total annual mortality rate, A (Table 8.2, ages 8–13).

TABLE 8.2.   Comparison of the biostatistical and true rates of exploitation (C/V and C/N) in a population in which rate of fishing (F) increases from age 3 to age 8 and remains constant thereafter; while the instantaneous rate of natural mortality, M, remains constant throughout. (The population, column 7, is rounded to the nearest integer, but fractions were retained in calculating it back from one fish surviving at age 16).

Age	F	M	Z	A	S	Population N	Total deaths	Catch C	Utilized stock V	C/V = u (max)	C/N = u
3	.005	.2	.205	.185	.815	4210	780	19	1723	.011	.0045
4	.03	.2	.23	.206	.794	3430	707	93	1704	.055	.027
5	.1	.2	.3	.259	.741	2720	706	235	1611	.146	.086
6	.3	.2	.5	.394	.606	2020	794	478	1376	.35	.236
7	.5	.2	.7	.503	.497	1223	615	440	898	.49	.359
8	.6	.2	.8	.551	.449	608	335	252	458	.55	.413
9	.6	.2	.8	.551	.449	273	151	113	206	.55	.413
10	.6	.2	.8	.551	.449	123	68	51	93	.55	.413
11	.6	.2	.8	.551	.449	55	30	23	42	.55	.413
12	.6	.2	.8	.551	.449	25	14	10	19	.53	.413
13	.6	.2	.8	.551	.449	11	6	5	9	.56	.413
14	.6	.2	.8	.551	.449	5	3	2	4	.50	.413
15	.6	.2	.8	.551	.449	2	1	1	2	.50	.413
16	.6	.2	.8	.551	.449	1	1	1	0	1.00	.413
						0			0		

185

TABLE 8.3. Comparison of the biostatistical and true rates of exploitation (C/V and C/N) in a population in which rate of fishing decreases after age 7. (The population is calculated back from 20 fish at age 18, to reduce error in rounding to the nearest integers.)

Age	F	M	Z	A	S	Population N	Total deaths	Catch C	Utilized stock V	C/V = u (max)	C/N = u
5	.6	.2	.8	.551	.449	42100	23200	17360	31360	.55	.413
6	.6	.2	.8	.551	.449	18860	10400	7800	14000	.56	.413
7	.6	.2	.8	.551	.449	8470	4670	3500	6200	.56	.413
8	.55	.2	.75	.528	.472	3810	2010	1473	2700	.54	.387
9	.50	.2	.70	.503	.497	1796	905	645	1227	.53	.359
10	.45	.2	.65	.478	.522	894	427	296	582	.51	.331
11	.40	.2	.60	.451	.549	467	210	141	286	.49	.301
12	.35	.2	.55	.423	.577	257	108	69	145	.48	.269
13	.30	.2	.50	.394	.606	148	58	35	76	.46	.236
14	.25	.2	.45	.362	.638	90	32	18	41	.44	.201
15	.20	.2	.40	.330	.670	57	19	9	23	.39	.165
16	.15	.2	.35	.295	.705	38	11	5	14	.36	.126
17	.10	.2	.30	.259	.741	27	7	2	9	.22	.086
18+	.10	.2	.30	.259	.741	20	13	7	7	......	......

TABLE 8.4. Comparison of the biostatistical and true rates of exploitation (C/V and C/N) in a population in which natural mortality rate increases after age 7. (The population is calculated back from 10 fish at age 16, to reduce error in rounding off).

Age	F	M	Z	A	S	Population N	Total deaths	Catch C	Utilized stock V	C/V = u (max)	C/N = u
3	.03	.2	.23	.206	.794	68380	14090	1838	31623	.06	.027
4	.1	.2	.3	.259	.741	54290	14058	4686	29785	.16	.086
5	.3	.2	.5	.394	.606	40232	15852	9511	25099	.38	.236
6	.4	.2	.6	.451	.549	24380	10995	7330	15588	.47	.301
7	.4	.2	.6	.451	.549	13385	6037	4025	8258	.49	.301
8	.4	.25	.65	.478	.522	7348	3512	2162	4233	.51	.294
9	.4	.3	.7	.503	.497	3836	1930	1103	2071	.53	.287
10	.4	.35	.75	.528	.472	1906	1006	537	968	.55	.282
11	.4	.4	.8	.551	.449	900	496	248	431	.58	.276
12	.4	.45	.85	.573	.427	404	232	109	183	.59	.270
13	.4	.5	.9	.593	.407	172	102	45	74	.61	.264
14	.4	.55	.95	.613	.387	70	43	18	29	.62	.258
15	.4	.6	1.0	.632	.368	27	17	7	11	.64	.253
16+	.4	.65	1.05	.650	.350	10	10	4	4		.253

3. When natural mortality, M, increases with age, and rate of fishing is constant, C/V tends to increase with age whereas u decreases with age (Table 8.4). The net result is an increasing discrepancy between C/V and u, so that the former can become 2 or 3 times the latter. However, C/V is close to A in such circumstances, being slightly greater than A.

4. Within age-groups in which recruitment occurs, i.e. when F is increasing with age, C/V exceeds u by a much greater fraction of the latter than during a series of years when F does not vary with age. However, during such years C/V is less than A (Table 8.2, ages 3–7).

5. If rate of fishing (F) and thus rate of exploitation (u) decrease with age after reaching some maximum, while M is constant, C/V decreases with age but not as rapidly as u does; thus C/V considerably exceeds the total mortality rate, A, over the years concerned (Table 8.3, ages 8–17).

6. If the age sample taken from the catch (and which is applied proportionally to the whole catch to represent the age composition of the latter) is biased so that older fish tend to appear more frequently than their true abundance warrants, C/V is somewhat less than it would otherwise be, but is not very seriously changed. However, if older fish appear disproportionately only in occasional years, important bias is introduced (Example 8.2).

7. It is possible to use C/V for a whole stock (including recruits) as an "arbitrary" index that reflects the direction and magnitude of changes in fishing intensity over a period of years. *However, it is important that the same minimum age be used in all years.*

8. Considering the utilized population for two successive ages, V_1 and V_2, the relationship in paragraph 2 above can be written:

$$\frac{C_1}{V_1} = \frac{V_1 - V_2}{V_1} = A_1 = 1 - S_1; \text{ thus } S_1 = \frac{V_2}{V_1}$$

This estimate of S, like A, is consistent only when M and F do not vary with age or, more exactly, when F/Z does not change during the remaining life of the year-class. When this is not so, bias in S will be opposite to that indicated for A in paragraphs 3–5.

9. Space does not permit an evaluation of effects of secular trends in F or M upon statistics computed from utilized stocks (as was done for catch curves in Chapter 2) but these effects should be examined in experiments where they might be significant. Table 2 of Ricker (1971b) is of this sort. It indicates that moderately large changes in F from year to year have little effect on C/V, provided only fully-recruited ages are included in V.

EXAMPLE 8.2. UTILIZED STOCKS (VIRTUAL POPULATIONS) AND BIOSTATISTICAL RATE OF EXPLOITATION OF OPEONGO TROUT. (From Ricker 1958a, after Fry 1949.)

The estimated age composition of the catch of Opeongo lake trout (*Salvelinus namaycush*) for 1936–47 was tabulated in Table 2.8 of Example 2.7. The minimum number of survivors of each brood at each age is obtained by summing the table diagonally from upper left to lower right; results are shown in Table 8.5. (The figures in parentheses in the lower right corner are the average of previous entries, since catches for years later than 1947 would be needed to supply actual data.)

Contributions of all ages to the fishery are available for the year-classes 1934–37; these comprise the utilized population at age 3, for those broods. (Fry points out that the average contribution of a year-class is less than 1 fish per 4 hectares of lake — indicating the sparseness of this population.)

Total utilized stock of all ages is found by summing entries of Table 8.5 diagonally from lower left to upper right; this gives a result corresponding to the summing of column 5 of Table 8.1 above (in the sturgeon example). Thus, at the start of the 1936 fishing season there were *at least* 10,129 fish of age 3 and older in the lake; there were at least 8640 in 1937, 7210 in 1938, 6959 in 1939, and 6599 in 1940.

The biostatistical rate of exploitation at each age and in each year is now estimated as the ratio of catch to utilized population. A random example: 104 age-9 trout were caught in 1941 out of at least 205 present; hence $C/V = 51\%$.

Average values of C/V are plotted in Fry's figure 2. These suggest that after the increase in vulnerability during the recruitment phase there is a decrease in C/V at older ages, from 0.53 at age 8 to 0.26 at age 13. According to Table 8.3 above, in an equilibrium fishery this would reflect an even greater decrease in true rate of exploitation, u, with age. However, the apparent decrease should be at least partly discounted. It depends very heavily on an exceptionally large estimated catch of fish older than age 13 in one year, 1945, which in turn seems to be based on only 7 actual specimens. Considering only the first four years of the fishery, the series of biostatistical rates of exploitation is as follows:

Age	C/V	Age	C/V	Age	C/V
7	0.37	10	0.68	13	0.52
8	0.60	11	0.52	14	0.60
9	0.66	12	0.55	15	0.61

Here the apparent maximum vulnerability is at age 10 rather than age 8. The small decrease in C/V beyond age 10 indicated by this series may well reflect a real decrease in the true rate of exploitation, though for a complete analysis the possible effects of the secular changes in F should be examined.

8.3. ESTIMATION OF ACTUAL POPULATION FROM UTILIZED POPULATION AND RE-
     COVERY OF MARKED FISH — FRASER'S METHOD

Fraser (1955, p. 172) showed how a utilized (virtual) population estimate can be converted to an estimate of actual population when combined with results of a

TABLE 8.5. Utilized stocks of Opeongo lake trout, arranged by age and brood. (Adapted from Fry 1949.)

Year-class	\	\	\	\	\	A g e	\	\	\	\	\	\	\	\	\	Total utilized stock	Year
	3	4	5	6	7	8	9	10	11	12	13	14	15	16	17		
1919															11	10129	(1936)
1920														15	0	8640	(1937)
1921													10	0	0	7210	(1938)
1922												29	4	4	4	6959	(1939)
1923											34	15	7	4	0	6599	(1940)
1924										69	12	4	4	0	0		
1925									140	22	18	15	0	0	0		
1926								326	66	28	19	6	0	0	0		
1927							685	207	31	28	4	4	4	4	4		
1928						1396	731	176	86	39	33	22	20	20	16		
1929					1861	1387	362	167	77	63	52	50	50	41	4		
1930				1665	1432	782	343	120	74	63	54	48	26	10	3		
1931			1371	1243	1045	625	232	112	90	82	58	49	6	6	(3)		
1932		1294	1199	1165	890	569	205	101	73	73	42	21	14	(8)	(3)		
1933	1223	1193	1189	1062	841	407	207	154	112	68	31	20	(11)	(8)	(3)		
1934	1129	1122	1059	939	715	480	359	226	133	49	23	(20)	(11)	(8)	(3)		
1935	1277	1265	1229	1147	872	655	444	246	87	40	(28)	(20)	(11)	(8)	(3)		
1936	1388	1349	1265	1121	1004	732	530	157	64	(46)	(28)	(20)	(11)	(8)	(3)		
1937	1194	1174	1095	1049	928	731	379	162	(86)	(46)	(28)	(20)	(11)	(8)	(3)		

marking experiment. The utilized population, V, at the start of a given year, (call it year 1), is the part of the then number of recruited fish which will subsequently be caught in all future years. The total recoveries from M fish marked at the start of year 1, in successive years of their appearance in the fishery, is $R_1 + R_2 + R_3 + \ldots = R$. The ratio R/M is an estimate of the exploitation ratio, $E = F/Z$, provided no tags are lost from the fish and all are reported when caught. But the exploitation ratio is the ratio of fish caught from a year-class (V) to its total abundance at recruitment (N); thus $V/N = E = R/M$, and the estimate of N becomes:

$$N = \frac{VM}{R} \qquad (8.5)$$

This estimate is analogous to a Petersen estimate (expression 3.5), but it has some advantages. One is that it is free from the bias which can occur in Petersen estimates because of differences in vulnerability of different sizes of fish (Section 3.7). In addition, the Fraser estimate is less affected by systematic error arising from aberrant behaviour or aberrant vulnerability of fish immediately after marking. If marking makes a fish relatively invulnerable for some weeks or months, the reduced recoveries during that season will largely be compensated by increased recaptures the following season (unless natural mortality is large). Equally, if marking increases vulnerability temporarily, the excess recaptures in the first year are mostly compensated by fewer recaptures later. For a similar reason, with Fraser's method it is not so important to do the marking exactly at the start of year 1: the period of marking may be extended some days or weeks into the fishing season of year 1 without much effect upon the estimate of N.

On the debit side, a population estimate from (8.5) involves the delay and (usually) the sampling error inherent in any computation of utilized population. It also implies the use of a tag or mark which will not become progressively lost or indistinguishable *over the whole vulnerable life-span* of the fish, rather than for one or two years only.

8.4. COMBINATION OF UTILIZED STOCK ESTIMATES WITH THE BEVERTON–HOLT ITERATIVE PROCEDURE

At the 1957 Lisbon conference on fishing, several contributors apparently suggested using virtual populations to obtain separate estimates of rate of fishing and natural mortality rate, by methods similar to the Beverton–Holt procedure of Section 7.4 above. Most of these contributions seem never to have been published; but shortly afterward Paloheimo (1958) and Bishop (1959) described the method, and Bishop made an extensive analysis of the magnitudes of errors of the statistics estimated under different conditions. She found that estimates of natural mortality (M) would be too small and those of catchability (q) too large when fishing effort tended to increase over a period of years or when it fluctuated without any trend; the opposite biases occurred when there was a trend of decreasing effort.

. The method appears to be of limited usefulness so is not described here. Both Paloheimo and Bishop used ordinary predictive regression lines for their estimates, so these have an additional and usually unknown bias (M too large, q too small) insofar as there is uncertainty in the measurement of effective fishing effort.

8.5. RATES OF NATURAL AND FISHING MORTALITY, GIVEN CONSTANT RECRUITMENT AND NATURAL MORTALITY AND TWO OR MORE LEVELS OF STABLE CATCH — TESTER–GULLAND METHOD

When the history of a fishery shows two or more periods of (different) stable catch, C, which have the same natural mortality rate, M, and the same absolute level of recruitment, R, Tester (1955) showed that an estimate of M and R can be made. The type of calculation depends on the relative timing of natural and fishing mortality and recruitment (Section 1.5). The most manageable situations are:

1. Type 1A populations, with recruitment immediately before the fishing season.

2. Type 2A populations, with "instantaneous" recruitment at the start of the year.

3. Type 2B populations, with recruitment occurring throughout the year along with fishing and natural mortality.

The information needed is statistics of catch (C) in numbers and an estimate of instantaneous mortality rate (Z) for each period, based on age structure (Chapter 2).

8.5.1. TYPE 2A POPULATIONS. With arguments similar to those which lead to expressions (1.13) and (1.18) of Chapter 1, Tester developed the relationship:

$$Z = M + \frac{ZC}{R}$$

Gulland (1957) showed that for the purpose of estimating M and R it would be better to have Z appear only once. For an equilibrium situation we can write:

$$C = Nu = \frac{NAF}{Z} = \frac{RF}{Z} = \frac{R(Z - M)}{Z} = R - \frac{MR}{Z}$$

Transposing and dividing by MR:

$$\frac{1}{Z} = \frac{1}{M} - \frac{C}{MR} \tag{8.6}$$

Thus the regression of $1/Z$ against C is an estimate of $-1/MR$, and its Y-axis intercept is an estimate of $1/M$ (Gulland inadvertently quoted the reciprocals of these quantities), from which both M and R can be obtained. Since catch data are usually much more accurate than estimates of mortality rate, the ordinary predictive regression is the best one to use for this estimate.

If only two periods of stable catch are available, (8.6) can be developed into an expression explicit for R:

$$R = \frac{C_1 Z_1 - C_2 Z_2}{Z_1 - Z_2} \tag{8.7}$$

Indeed it is probably better to solve the data by pairs of periods, even if more than two are available.

8.5.2. TYPE 2B POPULATIONS. When recruitment is continuous and is balanced by mortality, we have, from (1.17) and (1.25):

$$C = F\bar{N} = NAF/Z = RF/Z$$

so that the expressions (8.6) and (8.7) and the procedure developed for Type 2A are applicable.

8.5.3. TYPE 1A POPULATIONS. Of N fish present at the start of the year, Nu are caught; also $u = m$, since no natural mortality occurs during the time of fishing. The recruitment, R, is however still equal to NA: initial population times total annual mortality rate. Thus:

$$C = Nu = Nm = Rm/A; \quad \text{or } m = CA/R$$

Combining (1.6) and (1.3) with the above, and taking logarithms:

$$Z = M - \log_e(1 - CA/R) \tag{8.8}$$

To fit the best straight line to (8.8), it is necessary to use trial values of R, and to continue fitting until the best fit is obtained. However, the specification of what constitutes the best fit presents difficulty, so it is better to compare situations by pairs, using a rearrangement of (8.8), as follows:

$$R = \frac{C_1 A_1 e^{Z_1 - Z_2} - C_2 A_2}{e^{Z_1 - Z_2} - 1} \tag{8.9}$$

In applications of the Tester–Gulland method the principal point will be whether or not average level of recruitment can be considered constant over the times involved. Tester's (1955) account should be consulted for this, and also for his method of computing catch-curve mortality rates so they refer to catches for the appropriate series of years.

8.6. SEQUENTIAL COMPUTATION OF RATE OF FISHING AND STOCK SIZE (COHORT ANALYSIS)

8.6.1. GENERAL AND HISTORICAL. If the natural mortality rate or rates in a stock be known or assumed, computations can be made of the fishing mortality rate experienced by a year-class at successive ages, using its catch at each age as obtained from catch statistics and yearly age-composition data. If is also necessary to know or to assume a value for rate of fishing F (or u or Z or S) for one age as a starting point for the computation. The age chosen should be the oldest, or one of the oldest, to which the computation is applied, because estimates of F computed for younger ages will converge asymptotically to their true values for the given M, whereas estimates for older ages will diverge progressively (unless the initial trial value of F happens to be correct).

The mechanics of a sequential computation based on Baranov's catch equation were described by Ricker (1948) in an application to the incompletely-vulnerable ages of a halibut stock (Example 5.6 above). Jones (1961) constructed an artificial example of a sequential computation for successive ages of a year-class, starting with the oldest, and he was the first to demonstrate the convergence and divergence mentioned above. Jones used Paloheimo's (1958) approximate formula for estimating fishing effort, but this seems to have no advantage over the original Baranov relationship in this context. Murphy (1965) proposed computations similar to Ricker's, i.e. using the Baranov equation, and developed a computer program for the calculations. He and Tomlinson (1970) have compared "forward" with "backward" calculations and confirmed the superiority of the latter. Gulland (1965) also employed the Baranov equation, but used as control data the utilized or virtual populations found by summing catches (Section 8.1). The theory of this procedure requires that $V = FN/Z$ or $C/V = A$ at each age, which is true only if F/Z remains constant throughout the remaining life of the year-class (see Section 8.2 for the effect of variation in F on the ratio C/V). Thus using virtual populations seems merely to make the calculation less exact and a little more complicated. Jones (1968) returned to the method of Ricker and Murphy, putting the computations into a slightly different form (described below). Schumacher (1970) was the first to publish a "work table" of $SZ/FA = S/u$ against F (for $M = 0.2$). Pope (1972) suggested an approximation which is easier to use; he called it "cohort analysis," but this term had earlier been applied to sequential computations in general, and is so used here.

8.6.2. RELATIONSHIPS. Considering a Type 2A fishery (Section 1.5.1), Baranov's equation for catch in numbers (expression 1.17) can be written as:

$$C = \frac{FAN}{Z} = uN \tag{8.10}$$

where N is the population at the beginning of a year. Consider any two successive ages in a cohort, called 1 and 2 for convenience. From (8.10):

$$N_2 = \frac{C_2}{u_2} \tag{8.11}$$

By definition:

$$N_2 = N_1 S_1 \tag{8.12}$$

From (8.10) and (8.12):

$$\frac{C_1}{N_2} = \frac{u_1 N_1}{N_1 S_1} = \frac{u_1}{S_1} = \frac{F_1 A_1}{S_1 Z_1} \tag{8.13}$$

Expression (8.13) is the same as (5.37). It can be solved by trial as described in Section 5.7.3, though for a long series of ages this becomes tedious. Computer programs have been written for it several times; two such are MURPHY by P. K.

194

Tomlinson (Abramson 1971), and COHORT by K. R. Allen (Fisheries Research Board of Canada, Nanaimo, B.C.). The latter permits use of different values of M at different ages.

Selection of the initial trial value of F should be considered carefully. Earlier we noted that computed values of F converge toward the correct value as a sequential computation moves to progressively younger ages, whereas they diverge when moving toward older ages. Nevertheless, it is sometimes undesirable to use the oldest age as a starting point, either because the estimate of its catch is subject to large sampling error, or because it is possible that natural mortality rate may increase rapidly among the oldest fish, so that the usual assumption of constant M is invalid. When there is semiquantitative information about fishing effort, an excellent starting point, if it can be found, is two successive (older) ages for which fishing effort is the same or nearly the same. Then C_2/C_1 will be an estimate of both S_1 and S_2, from which (plus M) F_2 can readily be calculated and used to start sequential computation. If the efforts of the two years are close but not exactly the same, a good approximate estimate of S_1 can be obtained from:

$$S_1 \sim \frac{C_2 f_1}{C_1 f_2} \tag{8.14}$$

When the F_2 used to start the sequential series refers to some age other than the oldest catch available, the computation can also be continued forward to the last age represented.

8.6.3. POPE'S METHOD. Still considering Type 2A fisheries, Pope (1972) developed an approximate expression for N_1 that can be used for a more direct analysis. For any two successive ages, 1 and 2, he writes:

$$N_1 \sim N_2 e^M + C_1 e^{M/2} \tag{8.15}$$

The rationale of this expression is obvious. Approximation is involved only in the second term, which would be exact if all of the catch C_1 were taken at the mid-point of the time period under consideration (usually a year). If the fishing mortality rate F is distributed uniformly throughout the year, more fish of any fully-recruited age are caught during the first half of the year than during the second half; thus (8.15) tends to overestimate N_1 somewhat, but by only a few percent for ordinary values of F and M. In any event most fisheries are concentrated seasonally, so using (8.15) may be no more approximate than using (8.13) — often less so. Also, during years of recruitment (8.15) is likely to be superior to (8.13), if fish are growing and becoming more vulnerable throughout the fishing season.

The great advantage of (8.15) is that it makes the computations easy without a computer. N_1 is calculated without iteration; $S_1 = N_2/N_1$, and thus Z_1 and F_1 are calculable. Expression (8.15) also makes it easier to determine effects of systematic and random errors in a sequential computation; Pope (1972) has examined these in detail.

EXAMPLE 8.3. SEQUENTIAL COMPUTATIONS OF RATES OF FISHING FOR PERUVIAN ANCHOVIES. (Data from Burd and Valdivia 1970.)

Burd and Valdivia did a sequential analysis, at monthly intervals, of a series of cohorts of Peruvian anchovies (*Engraulis ringens*), based on the catch equation and using a computer program. Table 8.6 here shows their catch data for cohort 1969–1 in the central region (i.e. the fish that first became vulnerable in important quantities in the early months of 1969), grouped by 2-month intervals. Analysis of Table 8.6 is done using Pope's method (expression 8.15), with $M = 0.2$ per 2-month period. Column 2 is the catch in individuals/10^7. Column 3 is the catch times 1.1052 ($= e^{M/2}$). Column 5 shows *initial* populations for the period indicated; thus the quotient of any two successive entries (older/younger) is the survival rate for the interval in which the denominator stands. These are shown in column 6. Figures in column 4 are 1.2214 ($= e^M$) times the figure in column 4 *for the next older interval*.

To get the computation started a figure for rate of fishing has to be assumed for the last catch interval — in this case $F = 0.1$. Adding $M = 0.2$ gives $Z = 0.3$, and the corresponding $S = 0.741$ is found from Appendix I and entered in column 6. By trial we find the number of survivors in period 3 of 1972 which, when multiplied by e^M and added to the catch of period 2, provides an estimate of S as close as possible to 0.741. In this case 94 is correct: $94 \times 1.2214 = 115$, $115 + 12 = 127$, and $94/127 = 0.740$, the closest available value to 0.741. For the next step 127 is multiplied by 1.2214 to get 157, which is entered in column 4 of the next youngest interval, and as there is no catch, 157 is entered in column 5 also. Then 157 is multiplied by 1.2214 to get 192, and 128 is added from column 3 to give 320, which is an estimate of initial population in period 1971(6). For the same period the quotient $157/320 = 0.491 = $ S; $Z = -\log_e S = 0.71$, and $F = 0.71 - 0.2 = 0.51$. The same procedure is repeated for each pair of intervals back to the first.

In column 9, rates of fishing are added by years; the grand total is 3.38. To this can be added a total natural mortality rate of $M = 21 \times 0.2 = 4.20$, so that total mortality is $Z = 7.58$. This can be checked by calculating the overall figure $Z = -\log_e (94/189244) = 7.61$, which agrees within the limits of rounding error used.

F values found in Table 8.6 are influenced by the natural mortality rate used, and by the initial value of F. Table 8.7 compares estimates obtained using several different values of these parameters. (1) For any given M, the larger the value of initial F, the greater the total F. However, most of the difference occurs in the later part of life when fish are scarce; the discrepancy is not great during the main period of the catch. (2) At any given initial F, the natural mortality rate used makes a major difference to the absolute level of fishing effort computed. The larger M is, the smaller is the computed F at all intervals, and increasingly so (relatively) among younger fish. However, the larger M is, the larger is the estimated total mortality throughout life. In actual fact, it is likely that M would vary with age. Natural mortality may increase as fish become senile, and at the other end of the size range small fish may be more vulnerable to predators. In addition, the smallest fish probably suffer important attrition from catching operations, which appears as "natural" mortality:

196

TABLE 8.6. Sequential computation of rate of fishing by Pope's method, for a cohort of Peruvian anchovies. Natural mortality is held constant at $M = 0.2/2$ months, and initial $F = 0.1$. C — catch $(\times 10^{-7})$; N — initial population; S — survival rate; Z — total mortality rate; F — rate of fishing. Column 1 indicates the six 2-month periods into which each year is divided.

1	2	3	4	5	6	7	8	9
	C	$Ce^{M/2}$	Ne^M	N	S	Z	F	F
1968								Totals
6	823	910	188,334	189,244	.815	0.20	0.00	0.00
1969								
1	12,006	13,269	140,926	154,195	.748	0.29	0.09	
2	16,858	18,631	96,750	115,381	.687	0.38	0.18	
3	2,138	2,363	76,850	79,213	.794	0.23	0.03	
4	0	0	62,919	62,919	.819	0.20	0.00	
5	2,186	2,416	49,098	51,514	.780	0.25	0.05	
6	741	819	39,379	40,198	.802	0.22	0.02	0.37
1970								
1	739	817	31,424	32,241	.798	0.23	0.03	
2	1,556	1,720	24,008	25,728	.764	0.27	0.07	
3	642	710	18,946	19,656	.789	0.24	0.04	
4	0	0	15,512	15,512	.819	0.20	0.00	
5	4,331	4,787	7,913	12,700	.510	0.67	0.47	
6	2,722	3,008	3,471	6,479	.439	0.82	0.62	1.23
1971								
1	0	0	2,842	2,842	.819	0.20	0.00	
2	1,116	1,233	1,094	2,327	.385	0.95	0.75	
3	129	143	753	896	.688	0.37	0.17	
4	0	0	616	616	.818	0.20	0.00	
5	102	113	391	504	.635	0.45	0.25	
6	116	128	192	320	.491	0.71	0.51	1.68
1972								
1	0	0	157	157	.809	0.20	0.00	
2	11	12	115	127	.741	0.30	0.10	0.10
3	0	0		94				
Totals	46,216					7.58	3.38	3.38

many escape the net in damaged condition, and others simply disintegrate in the course of the two pumping operations that intervene between the closed seine and the arrival of the anchovies at the plant.

Burd and Valdivia (1970) did their sequential analysis over a period of years. Comparing successive estimates of F with the fishing effort, they found that computed catchability tended to increase by about 15% per year. This implied a rather rapid increase in efficiency of fishing operations that was not reflected in the unit of fishing effort used (based on vessel tonnage and time fished).

197

TABLE 8.7. Results of sequential computations of the data of Table 8.6 using different values of M and of initial F.

M	0.1	0.1	0.2	0.2	0.2
Initial F	0.1	0.2	0.1	0.2	1.0
F-values					
1968–9	0.79	0.82	0.37	0.37	0.37
1970	1.49	1.64	1.23	1.30	1.36
1971	2.15	2.69	1.68	2.25	3.45
1972	0.10	0.20	0.10	0.20	1.00
Total F	4.53	5.35	3.38	4.12	6.18
Total M	2.10	2.10	4.20	4.20	4.20
Total Z	6.63	7.45	7.58	8.32	10.38
Computed Z	6.73	7.46	7.61	8.33	10.33
Initial stock ($\times 10^{-9}$)	82.5	83.8	1892	1859	1832

8.6.4. NATURAL MORTALITY. Unless the natural mortality rate used in a sequential computation is based on objective information, it is desirable to try several values (or sets of values) of M to assess effects of differences in this parameter.

Lassen (1972) proposed using the catch data themselves to obtain an estimate of M (the same for all ages) by successive trials, choosing the one which minimizes the sum of squares of the differences between calculated and observed catches. The necessary computer program is FMESTIM of Danmarks Fiskeri- og Havundersögelser, Charlottenlund. However this minimum sum of squares is not very sensitive to changes in M, and it is easy to obtain "best" estimates of M that are at variance with other known characteristics of a stock.

8.7. SEQUENTIAL COMPUTATIONS WHEN FISHING PRECEDES NATURAL MORTALITY

In Type 1 fisheries fishing precedes natural mortality and it is evident that:

$$N_2 = (N_1 - C_1)e^{-M} \qquad (8.16)$$

This can be rearranged in a form similar to Pope's basic expression (8.15 above):

$$N_1 = N_2 e^M + C_1 \qquad (8.17)$$

Thus a cohort analysis can be made for Type 1 fisheries using the same sequence of operations as in Pope's method, the only difference being that no approximation is involved.

8.8. Population Estimates by "Change of Composition" or "Dichotomy" Methods

If a population is classifiable in two or more ways, and harvesting from it is selective with respect to this classification, it is possible to make a population estimate from knowledge of original composition, final composition, and composition of the harvested catch. Classification might be by age, size, colour, sex, etc. To date the procedure has been used mostly with game birds or mammals, for which classification by sex is often easy and the kill is frequently very selective — whether because of legal restrictions or the habits of the animals. Chapman (1955) found it difficult to locate the origin of this method in space or time, but he has himself given it the most complete treatment to date.

Designating two classifications of individuals by X and Y, the information available is:

n_1, n_2 size of samples taken at the beginning and end of the "harvest" period (times 1 and 2)

x_1, x_2 number of X-items in samples n_1, n_2

y_1, y_2 number of Y-items in samples n_1, n_2

$p_1 = x_1/n_1; p_2 = x_2/n_2$

C_x number of X-individuals caught during the harvest period (the period between times 1 and 2)

C_y number of Y-individuals caught

$C = C_x + C_y$ total catch

We wish to know:

N_x number of X-items at time 1

N_y number of Y-items at time 1

$N = N_x + N_y$

The maximum likelihood estimates of N_x and N are, after Chapman:

$$N_x = \frac{p_1(C_x - p_2 C)}{p_1 - p_2} \qquad (8.18)$$

$$N = \frac{C_x - p_2 C}{p_1 - p_2} \qquad (8.19)$$

N_y is obtained by difference. These formulae assume there is no natural mortality, nor any other unaccounted mortality, during the time of the kill or harvest. During the harvest the two kinds of individuals in the population must be *un*equally vulnerable; however, during the sampling done at time 1 in order to determine p_1, the X-type and Y-type individuals should in general have the same vulnerability to the sampling apparatus; and similarly for the sampling at time 2.

An exception to this latter condition occurs when the Y-type (say) is not caught at all during the harvest period ($C_y = 0$). If so, it is only necessary that the X-type and Y-type have the same *relative* vulnerability at times 1 and 2, in order to obtain from expression (8.18) an estimate of N_x (i.e. the ratio of the vulnerability of X to the vulnerability of Y at time 1 should be the same as this ratio at time 2). However (8.19) is not applicable in that event, so that no estimate of the Y-type population is then possible. In this situation it would be possible, as Chapman points out, to use a sport fish as the X-type and a trash fish as the Y-type, though if times 1 and 2 are far apart the postulate of unchanged relative vulnerability of the two species may become risky.

Chapman's account can be consulted for estimates of asymptotic variances of these estimates under different conditions and for other pertinent information. In practice, the size of C_x may have to be estimated by sampling the catch, thus increasing the variability, but this is not serious as long as a fair-sized and representative sample is taken.

Since a pre-harvest sample has to be taken for the dichotomy method, it will often be easy to mark the fish involved and try to obtain, concurrently with the dichotomy estimate, Petersen or Schnabel estimates (of the X- and Y-types of fish separately). This would provide the check which is always so desirable in population estimation.

An advantage of the dichotomy method over marked-fish methods is that it avoids potential mortality or distortion of vulnerability which are apt to be inherent in handling and marking fish. However, conditions appropriate for using the method do not seem to occur very frequently.

8.9. ESTIMATE OF SURVIVAL FROM DIFFERENCES BETWEEN THE SEXES IN AGE AT MATURITY — MURPHY'S METHOD

8.9.1. RELATIONSHIPS. Murphy (1952) used the age composition of individual year-classes of coho salmon (*Oncorhynchus kisutch*) returning to a fishway for an estimate of survival during the last year of sea life. In the southern part of their range cohoes mature at age 2 and age 3; among mature age-2 fish males are in excess, whereas females are usually in excess at age 3. Let there be M' males and F' females approaching the end of their second year of life. Let x be the fraction of age 2 males which matures, y the fraction of age 2 females which matures, and S the survival rate of non-maturing 2's of both sexes up to the time they approach maturity as 3's. Then the expected numbers in each category are as below, and can be equated to observed numbers A, B, C, and D:

		Age 2		Age 3
	Total	Maturing	Not maturing	Maturing
Males	M'	$M'x = A$	$M'(1 - x)$	$M'S(1 - x) = C$
Females	F'	$F'y = B$	$F'(1 - y)$	$F'S(1 - y) = D$

For age-2 matures, males exceed females and the difference is $A - B = M'x - F'y$. For age-3 matures, females usually exceed males and the difference is $D - C = S(M'x -$

200

$F'y + F' - M'$). If we know the ratio of the two sexes just before any mature at age 2, $M'/F' = a$ say, these expressions can be developed algebraically into estimates of S and F' (thus also M'):

$$S = \frac{aD - C}{A - aB} \tag{8.20}$$

$$F' = \frac{AD - BC}{aD - C} \tag{8.21}$$

If a fishery attacks age-3 individuals near the end of their life span, the method can still be used if the two sexes are equally vulnerable *or* if the catch of each sex can be added in to C and D. In that event fishing mortality is included in the estimate of total mortality rate $(1 - S)$. If a significant number of age-2 fish are caught in salt water, as is true today in sport fishing areas, the method will fail unless estimates of number and sex of these removals are obtained and brought into the equations.

Murphy's formulae were simpler than the above because he assumed that $F'/M' = 1$, and he could put $B = 0$ because mature age-2 females are almost unknown among cohoes (and he observed none). However, the method is sensitive to quite small deviations from a 50:50 sex ratio among age-2 stock, and the assumption of equality may not be valid. Some data are available on sex ratios of salmon smolts at the time they go to sea, though apparently not for cohoes. Jensen and Hyde (1971) found 490 males and 480 females among hatchery chinook fingerlings in 1968 — a difference which can be of interest only by comparison with other data. Suggestive and in one case significant deviations from 50:50 were found among sockeye smolts at Cultus Lake in some years (Foerster 1954b), indicating a small excess of males among the yearlings which predominated in the runs. Foerster also showed that females tend to predominate during the early part of the run and males in the later part, so that single samples, such as some of those listed by Clutter and Whitesel (1956, table 63) are not necessarily representative of a complete run. However, the largest samples of sockeye yearlings in Clutter and Whitesel's table all show an appreciable excess of males (Rivers Inlet 1915 and 1916, Chilko 1954). Also, in the one year (1953) in which Dombroski (1964) sampled Babine sockeye smolts proportionally throughout the season there was an excess of males, as there was among total smolts taken in all 4 years. Thus there are indications that among Pacific salmon, as among most other vertebrates where the subject has been studied, there is a small excess of males early in life. Possibly this excess would be somewhat reduced after the first growing season in the sea, but absence of direct information on the sex ratio at that time is evidently a deficiency of the estimates obtained by Murphy's method.

Even apart from uncertainty about sex ratios, any attempt to apply the Murphy method to species having more varied life-histories seems out of the question at the present time. For example, in British Columbia the sockeye have ages 3, 4, and 5 all represented by both sexes (although age 3 females are uncommon); also, a selective fishery seriously distorts the sex composition of maturing 4's and 5's, so that accurate information about the number and sex of fish captured at each age would be needed, as well as of the escapement.

8.9.2. ILLUSTRATION. A Murphy-type experiment provides the following numbers of mature fish (including those caught near the end of life):

Number of age 2 males		A = 100
,, ,, ,, 2 females		B = 10
,, ,, ,, 3 males		C = 50
,, ,, ,, 3 females		D = 75
Ratio of males to females before any mature at age 2		a = 1.08

From (8.20) and (8.21) we compute:

Survival rate $= S = (81 - 50)/(100 - 11) = 0.348$
Number of females at age 2 $= (7500 - 500)/(81 - 50) = 226$
Number of males at age 2 $= 226 \times 1.08 = 244$

If the ratio a were taken as 1, the estimates are $S = 0.278$, $F = M = 280$.

CHAPTER 9. — GROWTH IN LENGTH AND IN WEIGHT

9.1. ESTIMATION OF AGE AND RATE OF GROWTH

9.1.1. GENERAL AND HISTORICAL. Techniques for determining the age of fish are diverse and have been amply reported. Few kinds of fish in temperate waters can hope to conceal their age from a persistent investigator: length frequency distributions, tagging experiments, scales, otoliths, opercular bones, vertebrae, fin rays, etc., can all be called on.

The first known account of reliable age determinations of fishes is by the Swedish clergyman Hans Hederström (1959; original version 1759). By counting rings on the vertebrae he obtained the age of pike (*Esox lucius*) and other species, and his growth rates are similar to modern estimates. However, the art of age determination had to be rediscovered toward the end of the 19th century, and the history of this period has been reviewed by Maier (1906), Damas (1909), and others. This time the first method to be applied was that of length frequencies, by C. G. J. Petersen (1892) — Fig. 9.1 here. Scales were first used for age reading by Hoffbauer (1898), otoliths by Reibisch (1899), and various other bones by Heincke (1905). Much early work by D'Arcy Thompson and others, using Petersen's method, was later shown to be inaccurate because a succession of modes had been treated as belonging to successive year-classes, whereas they actually represented only *dominant* year-classes, separated by one or more scarce broods. More recently length-frequency analysis has been made easier and more reliable by several statistical and mechanical aides: Harding (1949), Cassie (1954), and others used probability paper; Tanaka (1956) devised the method of fitting parabolas to the logarithms of frequencies; and Hasselblad (1966) fitted successive approximations to normal curves using a computer. Direct determination of rate of growth from successive measurements of tagged fish has sometimes been possible, but frequently the capture or tagging operation affects growth rate. There can also be an appreciable difference between the length of a fish when alive and some hours after it has died, which may partly explain why a net *decrease* in size of tagged individuals, even after many months at large, has been observed both in freshwater and marine fishes (cf. Holland, 1957).

The various techniques of age determination have been reviewed by Rounsefell and Everhart (1953) and Tesch (1971), and a number of comprehensive works justify their applicability in general or their application to particular species (e.g. Creaser 1926; Graham 1929a,b; Le Cren 1947; Van Oosten 1929; Chugunova 1959). However, no one claims that *all* his age determinations are infallibly accurate, and older fish often present considerable difficulty.

As well as telling the current age of the fish, markings on the hard parts (usually scales) are regularly used to compute fish length at the end of previous growing seasons, as indicated by the spacing of the "annuli". Again an extensive literature exists concerning methods of making this computation, the most suitable method being different for different fishes. In anadromous fishes the scales also reveal the length of time spent in fresh water and in the ocean, respectively. Finally, in some species the scale, otolith, or fin ray indicates at what age the fish first spawned (Rollefsen 1935; Monastyrsky 1940) and, in sturgeons, the sequence of years between spawnings (Derzhavin 1922, Roussow 1957).

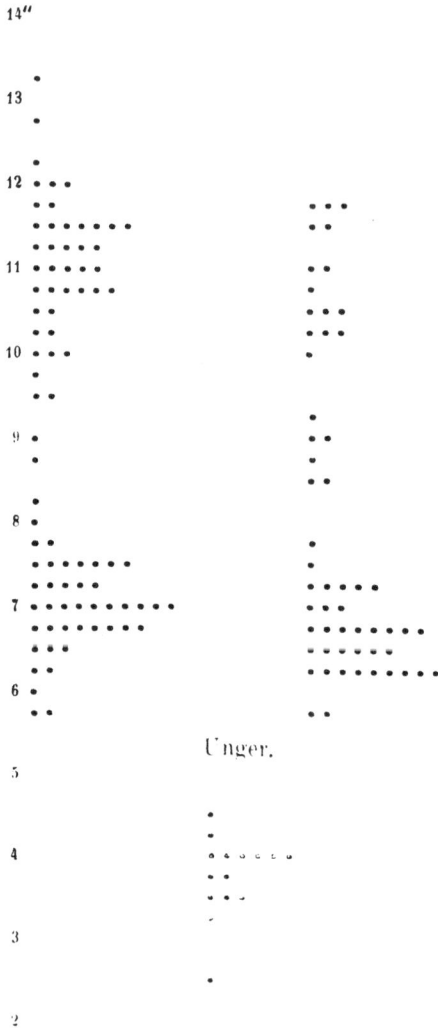

FIG. 9.1. Reproduction of the first length-frequency table used to indicate age in a fish population (Petersen 1892, p. 124). Measurements are of *Zoarces viviparus*, in Danish inches (= 26.15 mm).

Suppose a sample of a fish population has been taken and the age of each fish in it has been determined. The average size of fish at each age is then computed. A plot of these average sizes can be used directly to estimate rate of increase in size from year to year, provided (a) that there is no difference between year-classes in respect to rate of growth at any given age; and (b) that the fish taken constitute a random sample of each of the age-classes involved (not necessarily a random sample of several age-classes simultaneously, as was desirable in estimating mortality rate).

As far as they have been intensively studied, fishes apparently all exhibit an initial period of increasingly rapid absolute increase in length, followed by a decrease. The initial increasing phase is usually completed within the first two years of life, and if so, it may not appear at all in a graph of yearly increments. However, it is frequently exhibited in centrarchids (Fig. 9.3A).

The changeover from increasing to decreasing length increments may be so slow and protracted as to make the age-length relationship effectively linear for almost the whole of the fish's life, or for the part covered by the available data. This approximation has been used successfully in some production computations (Example 10.4). More commonly, a decrease in yearly increment of length is quite evident as the fish grow older.

9.1.2. DIFFERENCES BETWEEN YEAR-CLASSES. Differences between successive broods in rate of growth can be tested easily by taking samples in two or more successive years and comparing fish of the same age. If only one year's data are available, such differences will show up as irregularities in the line of plot of length against age, which can be adjusted to some extent by smoothing. Although differences in the rate of growth of successive broods in a population are fairly common, particularly when broods vary greatly in abundance, they are not often a serious obstacle to obtaining a picture of the *average* growth pattern by this method.

9.1.3. LIMITED REPRESENTATIVENESS OF SAMPLING. If only one sampling method is used, it is unlikely to be representative for all ages included. If it is most efficient for fish of intermediate size, it tends to select more of the larger members of the younger age-groups, and more of the smaller members of the older age groups (Fig. 9.2, Curve A). If this is not considered, and the sample is treated as representative, the growth rate obtained will always be *smaller* than the actual. The same is true whether the gear's maximum efficiency is for the smallest fish, or for the largest (Fig. 9.2, Curve B). The most direct way to avoid this trouble is to use several different kinds of sampling apparatus, all of which will probably be selective for size to some degree, but which will select different size ranges. In sampling some lake fish, for example, shore seining might take age 0 and even age 1 representatively; minnow traps might cover ages 1 and 2; fykenets, ages 4–7; and angling, ages 5–10. This would leave age 3 in doubt, but from specimens taken in traps and nets a fairly accurate average value might be obtained; or, that point could be interpolated. For further discussion of gear selectivity see Section 2.11.

9.1.4. AGE–LENGTH KEYS. Sometimes the length composition of catches taken from a stock is available from many localities, whereas age determinations are available from only part of them, or from another source. It is desired to estimate the age distribution in the total length sample. Subject to the conditions below, this can be

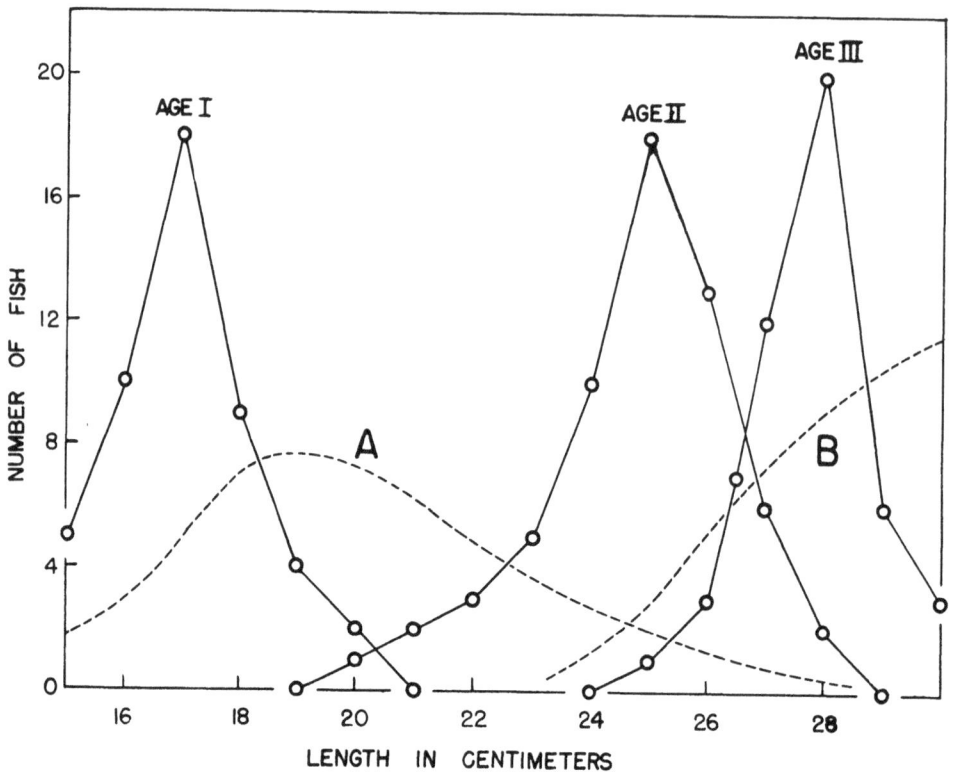

FIG. 9.2. Illustration of the effect of selectivity of gear. Three age-frequency polygons (adapted from those for Clear Lake ciscoes, Hile 1936, table 24) are fished by two hypothetical gears whose relative powers for taking different sizes of fish are as shown by curves A and B.

done by constructing an *age-length key* from a representative sample of fish from the population. This is a double frequency table, usually with age in the columns and lengths in the rows. A similar table is then constructed giving the percentage of each age among fish of a given length, and this is used to convert any observed length distribution to age[1].

In using an age-length key, one must remember that *the fish used for age determination must be taken from the same stock, during the same season, and using gear having the same selective properties as that used to take the length-frequency samples.* Above all, an age-length key cannot be applied to length samples of any year except the one from which it was derived, unless the year-classes represented always have the same initial abundance and are subjected to the same fishing experience — a condition seldom encountered.

---

[1] Allen (1966a) developed a computer program to perform these computations and print the resulting age distribution in numbers and percentages; it can be obtained from the Fisheries Research Board of Canada at Nanaimo, B.C., or St. Andrews, N.B.

As a rule the above restrictions mean that the increase in accuracy achieved by combining a length sample with a smaller age sample is not very great. Usually it is more profitable to put all available resources into increasing the size of the age sample, rather than mounting a massive length-sampling program, unless the latter also serves some other purpose.

9.1.5. GROWTH STANZAS. In their early development, fish typically pass through several distinct stages or *stanzas* of growth (Vasnetsov 1953), between which there occurs a rather abrupt change in structure or physiology. In extreme cases, such as eels, this may involve a metamorphosis comparable to what occurs in the higher insects; more commonly the stanzas are separated merely by a change in body form which shows up in the weight: length relationship, or sometimes simply by a sudden change in rate of growth. Usually the final growth stanza begins sometime during the first year of life; for example, Tesch (1971, p. 117) illustrates an abrupt change in the weight:length relationship of brown trout at 42 mm and about 1.1 g. Among salmon and other anadromous species there is typically a sudden change in rate of growth when the fish go to sea, which may occur at ages from $0+$ to $4+$ or more.

9.2. TYPES OF GROWTH RATES

Growth may be described in terms of length (l) or weight (w), and for each we may distinguish:

(1) *absolute increase* (increment) in a given year: $l_2 - l_1$ or $w_2 - w_1$;

(2) *relative rate of increase*: $\dfrac{l_2 - l_1}{l_1}$ or $\dfrac{w_2 - w_1}{w_1}$ (usually expressed as a percentage);

(3) *instantaneous rate of increase*: $\log_e l_2 - \log_e l_1$ or $\log_e w_2 - \log_e w_1$.

Figure 9.3 presents a typical population growth curve in several forms.

Relative and instantaeous rates are used mostly in connection with weight. Instantaneous rate of length increase and instantaneous rate of weight (G) increase are similar statistics, differing only by a constant. Anticipating expression (9.3) of the next section, this is evident from the following:

$$\begin{aligned} G &= \log_e w_2 - \log_e w_1 \\ &= \log_e a + b(\log_e l_2) - \log_e a - b(\log_e l_1) \\ &= b(\log_e l_2 - \log_e l_1) \end{aligned} \qquad (9.1)$$

This provides a convenient method of estimating G from length data, provided b is known.

9.3. WEIGHT–LENGTH RELATIONSHIPS

9.3.1. BASIC RELATIONSHIP. It has been found that, within any stanza of a fish's life, weight varies as some power of length:

$$w = al^b \qquad (9.2)$$

$$\log w = \log a + b(\log l) \qquad (9.3)$$

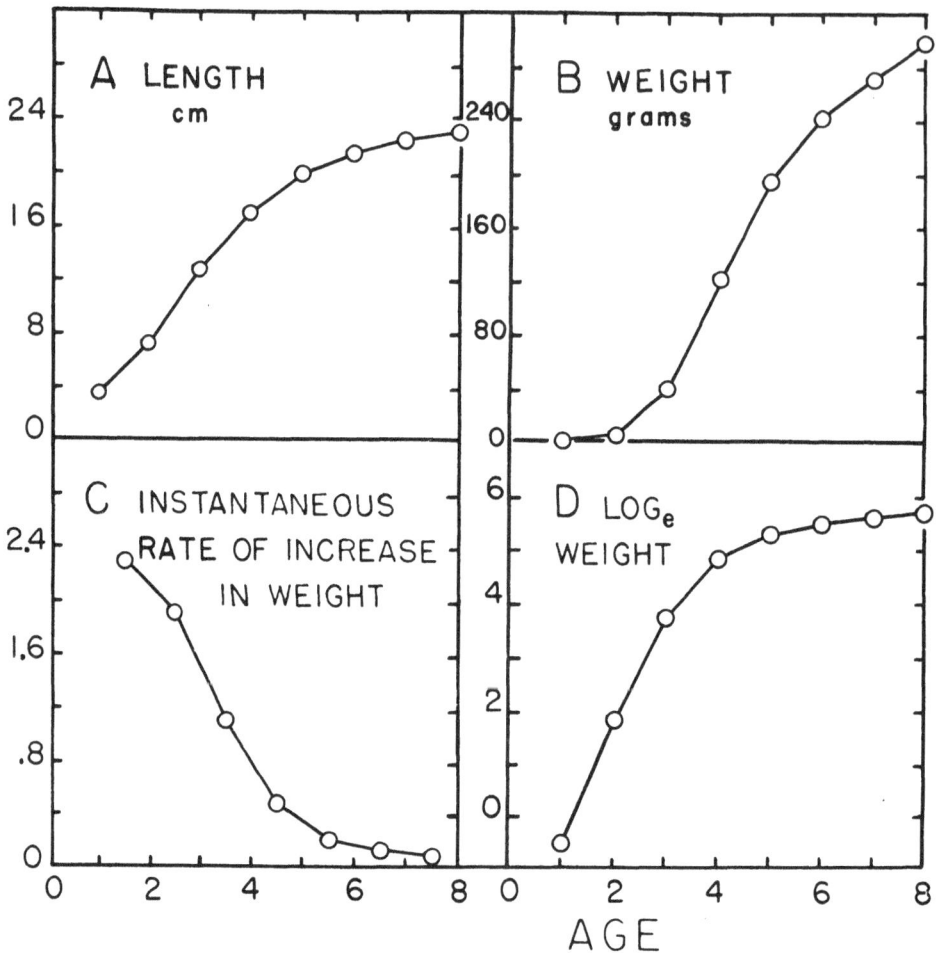

FIG. 9.3. Example of different measures of growth (for *Lepomis macrochirus* in Spear Lake, Ind.). (A) Fork length (cm); (B) Weight (g); (C) Instantaneous rate of increase in weight; (D) Natural logarithm of weight.

These expressions would apply best to an individual fish that was measured and weighed in successive years of its life. This of course is rarely possible. The value of b is usually determined by plotting the logarithm of weight against the logarithm of length for a large number of fish of various sizes, the slope of the fitted line being an estimate of b. The GM functional regression should be used, rather than the predictive regression which has commonly been employed in the past (Appendix IV).

9.3.2. GROUPING AND SELECTION OF DATA. Extensive data have sometimes been grouped into short length-classes, and the mean length and weight of each class have been used as the primary data in computing the regression line. The motive was to speed up computations, but with modern equipment this is no longer necessary, and when using means it is impossible to compute a representative standard error, nor can the functional regression be obtained.

A more legitimate reason for grouping data is to distribute observations more evenly among the range of sizes present, and so get a more representative relationship. This is best done by measuring some fixed number within each of a series of short length intervals and weight intervals. The intervals should preferably be in logarithms. Also, to obtain a representative functional regression one must select half the total sample on the basis of length and half on the basis of weight — otherwise there will be bias (Ricker 1973a).

When a general weight–length relationship for a population is desired, every effort should be made to obtain fish of a wide range of sizes, down to and including age 0 (unless of course the younger fish belong in a different growth stanza — see Section 9.1.5). When only the older, commercial-sized, fish are available, the parameters estimated can deviate importantly from their true values simply from sampling variability, and partial representation of younger ages frequently creates bias (Example 9.1).

9.3.3. ISOMETRIC AND ALLOMETRIC GROWTH AND CONDITION FACTORS. The functional regression value $b = 3$ describes *isometric growth*, such as would characterize a fish having an unchanging body form and unchanging specific gravity. Many species seem to approach this "ideal," though weight is affected by time of year, stomach contents, spawning condition, etc. On the other hand, some species have b-values characteristically greater or less than 3, a condition described as *allometric growth*. There are sometimes marked differences between different populations of the same species, or between the same population in different years, presumably associated with their nutritional condition. (The term allometry applies also to changes in the ratios of linear measurements of the fish).

To compare weight and length in a particular sample or individual, condition factors are employed. The commonest is *Fulton's condition factor*, equal to w/l^3, often considered to be *the* condition factor (Fulton 1911). It is the parameter a in expression (9.2) when $b = 3$. The heavier a fish is at a given length, the larger the factor, and (by implication) the better "condition" it is in. Fulton's condition factor is suitable for comparing different individual fish of the same species; it will also indicate differences related to sex, season, or place of capture. It is most useful if, under average or standard conditions, the exponent b in (9.2) is actually equal to 3 for the species in question. Fulton's factor can also be used to compare fish *of approximately the same length* no matter what the value of b.

The *allometric condition factor* is equal to w/l^b, where b is given a value determined for the species under standard conditions. As it is usually difficult to decide what conditions are standard, and as there is usually considerable error in estimates

209

of *b*, this factor has been much less used than Fulton's. Often Fulton's factor is used as an approximation even when the allometric factor is theoretically more appropriate.

9.3.4. MEASUREMENTS OF LENGTH AND WEIGHT, AND CONVERSION FACTORS. Condition factors have been computed using different kinds of lengths and weights. Among other ways, fish length may be measured, at the head end, either from the tip of the snout, from the end of the lower jaw (if it protrudes past the snout), from the middle of the orbit of the eye, or from the hind margin of the orbit. At the tail end, the measurement may be made to the end of the vertebral column, to the margin of the median rays of the tail fin, to the end of the longest caudal rays when held in a natural position, to the end of the longest rays when squeezed to an extreme position, to the longest ray of the upper lobe of the tail when held at its position of maximum extension, or in several other ways. In addition, it makes a difference whether length is measured on live fish, freshly killed fish, fish held in cool storage, or fish preserved in formalin or alcohol. Any length may be measured either "on the contour," by holding a flexible tape from snout to tail, or by laying the fish on a flat ruled measuring board. Weight may be of the whole fish, or of the fish less stomach contents, less gonads, less all entrails, or less entrails and gills.

Whatever length and weight are used, they should be specified in detail. The combination most frequently used in fishery work is fresh whole weight and length measured on a board from the most anterior point (tip of the snout or of the more protruding jaw when the mouth is closed) to the end of the median rays of the tail (= "fork", Smitt or median length), whenever these two points can be ascertained readily. These are the standard measurements for computing Fulton's condition factor. But when tails or snouts are frequently damaged, other reference points must be used. The combination of weight less entrails divided by fork length cubed is known as *Clark's condition factor*.

The variety of lengths and weights in use makes it necessary to obtain conversion factors by measuring the same fish in two or more different ways. If one length is plotted against another, a fitted straight line determines the conversion equation. This should be the GM functional regression line (Ricker 1973a). For many purposes a line through the origin is more convenient and sufficiently accurate; indeed it is not often that the best fitted line differs significantly from one through the origin. A line through the origin should also be fitted functionally (Appendix IV).

9.3.5. WEIGHT–LENGTH RELATIONSHIP WITHIN AGE-GROUPS. It is sometimes interesting to examine the nature of the relation of weight to length within individual age-groups. This is a very different concept from the weight:length relationship associated with growth. For example, within an age-group the functional exponent *b* could conceivably be less than 3 at the same time as the same exponent for each individual fish, measured throughout its life, was greater than 3. An overall weight–length relationship computed from fish of several age-groups reflects mainly the change in form as the fish grow, rather than the intra-age-group relationship. However, it is essential that the functional rather than the ordinary regression be used to compare an intra-age-group with an overall exponent.

9.3.6. WEIGHT–LENGTH RELATIONSHIPS BETWEEN AGE-GROUPS. The functional a and b coefficients calculated from expression (9.3) may be called "individual-a" and "individual-b". It is sometimes useful to calculate analogous functional coefficients, call them a' and b', from the relation of the logarithm of average weight ($\bar{w}$) of a random sample of the fish *of a given age* to the logarithm of their average length ($\bar{l}$); that is:

$$\log \bar{w} = a' + b'(\log \bar{l}) \tag{9.4}$$

Trial calculations will show that the average weight of all fish of a given age ($\bar{w}$) is greater than the average weight of a group of individual fish whose lengths are each exactly equal to the average length of the age-group. The difference is commonly of the order of 5%. For example, Graham (1938b, p. 62) obtained 531 grams and 506 grams, respectively, for gutted age-2 North Sea cod. Pienaar and Ricker (1968) give a formula by which $\bar{w}$ can be computed for an age-group when its variability in length and the relationship (9.2) are known.

Because $\bar{w}$ exceeds w at any given length, one or both of the between-age-group parameters a' and b' must exceed the corresponding "individual" values a and b. The actual situation depends on amount of variability in fish size in a year-class at successive ages, and on how this variability changes with age. Trials show that if the standard deviation in length of a year-class increases exactly proportionally to mean length, b' is the same as b, but a' is greater than a. If standard deviation in length remains constant as length increases, and expression (9.2) applies, the relation between $\log \bar{w}$ and $\log \bar{l}$ is not strictly linear; but the best line will have a slope (b') *less* than the individual b, and a' becomes even greater. In practice, standard deviation in length of a year-class often tends to increase, during early years of life, approximately proportionally to mean length. Later, however, it increases less rapidly, becomes static, or even decreases at older ages.

The coefficients a' and b' and expression (9.4) are useful mainly for converting, to terms of weight, *average* lengths that have been calculated from scale annuli. This circumvents, with little loss of accuracy, the tedious calculation of weights at successive ages from each fish individually, using a and b.

EXAMPLE 9.1. WEIGHT-LENGTH RELATIONSHIPS IN A SAMPLE OF BRITISH COLUMBIA HERRING.

1. Figure 9.4 is a plot of logarithm of weight (log w) against logarithm of length (log l) for herring taken from a pre-spawning school, about 4 months prior to spawning. These fish were mostly in stage 3 of maturation. Data are shown in Table 9.1. The schedule below shows the calculation procedure using an ordinary calculator, but with modern equipment the statistics can be obtained quickly from a desk-size electronic computer.

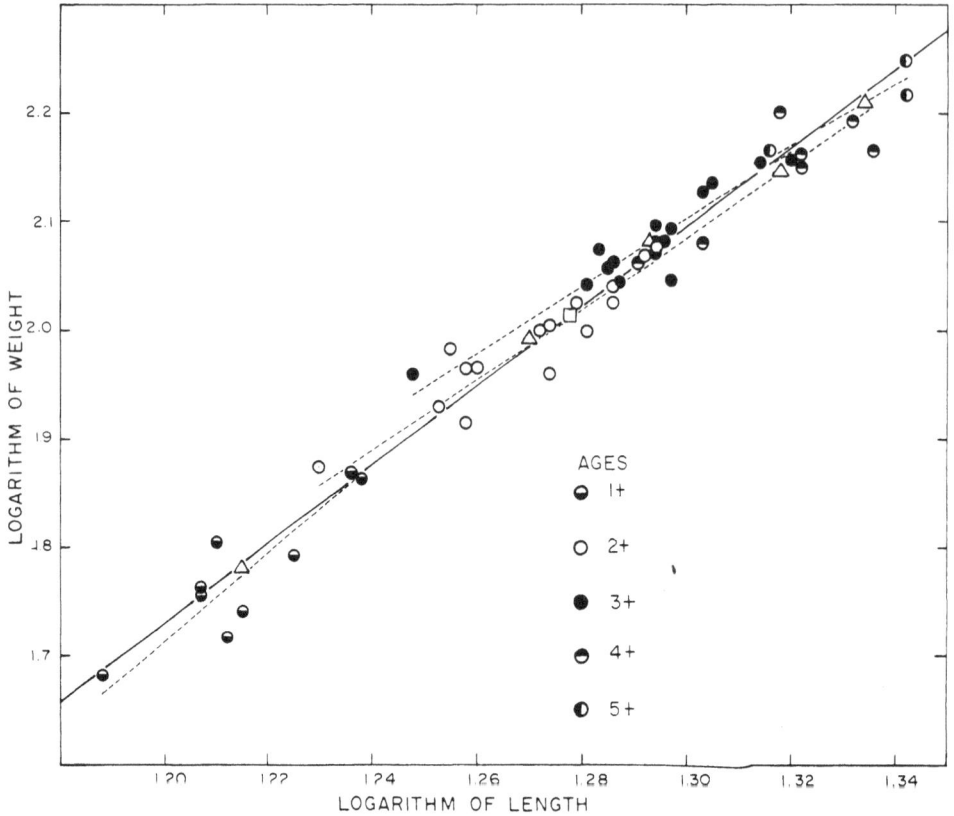

FIG. 9.4. Graph of logarithm of weight against logarithm of length, for the herring of Table 9.1. Solid line — functional regression line for all the points, passing through the grand mean (square point). Broken lines — functional regressions for individual ages. (These pass close to but not necessarily exactly through the triangular points, which represent the logarithms of the mean lengths and weights for each age-group.)

Sums and means:

    log l: 63.896   Mean: 1.2779

    log w: 100.712   Mean: 2.0142

Sums of squares and cross-products:

    log l 81.72599

    log w 203.80998

    (log l)(log w)   128.95911

Sums of squares and cross-products (measured from means):

Length	($= 81.72599 - 63.986^2/50$)	0.07201
Weight	($= 203.80998 - 100.712^2/50$)	0.95184
Products	($= 128.95911 - 63.896 \times 100.712/50$)	0.25723

212

TABLE 9.1. *Top*: Standard lengths (tip of lower jaw to end of scale covering) in centimeters, and whole weight in grams, for a sample of British Columbia herring taken from a purse seine catch at Porlier Pass, November 22, 1960. *Bottom*: Base-10 logarithms of the above lengths and weights. (from data supplied by Dr F. H. C. Taylor.)

	Age 1+		Age 2+		Age 3+		Age 4+		Age 5+	
	l	*w*	*l*	*w*	*l*	*w*	*l*	*w*	*l*	*w*
	17.2	74	19.7	119	19.3	115	20.8	159	22.0	177
	16.8	62	19.0	106	19.7	125	20.1	120	22.0	164
	16.2	64	18.8	101	19.2	118	19.5	115	20.7	146
	16.3	52	18.2	92	19.7	118	21.0	145		
	16.1	58	18.1	92	19.8	124	21.0	141		
	15.4	48	18.8	91	19.3	115	21.5	156		
	16.1	57	19.6	117	20.6	142	21.7	146		
	16.4	55	18.7	100	19.7	119				
	17.3	73	19.3	106	20.2	136				
			18.1	82	19.8	111				
			19.1	100	19.1	110				
			17.0	75	20.9	143				
			18.0	96	19.3	110				
			19.3	110	17.7	91				
			17.9	85	20.1	134				
					19.7	120				
Means	16.422	60.33	18.640	98.13	19.631	120.69	20.800	140.29	21.567	162.33
Log means	1.215	1.781	1.270	1.992	1.293	2.082	1.318	2.147	1.334	2.210

	log *l*	log *w*	log *l*	log *w*	log *l*	log *w*	log *l*	log *w*	log *l*	log *w*
	1.236	1.869	1.294	2.076	1.286	2.061	1.318	2.201	1.342	2.248
	1.225	1.792	1.279	2.025	1.294	2.097	1.303	2.079	1.342	2.215
	1.210	1.806	1.274	2.004	1.283	2.072	1.291	2.061	1.316	2.164
	1.212	1.716	1.260	1.964	1.294	2.072	1.322	2.161		
	1.207	1.763	1.258	1.964	1.297	2.093	1.322	2.149		
	1.188	1.681	1.274	1.959	1.286	2.061	1.332	2.193		
	1.207	1.756	1.292	2.068	1.314	2.152	1.336	2.164		
	1.215	1.740	1.272	2.000	1.294	2.076				
	1.238	1.863	1.286	2.025	1.305	2.134				
			1.258	1.914	1.297	2.045				
			1.281	2.000	1.281	2.041				
			1.230	1.875	1.320	2.155				
			1.255	1.982	1.286	2.041				
			1.286	2.041	1.248	1.959				
			1.253	1.929	1.303	2.127				
					1.294	2.079				
Means	1.2153	1.7762	1.2701	1.9884	1.2926	2.0791	1.3177	2.1440	1.3333	2.2090

213

Regressions and correlation:

Ordinary regression of log w on log l 3.5721
($= 0.25723/0.07201$)

Standard deviation from the regression line   0.025690

$$\left(= \left[\frac{203.810 - (128.959)^2/81.72599}{50 - 2} \right]^{\frac{1}{2}} \right)$$

Standard error of ordinary regression   0.09573
($= 0.025690/[0.07201]^{\frac{1}{2}}$)

Intercept on the log w axis   −2.5506
($= 2.0142 - 3.5721 \times 1.2779$)

Correlation coefficient   0.9825
($= 0.25723/[0.07201 \times 0.95184]^{\frac{1}{2}}$)

GM regression of log w on log l 3.636
($= 3.5721/0.9825$)

Standard error of functional regression   0.09573
(same as that of the ordinary regression)

Intercept on the log w axis   −2.634
($= 2.0142 - 3.636 \times 1.2779$)

The functional relation between log w and log l, as estimated by the geometric mean regression, is:

$$\log w = -2.634 + 3.636(\log l)$$
$$w = 0.002323 l^{3.636}$$

This functional line is drawn on Fig. 9.4.

The ordinary regression of log w on log l is estimated above as 3.572, differing from the functional regression 3.636 by only 0.064, less than the standard error of either. Obviously there is little difference between the two regressions in this case; however, a rather small change in the exponent makes a fairly large difference in a computed weight. In any event it is wise to avoid any kind of consistent bias, however small. Notice that the functional regressions in this situation should be used for prediction of w from l, or of l from w (Ricker 1973a).

2. Between-age-group regressions are computed similarly from the "log means" line of Table 9.1. They can be computed either directly from the five means, or by weighting each mean as the number of fish involved. The equations involving the functional regression (b' in 9.4) are:

Unweighted: $\log \overline{w} = -2.573 + 3.585(\log \overline{l})$
Weighted: $\log \overline{w} = -2.647 + 3.651(\log \overline{l})$

The weighted value for b' is very close to b: the line lies almost parallel to the line computed from the individual fish, but of course a little above it. The unweighted value of b' is smaller, because the means of fish weights for the two oldest, poorly-represented, ages happen to lie below the line for the individual fish, and they tilt the line down at the right end.

3. Regressions for individual age-groups are shown in Table 9.2. Values of predictive regressions in column 5 are all smaller than the predictive regression computed from fish of all ages combined, usually by a substantial amount. However, part of this difference is an artifact resulting from the shorter range of lengths available at any one age (Ricker 1973a); thus it does not necessarily indicate that the within-age-group weight:length relationship differs from the overall relationship. Functional regressions in column 6 do not have this bias; most are smaller than the overall value 3.636, but one is larger. Their weighted average is 3.310, so it is possible that the within-age relationship is different from the overall 3.636, but a much larger sample would be required to confirm this.

TABLE 9.2. Weight: length regressions (1) for individual age-groups, (2) for all ages combined, and (3) between age-group means, for the herring of Table 9.1. Column 7 is the standard deviation from the predictive regression line of log w on log l. Column 8 is the intercept of the functional relationship on the log w axis.

1	2	3	4	5	6	7	8
	No. of fish	Means		Regression			Functional intercept
		log l	log w	Ordinary	Functional b	s	log a
1+	9	1.2153	1.7762	3.557	4.037	0.02802	−3.130
2+	15	1.2701	1.9884	2.963	3.194	0.02021	−2.068
3+	16	1.2926	2.0791	2.872	3.059	0.01636	−1.875
4+	7	1.3177	2.1440	2.890	3.407	0.02649	−2.345
5+	3	1.3333	2.2090	2.596	2.819	0.01442	−1.550
Means (unweighted)	...	...	...	2.976	3.303	...	...
All ages	50	1.2779	2.0142	3.572	3.636	0.02569	−2.634

By age-group	No. of ages	Means		Regression			
		log l	log w	Ordinary	Functional b'	s'	log a'
Unweighted	5	1.2860	2.0424	3.578	3.585	0.00925	−2.573
Weighted	5	1.2780	2.0176	3.647	3.651	0.03478	−2.647

9.4. EFFECTS OF SIZE-SELECTIVE MORTALITY

9.4.1. LEE'S PHENOMENON. Frequently the larger fish in a year-class have a different mortality rate than the smaller ones: either greater or less, but usually greater. This can be detected when back-calculations of length at earlier ages are made from scales or otoliths, *using samples that are representative of the whole of each age-group involved.* When a larger fraction of the larger fish die, the result is "Rosa Lee's phenomenon"

215

(Sund 1911; Lee 1912)—a smaller estimated size for fish of younger ages, when calculated from scales of older fish, than the true average size at the age in question (see Table 9.3, below). A review and bibliography of this subject is given by Ricker (1969a).

1. *Natural* selection for size can bear more heavily on either larger or smaller fish. Faster-growing fish frequently tend to mature earlier and also become senile and die earlier than slower-growing fish of the same brood (Gerking 1957). This is the principal or perhaps the only cause of natural Lee's phenomenon in unfished populations. However, there are at least two possible situations that act in the opposite direction: (a) There is considerable evidence that during the first year of life slower-growing individuals are more susceptible to predation, for example among walleyes (Chevalier 1973). Such selective mortality during the first year cannot affect calculated growths differentially, because only after the first annulus is laid down can there be any back-calculation. However, if the same situation persists into the second or later years of life it means that, for example, the size at annulus 1 computed from fish of age 2 will tend to be *greater* than the same computed from fish of age 3. (b) The other situation occurs when fish of both sexes are sampled and analyzed together, but there are in fact sex differences in both growth rate and natural mortality rate. Among most flatfishes, for example, females grow faster and live longer than males; this makes calculated early growth increments *larger*, the older the fish from which they are computed, in samples where the sexes are not distinguished. The antagonistic action between situations such as these two and the earlier senility of fast-growing fish may result in the absence of detectable Lee's phenomenon, or even "reversed" Lee's phenomenon, in natural populations (Deason and Hile 1947).

2. Size-selection by a *fishery* is primarily the process of recruitment. The larger members of a year-class become vulnerable first, and it may be several years before the smallest ones are fully vulnerable. Obviously this can be a major cause of Lee's phenomenon. Also, the largest fish in a population may sometimes be less vulnerable than those of intermediate size, but in practice this is far less important.

9.4.2. SHAPE AND VARIABILITY OF LENGTH DISTRIBUTIONS. Size-selective mortality does not necessarily change the shape of the length distribution of a year-class. Jones (1958) showed that when the gradient of instantaneous mortality rates within a year-class is linear with respect to length, an originally normal length frequency distribution will remain normal no matter how severe the mortality gradient. In nature, of course, the length-mortality relation need not be exactly linear, but trials show that it requires a marked deviation from linearity to produce appreciable skewness in the derived length distribution. Jones (1958) also showed that with a linear size-mortality relation the variability of the frequency distribution is conserved, and trials show that even quite severe non-linear selection changes the variance too little to be detectable in practice (Ricker 1969a).

9.4.3. EFFECTS OF SELECTIVE SAMPLING. All methods of detecting selective mortality of fish of different growth rates presuppose that representative samples are taken. This is a vital condition, because non-representative sampling can produce Lee's phenomenon to a marked degree, whether the maximum catching power of the nets is for the largest fish, the smallest fish, or some intermediate size (Section 9.1.3; also (Ricker 1969a, p. 509). Nor does it matter whether the selection results from physical characteristics of the net, or from differing habits of fish of different sizes.

In addition, incorrect techniques of back calculation of size from annulus measurements can introduce an "artificial" Lee's phenomenon. For example, if scale annuli are taken as directly proportional to body length in a population where they are actually proportional to length less a constant quantity, the calculated first-year growth is always too small, and it becomes smaller, the greater the age of the fish from which it is calculated.

9.5. COMPUTATION OF MEAN GROWTH RATES

9.5.1. POPULATION GROWTH RATE AND TRUE GROWTH RATE. When there is size-selective mortality within a year-class, the true mean growth rate of the fish differs from the apparent or *population growth rate*, and it becomes necessary to distinguish between them. The population growth rate (G_x) is obtained simply by comparing the mean size of the surviving fish at successive ages. But if the larger fish of a year-class die more frequently than the smaller, this affects the mean size of the survivors, making it less than it would otherwise be. Accordingly it becomes necessary to estimate *true growth rate* from that of individual fish during the period involved. Usually the best available estimate of the growth rate of individual fish (G) comes from back-calculation of their length at the last two annuli on the scales;[2] the difference of the natural logarithms of these lengths, multiplied by the weight:length exponent, is the instantaneous rate of increase in weight for the year (expression 9.1). Figure 9.5 illustrates the difference between population growth rate (fine dotted line) and mean individual growth rate of the survivors at successive ages. The final, broken, segment of each sample line is the best available representation of the true mean growth rate $(\overline{G})$ of the fish present in the year indicated.

9.5.2. ESTIMATION OF TRUE GROWTH RATES. To estimate true mean growth rates the usual sequence of operations is as follows:

1. Determine age from scales, and make measurements to successive annuli.

2. Establish the relation of scale size to fish size.

3. For each fish, back-calculate the length at the start and close of the last complete year represented on the scale at each age.

4. Compute the mean intital and final length during this year for fish of each age separately.

---

[2] As there may be differential mortality within this interval and right up to time of sampling, the mean growth rate estimated from survivors at the time of sampling will be slightly less than a weighted mean value for all fish that were alive during the previous complete year. But this second-order effect can usually be disregarded.

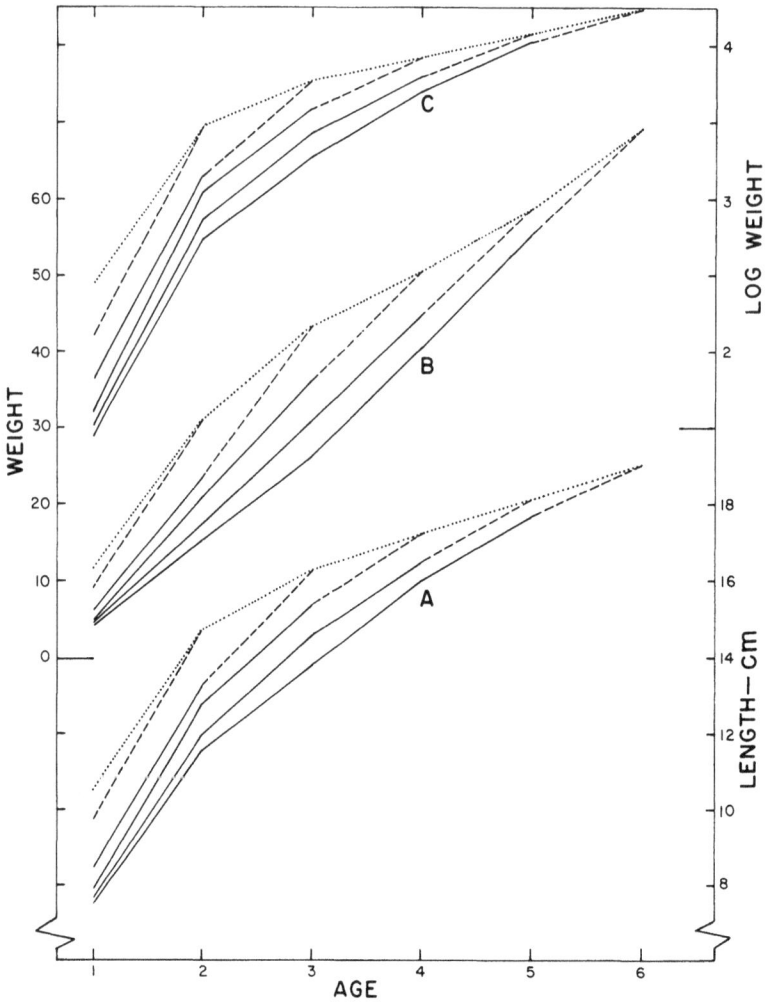

Fig. 9.5. Growth in an artificial cisco stock. (A) Standard length; (B) "Weight" (cube of length, which is approximately proportional to weight); (C) Natural logarithm of "weight". The broken segment of each sample line represents approximately the true mean growth rate of fish in the stock at successive ages. Dotted lines represent the population growth rates. (After Ricker 1969, fig. 1, slightly modified from the Silver Lake data of Hile 1936.)

5. Calculate the between-ages functional slope, b', from representatively-sampled ages (expression 9.4; see also part 2 of Example 9.1).

6. Take the natural logarithms of mean lengths in 4 above, and subtract the initial from the final for each age-group; this gives the instantaneous rate of increase in mean length at each age.

7. Multiply the above rates by b'; this gives $\overline{G}$ for each age.

218

An alternative and somewhat better procedure, following item 3 above, is as follows:

4. Calculate the functional slope b for individual fish (expression 9.3).

5. Take the natural logarithms of initial and final length for the last complete year of growth *of each fish*, and subtract; this gives the instantaneous rate of increase in length for each fish.

6. Average the instantaneous rates of increase in length for each age-group, and multiply by b; this gives the mean instantaneous rate of increase in weight, $\overline{G}$, at each age.

9.5.3. PONDERAL MORTALITY RATE. Size-selective mortality also complicates the concept of mortality rate when the latter is applied to weight. Ricker (1969a) defined an instantaneous *ponderal mortality rate* (Z_w). When mortality is not selective by size this is the same as the numerical mortality rate, Z. But if the larger fish of a year-class die more readily than the smaller, the rate of weight loss is greater than the corresponding decline in numbers, i.e. $Z_w > Z$; and vice versa. Ricker (p. 495) showed that:

$$Z_w - Z = \overline{G} - G_x \qquad (9.5)$$

This provides a method of estimating Z_w when Z, $\overline{G}$, and G_w are known.

EXAMPLE 9.2. COMPUTATION OF GROWTH RATES FOR SILVER LAKE CISCOES. (Modified from Ricker 1958a.)

Table 9.3 shows the back-calculated lengths of a population of ciscoes. We assume, for purpose of illustration, that age-group-b is equal to 3 exactly, and that sampling is non-selective except at age 1 (which is therefore disregarded). Age 7 is also disregarded because only one fish is available.

Columns 2–4 of Table 9.4 show the computation of population growth rates. The length differences in column 2 are from the lowermost diagonal of Table 9.3. Differences between the natural logarithms of length are shown in column 3, and these are multiplied by $b = 3$ to give the instantaneous rate of growth in weight, in column 4.

Columns 5–7 show the similar computation of true mean growth rates. In this case the lengths used are the last pair of each line of Table 9.3.

The true growth rates are all considerably greater than the population growth rates, reflecting the rather severe selective mortality among these fish.

9.6. MATHEMATICAL DESCRIPTION OF INCREASE IN LENGTH — THE BRODY–BERTA-
LANFFY PROCEDURE

9.6.1. BRODY'S EQUATIONS. Brody (1927, 1945) observed that in domestic animals a plot of length against time usually gives an S-shaped growth curve. To treat this

TABLE 9.3. Calculated standard lengths in millimeters of ciscoes from Silver Lake, Wisconsin, taken during the summer of 1931. The fish of age 1, only, are believed affected by net selectivity. (Data from Hile 1936, tables 5 and 9.)

Age	No. of fish	Length at capture	Calculated lengths at successive annuli						
			1	2	3	4	5	6	7
7	1	201	77	111	135	158	172	186	196
6	21	194	78	119	142	158	177	189	
5	108	188	80	126	150	168	182		
4	102	183	80	132	159	176			
3	61	177	83	137	166				
2	19	171	104	151					
1	66	141	105						

TABLE 9.4. Computation of growth rates for the ciscoes of Table 9.3.

1	2	3	4	5	6	7
	Population growth			Mean individual growth		
Age interval	Length interval mm	Difference of natural logarithms	Instantaneous growth rate Gx	Length interval mm	Difference of natural logarithms	Instantaneous growth rate G
2–3	151–166	0.094	0.282	137–166	0.192	0.576
3–4	166–176	0.059	0.177	159–176	0.102	0.306
4–5	176–182	0.033	0.099	168–182	0.080	0.240
5–6	182–189	0.038	0.114	177–189	0.065	0.195

mathematically he divided it at the inflexion point and fitted the two halves with separate curves. For the parts having increasing and decreasing slope, respectively, he used:

$$l_t = ae^{K't} \tag{9.6}$$

$$l_t = b - ce^{-Kt} \tag{9.7}$$

Here l is length and t is age; a, b, and c are constants (parameters) having the dimensions of length; while K' and K are constants determining the rate of increase or decrease in length increments.

Expression (9.6) may apply to the earlier growth stanzas in a fish's life; other expressions have been proposed by Hayes (1949), Allen (1950, 1951), and others. However, fish population studies have been concentrated mainly on the final stanza of growth, to which expression (9.7) has been found to be applicable in many, but not all, populations. Sometimes it describes growth from age 1 onward, but more commonly a good fit is obtained only from a somewhat greater age.

Fabens (1965) has an excellent exposition of the significance of the terms of (9.7). As t increases indefinitely, $l_t \rightarrow b$; thus b is the mean asymptotic length, usually represented by L_∞. This is the length that an average fish would achieve if it continued to live and grow indefinitely according to the pattern of expression (9.7). When $t = 0$, $l_t = b - c = L_\infty - c$, which represents the (hypothetical) size that the fish would have been at $t = 0$, if it had always grown according to (9.7); frequently $L_\infty - c$ is negative. A rearrangement of (9.7):

$$L_\infty - l_t = ce^{-Kt} \qquad (9.8)$$

illustrates that the difference between the asymptotic size and the actual size of a fish decreases exponentially at rate K. This difference decreases to half in time $(\log_e 2)/K = 0.693/K$ years, to a quarter in $1.386/K$ years, and so on. Obviously the larger K is, the more rapid is the decrease. Or, for any given initial fish size (at the time decreasing exponential growth starts), a larger K means a smaller L_∞ and thus slower growth from that time onward. Thus it is misleading to refer to K as a growth rate; a better name is the *Brody growth coefficient*.

9.6.2. VON BERTALANFFY'S EQUATION. Suppose the curve of expression (9.7) is extrapolated down to the time axis, and call this time t_0. Then expression (9.7) can be rearranged algebraically into the form used by von Bertalanffy (1934, 1938):

$$l_t = L_\infty(1 - e^{-K(t-t_0)}) \qquad (9.9)$$

The parameter c is replaced by the new parameter t_0, the relation between them being:

$$t_0 = \frac{\log_e(c/b)}{K}; \quad c = be^{Kt_0} \qquad (9.10)$$

In (9.9), instead of using age as measured in years from a conventional time zero (usually the beginning of the year in which the fish hatch), we in effect start from the hypothetical age, t_0, at which the fish would have been zero length if it had always grown in the manner described by the equation. Thus t_0 can be either positive or negative.

9.6.3. KNIGHT'S EQUATION. Knight (1969) has criticized the form of equation (9.9) on several grounds, but particularly because "although this is called a growth curve, no symbol in it has the dimensions of growth, length per unit time." He proposes an alternative equivalent form:

$$l_t = A + B(t - \bar{t}) + \frac{B}{K}[1 - K(t - \bar{t}) - e^{-K(t-\bar{t})}] \qquad (9.11)$$

$\bar{t}$ an arbitrary reference time, which can most usefully be either the middle of the range of ages represented, or the mean age

221

K   the Brody growth cœfficient

A   a parameter representing the length of the fish at time $\bar{t}$

B   a parameter representing the rate of growth of the fish at time $\bar{t}$

Transformations to the symbols of expression (9.9) are:

$$L_\infty = A + \frac{B}{K} \tag{9.12}$$

$$t_0 = \bar{t} - \frac{\log_e(1 + KA/B)}{K} \tag{9.13}$$

Conversely:

$$A = L_\infty(1 - e^{-K(\bar{t}-t_0)}) \tag{9.14}$$

$$B = KL_\infty e^{-K(\bar{t}-t_0)} \tag{9.15}$$

9.6.4. FORD'S EQUATION. Another form of (9.7) can be obtained by duplicating (9.9) using $t + 1$ for t, and subtracting the resulting equation from (9.9). Using *Ford's growth coefficient* k ($= e^{-K}$), this relationship is:

$$l_{t+1} = L_\infty(1 - k) + kl_t \tag{9.16}$$

This expression was developed empirically by Ford (1933) and later by Walford (1946); it has also been treated by Lindner (1953) and Rounsefell and Everhart (1953). It describes growth in which each year's increment is less than the previous year's by the fraction $(1 - k)$ of the latter, starting from a hypothetical initial length $L_\infty(1 - k)$ at "age 0" (the latter being 1 year prior to the time the age-1 sample was taken, rather than the mean time of egg deposition or of hatching). The relation between increments in successive years is clearer in the derived expression:

$$l_{t+2} - l_{t+1} = k(l_{t+1} - l_t) \tag{9.17}$$

Thus the larger the value of Ford's cœfficient, the more slowly the increments decrease, thus the greater the rate of growth from any fixed starting point.

9.6.5. THE WALFORD LINE. Walford's (1946) graphical presentation of (9.16), with l_{t+1} plotted against l_t, is rather convenient (Fig. 9.7A). The slope of this line is equal to k, and the Y-axis intercept is $L_\infty(1 - k)$, from which L_∞ can be calculated. The asymptotic length, L_∞, is also the length (measured on the abscissa) at which the line (9.16) cuts the 45° diagonal from the origin.

A modification of (9.16) suggested by Gulland (1964a) is obtained by subtracting l_t from both sides and rearranging:

$$l_{t+1} - l_t = L_\infty(1 - k) + l_t(k - 1) \tag{9.18}$$

Thus a regression of $(l_{t+1} - l_t)$ on l_t has a slope $(k - 1)$; its ordinate intercept is $L_\infty (1 - k)$, and its abscissal intercept is therefore:

$$\frac{-L_\infty(1-k)}{k-1} = L_\infty \qquad (9.19)$$

An example is given by Dickie (1968, p. 122).

9.6.6. TYPES OF FORD–WALFORD RELATIONSHIP. The principal types of Walford graphs are shown in Fig. 9.6. The Siglunaes herring (Curve A) are fitted with the line which Ford (1933) computed, with a Y-axis intercept of 9.57 cm and an asymptotic size (L_∞) of 37.1 cm.

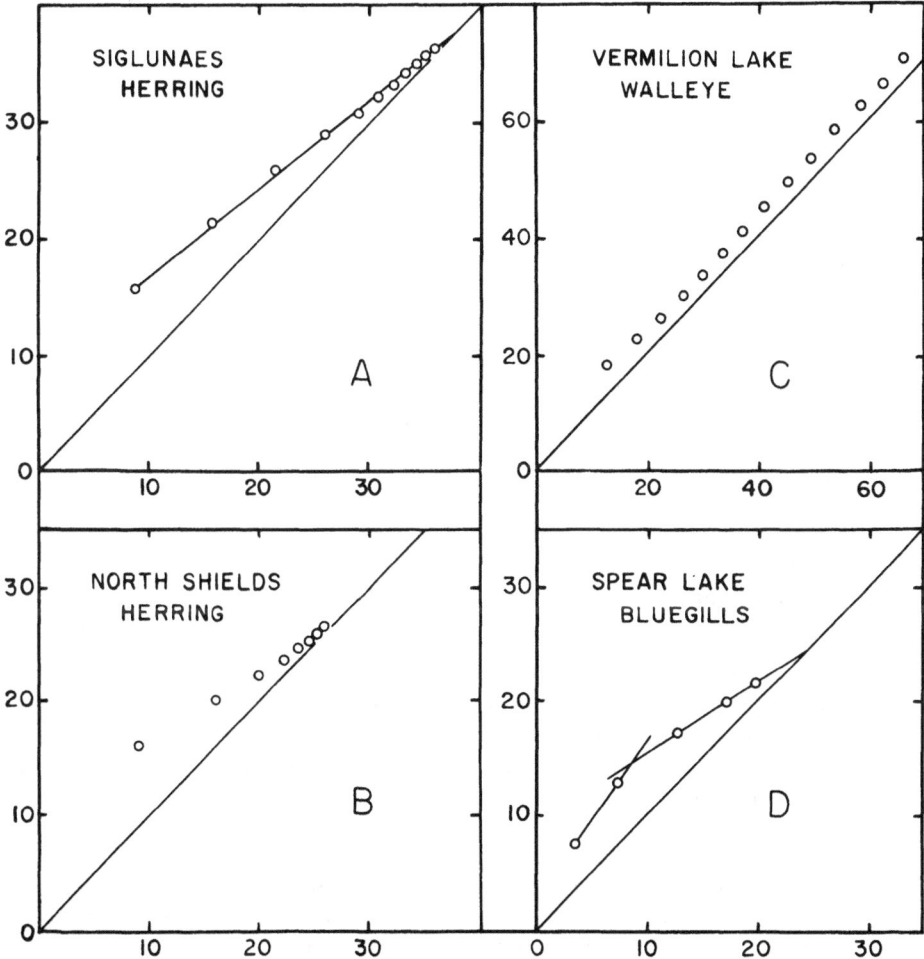

FIG. 9.6. Walford graphs, of length (cm) at age $t + 1$ against length at age t, for four fish populations. A and B, from Ford (1933) after data of Hjort; C, from Carlander and Hiner (1943); D, from Ricker (1955). In every instance the first point represents age 2 plotted against age 1, and later points proceed in sequence.

The North Shields herring (Curve B) are a population for which Ford found that k changed considerably, increasing from 0.56 to 0.77 among the older fish. A similar trend can perhaps be observed even in Curve A, and it appears in certain other well-studied populations such as Hile's (1941) rock bass (*Ambloplites rupestris*) of Nebish Lake or the Clear Lake walleyes (*Stizostedion vitreum*) of Carlander and Whitney (1961). Taylor (1962) points out that when Walford graphs are plotted for individual ages they commonly exhibit an increasing slope and increasing L_∞ with age, and that this effect can account for concavity of the overall line at the widely-spaced younger ages. However, this does not simplify the problem of using such lines, as Carlander and Whitney's paper illustrates.

In another walleye population (Curve C) a line practically parallel to the diagonal was obtained. This form describes uniform absolute increase in length with age, and it is approximated, up to a great age, by a number of long-lived freshwater and marine fishes inhabiting cool waters. Finally, a graph which increases in slope, then later decreases, has been found among centrarchids in the warmer parts of eastern North America; Curve D is an example. The same has been observed in several bivalve mollusc populations: see Weymouth and McMillin's data for razor clams as plotted by Rounsefell and Everhart (1953), or Stevenson and Dickie's (1954) data on growth of scallops.

9.6.7. SOURCES OF ERROR. It is most important, of course, to use truly representative measurements for Walford graphs. One common danger is selection for larger size among younger fish, and this leads to depression of the left end of the line. Age-groups so affected should not be included in the computation. A lesser danger is reading scales of old fish consistently too low. This causes some depression of the right end of the line, but the error has to be gross to produce any considerable effect.

In most or all fishery applications the Brody–Bertalanffy formulae have been fitted to the observed mean size at successive ages in the stock; thus they do not deal with the possibility that mean *individual* growth rates may be different (Section 9.5). That is, if there is differential mortality by size within year-classes, the true mean growth rate of the fish is greater than what is indicated by the Brody–Bertalanffy curve.

9.6.8. THEORETICAL CONSIDERATIONS. Von Bertalanffy (1934, 1938) tried to provide the relationship (9.9) with a theoretical physiological basis, and he and others have apparently considered it a generally-applicable growth law. However, one of the fundamental assumptions used is that anabolic processes in metabolism are proportional to the area of an organism's effective absorptive surfaces. This could seem reasonable if food were always available in excess, so that absorptive surface could actually be a factor limiting growth; and in the guppy experiments which are quoted in support of this relationship, food was actually provided in excess. In nature, fish are usually less fortunate; this is shown by the small average volume of food commonly found in their stomachs, and also by the great variability of their observed growth rates, both when we compare individual fish in the same environment and fish of the same stock living in different (but physically-similar) waters. Thus it seems unlikely that available absorptive surface is commonly a factor limiting growth of wild fish. Another complication is that some fish increase the relative surface area of their digestive tract more or less throughout growth, by increasing the convolutions of its inner surface.

Modifications of the Brody–Bertalanffy relationship can be obtained by using an additional parameter. Richards (1959) and Chapman (1961) describe such a curve, containing a parameter that controls the position of the point of inflexion. Taylor (1962) applied a 4-parameter curve to several sets of data, using more or less arbitrary values for one of the parameters; to obtain efficient estimates, lengthy iterations by computer are necessary (Paulik and Gales 1964, Pienaar and Thomson 1973 — Program WVONB). Paulik and Gales also say that the Chapman–Richards curve can be introduced into the Beverton–Holt model and evaluated by integrating the incomplete Beta function, but no examples have come to my attention. At the other extreme, many are opposed to using parameters that have no clear relation to reality, while Knight (1968) and others have questioned the reality of asymptotic growth in general.

Leaving theoretical considerations aside, observed growth curves are usually close enough to the Brody-Bertalanffy relationship to make the latter a useful empirical descriptive expression. This of course is the basis on which it was introduced by Brody. In fitting the curve, the main thing is to avoid including younger ages that do not conform to it; it is wise to be critical and reject any age that lies even slightly below the Walford line established by the older fish.

9.6.9. FITTING A VON BERTALANFFY CURVE — BEVERTON'S METHOD. To fit expression (9.9) to a body of data it is necessary to evaluate three parameters: L_∞, K, and t_0. The first two can be obtained from the GM functional regression line of l_{t+1} on l_t (Appendix IV). According to (9.16), its slope is equal to k, hence $K = -\log_e k$; also, the Y-axis intercept is equal to $L_\infty(1 - k)$, from which L_∞ can be calculated. The disadvantage in this procedure is that the first and last ages appear only once in the computations and all others twice. In any event further computation is needed to estimate t_0.

Beverton's (1954, p. 57) approach to estimating the parameters is to take a trial value of L_∞ (the one obtained from the regression of l_{t+1} on l_t is convenient) and use it in an expression derived from (9.9) by taking logarithms:

$$\log_e(L_\infty - l_t) = \log_e L_\infty + K t_0 - Kt \tag{9.20}$$

Thus a graph of $\log_e(L_\infty - l_t)$ against t should be straight, and this straightness is sensitive to changes in L_∞. A few trial plots will quickly yield the L_∞ which gives the best (straightest) line — which can usually be selected sufficiently well by eye. Finding the best L_∞ and corresponding line immediately determines K, which is the slope of the line; it also provides the value of t_0, since the Y-axis intercept of (9.20) can be equated to $\log_e L_\infty + K t_0$.

A somewhat better result can be obtained by weighting each age by the number of fish available at each, in fitting the line. The ordinary "predictive" regression is used in fitting (9.20), because the abscissal values (t) are known exactly.

9.6.10. FITTING A PARABOLIC APPROXIMATION TO THE BRODY–BERTALANFFY RELATIONSHIP — KNIGHT'S METHOD. Knight (1969) suggested fitting a predictive quadratic regression line, using standard statistical methods, as an approximation to the Brody relationship. Using the symbols of Section 9.6.3 the approximation is:

$$l_t = A + B(t - \bar{t}) - \frac{BK(t - \bar{t})^2}{2} \tag{9.21}$$

225

Knight also described a modification of (9.21) that gives a slightly better result. From (9.21) the statistics of Knight's expression (9.11) are available almost directly. Corresponding values for the Bertalanffy or Brody form of the equation can be found using expressions (9.12), (9.13), and (9.10).

9.6.11. FITTING BY COMPUTER. Several computer programs are available for fitting expression (9.11). Program BGC2 handles equally-spaced age intervals and BGC3, unequally-spaced ones (Abramson 1971). Program VONB of Allen (1966a, 1967) handles any age interval.

EXAMPLE 9.3. FITTING A FORD EQUATION AND BERTALANFFY CURVE TO LENGTH DATA FOR CISCOES OF VERMILION LAKE, MINNESOTA. (Modified from Ricker 1958a.)

The length column of Table 9.5 shows the mean length of ciscoes (*Coregonus artedii*) of ages 2 to 11 in a sample of 533 fish; they can be used to plot a growth curve of the "A" type of Fig. 9.3. Plotted on a Walford graph (Fig. 9.7A), the age 2 fish evidently do not conform to the linear series, possibly because of selection, so they have been omitted. Freehand fitting of a line to the Walford graph (discounting the last two points because based on so few fish) gave a Ford coefficient of $k = 0.70$ and an intercept on the diagonal of $L_\infty = 315$ mm. A first estimate of the Ford equation is therefore (from 9.16):

$$l_{t+1} = 93 + 0.70l$$

TABLE 9.5. Average weight and average standard length of ciscoes from Vermilion Lake, Minnesota, and data for fitting a Walford line to length. (Data from Carlander 1950.)

Age (t)	No. of fish	Weight (g)	Length (mm)	Using trial $L_\infty = 315$		Using final $L_\infty = 309$		Adjusted age (t−t₀)
				$L_\infty - l_t$ (mm)	$\log_e(L_\infty - l_t)$	$L_\infty - l_t$ (mm)	$\log_e(L_\infty - l_t)$	
2	101	99	172	....	....	....	....	1.76
3	14	193	210	105	4.66	99	4.60	2.76
4	136	298	241	74	4.30	68	4.22	3.76
5	52	383	265	50	3.91	44	3.78	4.76
6	67	462	280	35	3.56	29	3.37	5.76
7	81	477	289	26	3.26	20	3.00	6.76
8	54	505	294	21	3.04	15	2.71	7.76
9	20	525	302	13	2.56	7	1.95	8.76
10	6	539	299	16	....	10	....	9.76
11	2	539	306	9	....	3	....	10.76

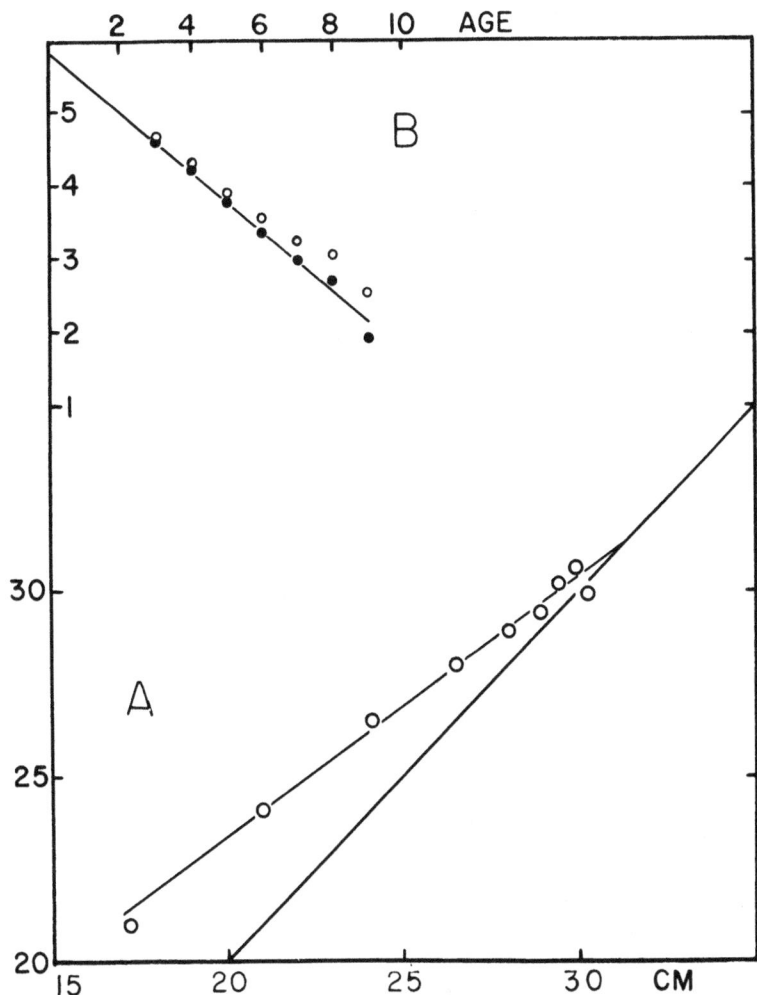

FIG. 9.7. (A) Walford graph for length of ciscoes of Vermilion Lake, Minn. The first point on the left is age 3 plotted against age 2. (B) $Log_e(L_\infty - l_t)$ plotted against age for trial values of $L_\infty = 309$ (open circles) and $L_\infty = 315$ (dots).

To obtain the Bertalanffy equation we use 315 mm as a trial value of L_∞ and $-log_e 0.70 = 0.37$ as a trial value of the Brody coefficient K. Trial values of $315 - l_t$ are computed (Table 9.5) and their natural logarithms are plotted against age for ages 3–9 (Fig. 9.7B, open circles). This line is somewhat curved: additional trials show that $L_\infty = 309$ mm gives the straightest plot (Fig. 9.7B, solid dots). For this value of L_∞ the slope of the natural log line is $K = 0.41$ (thus $k = e^{-0.41} = 0.66$), and the Y-axis intercept is 5.84. Equating the latter to $log_e L_\infty + K t_0$ in (9.20), with $log_e L_\infty = log_e 309 = 5.74$:

$$t_0 = \frac{5.84 - 5.74}{0.41} = 0.24$$

227

The Bertalanffy equation (expression 9.9) becomes:

$$l_t = 309(1 - e^{-0.41(t-0.24)})$$

The corresponding improved Ford equation (expression 9.16) is:

$$l_{t+1} = 105 + 0.66 \, l_t$$

The same series was fitted by means of Allen's computer program, VONB, using both the unweighted data and weighting each age by number of individuals present. Also, the two oldest ages were used, as well as ages 3–9. The results are:

Fitting	Ages	Weighting	K	L_∞	t_0
1. By eye	3–9	No	0.41	309	0.24
2. Computer	3–9	No	0.389	310	0.105
3. Computer	3–9	Yes	0.407	309	0.252
4. Computer	3–11	No	0.407	308	0.201
5. Computer	3–11	Yes	0.414	308	0.292

The preferred estimate here will be either no. 2 or no. 5, depending on whether differences between year-classes or random variability is the more important source of error. However, all the differences are small, and the eye-fitting by Beverton's method is completely acceptable; it has been used in Table 9.5 and Fig. 9.7.

9.7. Use of a Walford Line for Estimating Growth of Older Fish.

In general, the accuracy of age determinations tends to decrease among fish of larger sizes, and for really old fish they may be practically useless. Attempts to fill the gap by direct extrapolation along the curved line of an age–length or age–weight graph are usually unsatisfactory. A better result can usually be obtained by plotting the Walford line, provided available data extend into the region of decreasing increments. Proceeding from the oldest reliable age available, lengths at older ages can be read off the Walford line as far as desired; or expression (9.16) can be used.

9.8. Walford Lines and Brody–Bertalanffy Equations from the Growth of Marked Fish — Manzer and Taylor's Method

A Walford graph can also be drawn by plotting length at recapture against length at marking of marked or tagged animals (Manzer and Taylor 1947; Hancock 1965), though applicability to the wild population depends on the condition that the mark or tag does not retard growth. For fishes, the method will be useful chiefly when there are a number of recaptures made close to a year after marking, since use of intervals much shorter than a year would usually cause problems because of seasonal variations in growth rate. However, Lindner (1953) has applied the method to recaptures of marked shrimps made during successive 10-day intervals of the tagging season.

A computer program for fitting data of this type was proposed by Fabens (1965), who used the original Brody form of the equation (expression 9.7). The program has been modified as FABVB of Allen (1967) and BGC4 of P. K. Tomlinson (Abramson 1971).

Lines and curves obtained in this manner represent the growth of the surviving fish, hence can always be used to compute true growth rate, G, in the stock (subject to the caution about retardation of growth by the tag or mark used). Also, *provided there is no selective mortality by size within year-classes*, they can be used to reconstruct the length-age structure of the population, given the mean length at one age from another source. When selective mortality exists, however, such a reconstruction becomes impossible: the growth curve obtained by Manzer and Taylor's procedure then corresponds to the sum of the broken terminal segments of the ascending lines in Fig. 9.5, rather than the change in length with age in the population which is indicated by the dotted line of Fig. 9.5. Consequently, if a Walford line from tag recaptures is superimposed on one obtained from age-length data, it should lie *above* the latter if there is size-selective mortality within year-classes.

In an actual example for petrale sole by Ketchen and Forrester (1966, p. 87–88) the two lines cross: tag data indicate faster growth among young fish and slower growth among older fish than do the observed lengths at successive ages. The former difference might result from mortality selective by size. For the latter, Ketchen and Forrester postulate either an effect of the tag on rate of growth, or actual shrinkage: the fish are alive when measured at tagging, and dead when measured at recapture, and post-mortem shrinkage of about 5 mm has been observed in similar species. Both these effects could become important among older fish, whose normal length increments are small.

EXAMPLE 9.4. GROWTH OF ENGLISH SOLES FROM TAG RECAPTURES, USING A WALFORD LINE. (Modified from Ricker 1958a.)

Manzer and Taylor (1947) plotted length at recapture against length at tagging for female "English" or lemon soles (*Parophrys vetulus*) which had been at large approximately a year. (Tagging and recaptures were both done during the winter spawning season, when no growth was in progress, so the *exact* time interval was not important.) For the stock off Boat Harbour, Vancouver Island, the points determine a Walford line whose intercept with the diagonal indicates a mean asymptotic size of about 52 cm (Fig. 9.8). The GM functional regression is used to locate this line because the variability is mainly natural (Ricker 1973a).

The expected yearly increment of these soles for any initial length can easily be read from the trend line of Fig. 9.8. Their mean length at age 3 is known to be 29 cm. If we could assume no selective mortality within year-classes, lengths at successive older ages could be calculated from the line, as shown by the open circles in Fig. 9.8.

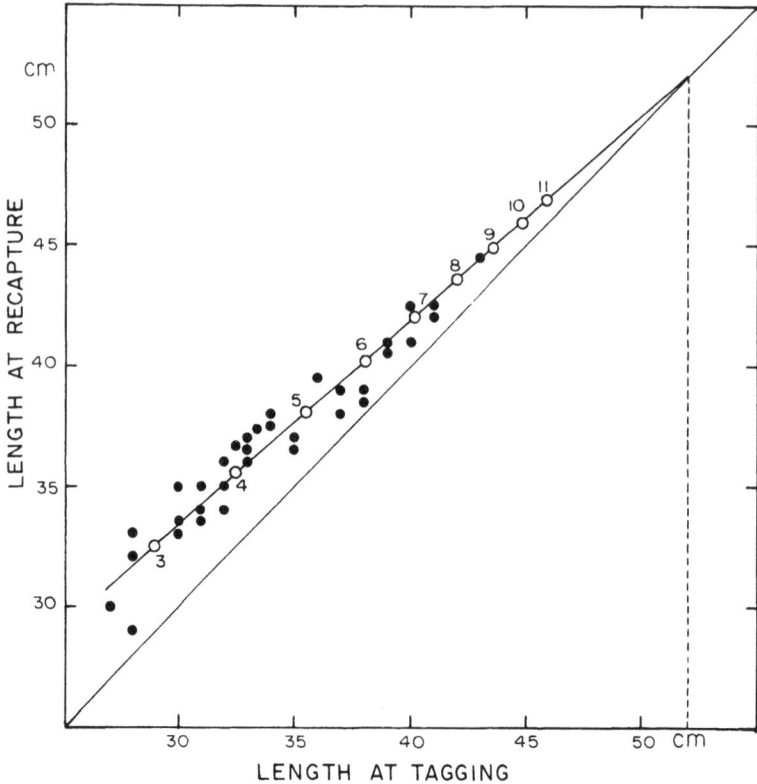

FIG. 9.8. Length at tagging plotted against length at recapture approximately a year later, for English soles. Open circles — computed lengths for ages indicated.

9.9. INCREASE IN WEIGHT WITH AGE

9.9.1. GENERAL. Graphs of weight against age resemble those of length in being usually S-shaped. The point of inflexion is at an older age than on the corresponding length graph (Fig. 9.3B). As with length, the two curves of Brody (9.6 and 9.7 above) can be used for the age-weight relationship, but *both* may be needed to describe the range of weights that are of interest in production calculations. Partial fits have been used, however: Thompson and Bell (1934) fitted an expression like (9.6) to part of a halibut age-weight graph, and Allen (1950) and Dickie and McCracken (1955) used the decreasing exponential for trout and flounders, respectively.

In terms of weight, (9.9) can be written:

$$w_t = W_\infty(1 - e^{-K(t-t'_0)}) \tag{9.22}$$

W_∞ is the mean asymptotic weight, corresponding to the asymptotic length L_∞; K is the Brody growth coefficient; and t'_0 is the hypothetical age at which weight (w_t) would have been zero if growth had always conformed to this relationship. That the Brody coefficient ideally has the same value when calculated from weight

230

as from length data was shown by Dickie and Paloheimo (in Ricker 1958a). However, for any given body of data, (9.22) will apply to a range of ages that starts at an older age than what can be used for (9.9). Also, the parameter t_0' of (9.22) will always be larger than the t_0 of (9.9).

9.9.2. FITTING A BERTALANFFY-TYPE RELATIONSHIP TO WEIGHT. Expression (9.22) can be fitted to weight data in the same way as the corresponding length equation (Section 9.6.9). A Walford-type graph is plotted, of w_{t+1} against w_t, and the value of W_∞ is obtained from the intersection with the 45-degree diagonal, or is calculated from the Y-intercept. As with length, the value of t_0' can then be obtained from trial graphs of $\log_e(W_\infty - w_t)$ against t, as indicated by the equation:

$$\log_e(W_\infty - w_t) = \log_e W_\infty + K t_0' - Kt \tag{9.23}$$

If a computer is available, the same programs as for length can be used, but deleting the younger ages to which (9.22) does not apply.

Allen (1967) also suggests fitting weight data by first converting each weight to a multiple of length. Taking the bth root of each side of (9.2) gives:

$$w^{1/b} = a^{1/b} l \tag{9.24}$$

so that $w^{1/b}$ is proportional to l. For $b = 3$ this transformation is an option in Allen's (1967) computer program FABVB.

EXAMPLE 9.5. FITTING A GROWTH CURVE TO WEIGHT DATA FOR THE CISCO POPULATION OF VERMILION LAKE, MINNESOTA. (Modified from Ricker 1958a.)

For most of the 533 ciscoes enumerated in Table 9.2, weight as well as length was recorded. Plotting these in a Walford graph (not shown here), neither age 2 nor age 3 falls on the trend established by the points for older years. Excluding these and the two oldest ages, free-hand fitting of a Walford line gave the trial values below:

$$k = 0.69; \quad K = 0.37; \quad W_\infty = 567 \text{ g}$$

Adjusting these in the same manner as for the length plot, the best fit is close to $W_\infty = 561$ g and $K = 0.40$ (compare $K = 0.41$ from Example 9.3.) For this line the natural log intercept is 3.80. Thus:

$$t_0' = \frac{3.80 - 2.99}{0.40} = 2.02$$

The weight equation, applicable to fish age 4 and over, is:

$$w_t = 561(1 - e^{-0.40(t-2.02)})$$

A computer fitting of the same data (unweighted and without transformation to length), using program VONB gives:

$$w_t = 565(1 - e^{-0.376(t-1.92)})$$

231

9.10. GOMPERTZ GROWTH CURVE

Another curve that can be used to describe fish growth is the Gompertz curve. This can be written in various ways (Fletcher 1973); Silliman (1967) uses:

$$w_t = w_0 e^{G(1-e^{-gt})} \qquad (9.25)$$

where w_t is weight at time t, measured from the conventional time t_0 when the fish would have had weight w_0; G is the instantaneous growth rate at time t_0; and g describes the rate of decrease of G. Expression (9.25) is S-shaped, with both a lower and an upper asymptote; the inflexion point is $1/2.718 = 0.368$ of the vertical distance from the lower to the upper asymptote. However, the lower asymptote is meaningless in describing growth, so only part of the curve is used.

The Gompertz curve will usually describe data on weight at age quite well, often including the early years of increasing increments. It can also be fitted to length data, in which case usually only the portion of the curve beyond the inflexion point is involved. Like the Brody–Bertalanffy curve, (9.25) has three parameters that must be estimated: w_0, G, and g. The easiest method of fitting is that developed by Ricklefs (1967).

The Gompertz curve has not been used much in fishery work because yield computations have usually employed the Brody–Bertalanffy. It is, however, more suitable than the latter for use with an analog computer, and Silliman used it in his analyses (Section 12.4.3).

9.11. GROWTH COMPENSATION

A by-product of the general type of growth in length and weight discussed above is the effect which has been called growth compensation. Although not necessary to our main theme, a brief description of its place in the growth picture seems desirable.

As it concerns length, growth compensation has been treated by a succession of authors, starting with Sund (1911) and Gilbert (1914). Some of the more comprehensive papers are by Watkin (1927), Hodgson (1929), Van Oosten (1929), Ford (1933), Hubbs and Cooper (1935), and Hile (1941). Growth in weight was brought into the picture by Scott (1949) and Ricker (1969a). The phenomenon these authors discuss is a correlation between increments in size in successive years of life among fish of a given year-class. Negative correlations indicate growth compensation, because they show that smaller fish tend to catch up with larger. Positive correlations have been called "reverse growth compensation," but a shorter term is "growth depensation" — adopting a word proposed in a different context by Neave (1954).

In a typical brood the fish vary considerably in size at the end of the first growing season, partly because of differences in time of hatching, partly from congenital physiological differences, and partly from differences in environment. Hodgson (1929) showed that growth compensation must occur among fish of any brood whose members increase in length by the same absolute amount at any given size, provided this increment decreases with size and provided the fish differ in size to begin with. Scott (1949) pointed out that growth compensation is associated with a decrease in the (absolute) yearly average increment of a year-class, whereas depensation is associated with an increase in yearly increment in the unit chosen (length or weight). As long as increases predominate in a year-class there is depensation; when decreases become more common, it shifts to compensation.

Considering weight first, the initially heavier fish of a brood usually tend to increase their advantage during the second year of life, and often continue to do so for one or more additional years. Eventually, however, the inflexion point of the weight-age curve is reached: the smaller fish start to catch up with the larger ones, and the correlation between increments in adjacent years shifts from positive to negative.

The course of length change is similar, but the shift from growth depensation to growth compensation occurs earlier in life; often compensation begins as early as the second growing season, so that the phase of positive correlation is omitted. This difference, between length and weight, in time of appearance of compensation and inflexion of the growth curve, is a necessary consequence of the fact that weight increases as a power of length. For example, if all fish of a brood were to increase in weight by the same absolute amount in a year, the smaller ones would be increasing more, in length, than the larger ones; thus growth in length would be compensatory.

These changes produce, or reflect, changes in variability in size of fish in a brood. Typically, standard deviation in weight or length of a brood increases early in life and decreases later, but the increasing phase lasts longer for weight than for length.

9.12. ESTIMATION OF SURVIVAL RATE FROM THE AVERAGE SIZE OF FISH CAUGHT

The average size of fish in a catch (above some minimum) is obviously related to annual mortality rate: the greater the mortality, the smaller the average size, provided recruitment is reasonably stable from year to year. Given some kind of expression for rate of increase in length or weight, a rough calculation of survival rate can be made from average size under these conditions. Appropriate formulae have been described by Baranov (1918, p. 94), Silliman (1945), and Beverton and Holt (1956, Appendix B). However, they have been little used, and it seems unnecessary to repeat the synopsis given by Ricker (1958a, Section 9H).

CHAPTER 10. — COMPUTATION OF YIELD
FROM A GIVEN RECRUITMENT

10.1. GENERAL CONDITIONS

The goal of most work on growth and mortality in fish populations has been assessment of the yield of a stock at different levels of fishing effort, or with different size limits for recruitment. In the computations of this chapter the yield calculated is that which will be obtained from whatever number of recruits are coming into the fishery. Regulation of recruitment is considered in Chapter 11; the calculations in this chapter are usually made in terms of yield *per recruit*, or per unit weight of recruits. Except in Section 10.9, equilibrium situations are postulated: that is, the situation which exists after the specified conditions have been in effect long enough to affect all ages for the whole of their exploited life.

An important prerequisite for these calculations is that instantaneous rates of natural mortality and of growth, at any given age, be constant over the range of conditions examined. The limited information available concerning *mortality* suggests that this may often be close to the truth over a fairly wide range of population densities, but the question needs constant re-examination. But *growth* has sometimes been found to vary markedly with change in population density — so much so that Nikolsky (1953) has suggested that rate of growth could be used, by itself, as an index of the degree to which a stock approaches its maximum productivity. Whether or not this is ever practical, variation in growth rate, when it occurs, sets strict limits to the range of stock densities over which useful predictions of yield can be made using the methods of this chapter, as Miller (1952) has emphasized. Fortunately, not all stocks react to exploitation by increasing growth rate: those which do seem usually to be the dominant species in the habitat from which they obtain the bulk of their food (cf. Ricker 1958c).

When the effect of a change in selectivity of gear or in minimum size limit *along with* a change in rate of fishing is being examined, the accompanying shifts in overall stock biomass need not be especially great, though age distribution will change dramatically. In such circumstances the methods of this chapter perform fairly well, provided the recruitment effect can be properly taken into consideration.

Subject to these conditions, computations of equilibrium yield per unit recruitment have been attempted by a number of authors. Generally the rate of growth in weight varies with age, and rate of fishing may also be different at different ages or sizes, particularly during the recruitment phase. The most direct approach breaks the population up into age-, size- or time-intervals sufficiently small that rates of growth and mortality can be considered constant within each, without important error; and yield statistics for whole populations are obtained by summing the results for all the intervals represented. Alternatively, we may attempt to obtain a single expression for yield by fitting mathematical expressions to growth and mortality, and combining the two.

We will confine the discussion here mainly to yield in weight of fish taken, with some reference to yield per unit of effort and the market value of the yield per kilogram. Particularly in sport fisheries, other characteristics of the take may sometimes be more important — size of fish caught, for example. Allen (1954, 1955) has dealt extensively with effects of size limits and bag limits upon various characteristics of catch and yield in sport fisheries.

Computations of the kinds described in the sections to follow can be quite time-consuming, particularly if charts are being prepared to show a number of complete surfaces of possible conditions — each of one of the general types presented by Baranov (1918, fig. 10), Thompson and Bell (1934, fig. 9), or Beverton (1953, fig. 2). For a large-scale operation a computer is almost indispensable. However, values in regions of special interest can be obtained quite rapidly with a desk calculator, and frequently nothing more is needed.

10.2. ESTIMATION OF EQUILIBRIUM YIELD — METHOD OF THOMPSON AND BELL

In Section 1.6.1 the technique was described of computing weight of a population by combining an age frequency distribution with the empirically determined average weight of fish of successive age-groups. Mean abundance ($\overline{N}$) at successive ages can be computed from appropriate series, for example those of expressions (1.18), (1.32), or (1.33); and each $\overline{N}$ is multiplied by the average weight ($\overline{w}$) of the fish at that age, which gives the corresponding mean biomass ($\overline{B}$). The biomass of fish caught at each age is then $F\overline{B}$, while the biomass which dies naturally is $M\overline{B}$ (expressions 1.40 and 1.41).

This is essentially the procedure adopted by Thompson and Bell (1934, p. 29) to compute the yield of halibut (*Hippoglossus stenolepis*) under equilibrium conditions for different combinations of fishing and natural mortality[1]. It can be used to compute, by successive trials, either or both of two pieces of information: (1) the value of F which produces maximum yield (in weight) for a given value of M, and (2) the value of F which produces maximum yield for a given value of Z.

This procedure ignores a direct effect of fishing upon the average weight of the fish taken at each age. Regardless of when recruitment occurs, if growth and fishing are concurrent, then the greater the rate of fishing, the smaller is the average size (during the fishing season) of a fish of a given age, because more are taken early in the season before much growth is made. However, the method is useful for orientation.

The earliest complete computations of this type are by Nesbit (1943), who portrayed the catch and population structure for striped bass (*Morone saxatilis*), halibut, and lake whitefish (*Coregonus clupeaformis*) for all combinations of a series

---

[1]However, Thompson and Bell divide the total mortality in the ratio of the conditional mortality rates *m* and *n*, instead of F and M, when computing separate shares for catch and natural mortality. In situations where F and M are nearly equal or where neither F nor M is really large, *m*:*n* is a fair approximation to F:M, but it is just as easy to use the correct ratio.

of values of natural mortality, fishing mortality, and age of recruitment. Tester's (1953) curves are also of considerable interest: they show the variation in equilibrium yield per unit weight of recruits that occurs with change in rate of fishing and change in natural mortality rate, for three different types of weight-age relationships. Also, Clayden (1972) recently used this yield model and incorporated it into a computer program to simulate changes in yields of Atlantic cod (*Gadus morhua*) in the North Atlantic.

EXAMPLE 10.1. COMPUTATION OF EQUILIBRIUM YIELDS AT DIFFERENT LEVELS OF FISHING EFFORT BY THOMPSON AND BELL'S METHOD. (From Ricker 1958a.)

A population is characterized by an instantaneous natural mortality rate of $M = 0.35$, and by the distribution of average individual weights at successive ages shown in column 2 of Table 10.1. If fishing occurs concurrently with natural mortality, what rate of fishing gives maximum yield?

The computation for a rate of fishing $F = 0.5$ is shown in columns 3–6 of Table 10.1. The sum of instantaneous natural mortality rate and rate of fishing is $0.35 + 0.50 = 0.85 (= Z)$, and this determines a survival rate of $S = 0.427$ (Appendix I). A stock of 1000 fish on hand at the start of age 3 is reduced to 427 in one year, to 182 the next year, etc. Total deaths each year are found by subtraction, and the fishery takes 0.5/0.85 or 58.8% of these (column 5). These numbers, multiplied by the average weight at each age, give the removals shown in column 6. The total is 2508 kg per 1000 recruits to age 3 — a result which is correct within about 25 kg.

TABLE 10.1. Computation of survivors and annual yield from an annual recruitment of 1000 fish at the start of age 3, under equilibrium conditions. The instantaneous rate of natural mortality is 0.35; of fishing, 0.50. Individual weights at each age are as shown in column 2.

1	2	3	4	5	6
Age	Average weight	Initial population	Deaths	Catch	Yield
	(kg)	(pieces)	(pieces)	(pieces)	(kg)
		1000			
3	1.86		573	337	626
		427			
4	5.53		245	144	798
		182			
5	8.80		104	61	535
		78			
6	10.96		45	26	286
		33			
7	12.28		19	11	136
		14			
8	13.60		8	5	68
		6			
9+	(14.5)		6	4	59
Totals		1745	1000	588	2508

Similar calculations for other rates of fishing give the results shown in Table 10.2. Maximum yield is apparently obtained with a rate of fishing slightly greater than 0.5. However, the important conclusion would be that there is a broad range of F-values over which yield varies little: from 0.35 to 0.8, no really significant change occurs.

Although the total weight taken does not change much, the biomass of the stock, and thus the catch per unit of effort and the average size of the fish caught, are very different at the different levels of F shown in Table 10.2 (see also Section 10.8). Either one of these might largely determine the most suitable type of regulation, depending on the kind of fishery involved and the commercial or esthetic value of fish of different sizes.

TABLE 10.2. Catch and yield per 1000 recruits for different rates of fishing of the population of Table 10.1. Also the average weight of a fish caught (in kilograms) and the yield (in weight) per unit of fishing effort, expressed on an arbitrary scale (effort is considered proportional to rate of fishing).

1	2	3	4	5
Rate of fishing	Catch	Yield	Avg weight of a fish	Yield per unit effort
(F)	(pieces)	(kg)	(kg)	
0.2	363	2090	5.76	100
0.35	500	2440	4.85	67
0.5	588	2510	4.27	48
0.65	650	2450	3.77	36
0.8	695	2390	3.45	29

10.3. ESTIMATION OF EQUILIBRIUM YIELD — RICKER'S METHOD

10.3.1. COMPUTATIONS. Mechanics of a direct balancing of growth rate against death rate, to compute net change in bulk of a year-class, were presented in Section 1.6.3. In applying the method to a whole population, the life of the fish must be broken down into periods such that neither growth rate nor mortality rate is changing very rapidly within any one of them.

The combined estimate of equilibrium yield (Y_E) under given conditions can be represented by the expression below, derived from (1.40):

$$Y_E = \sum_{t=t_R}^{t=t_\lambda} F_t \bar{B}_t \tag{10.1}$$

238

where t represents successive intervals or periods in the life of the fish (these not necessarily of equal length), t_R is the first period under consideration, and t_λ is the last period under consideration (usually the last period in which an appreciable catch is made). Other symbols are as in Section 1.3.

The easiest and most useful estimate of average weight of stock, $\overline{B}$, is the arithmetic mean of the initial and final value of B for each interval. Designating stock weight at the start and finish of interval t as B_t and B_{t+1}, this average is:

$$\overline{B}_t = \frac{B_t + B_{t+1}}{2} = \frac{B_t(1 + e^{G_t-Z_t})}{2} \tag{10.2}$$

Thus the yield is equal to:

$$Y_E = \sum_{t=t_R}^{t=t_\lambda} \left(\frac{F_t B_t [1 + e^{G_t-Z_t}]}{2} \right) \tag{10.3}$$

If a stock were to increase or decrease strictly exponentially, its average size during any interval t would not be the arithmetic mean (10.2), but rather expression (1.38), which may be written here as:

$$\overline{B}_t = \frac{B_t(e^{G_t-Z} - 1)}{G_t - Z_t} \tag{10.4}$$

Values of this expression can be obtained readily enough, because the factor $(e^{G_t-Z_t} - 1)/(G_t - Z_t)$ is available in Appendix I: column 5 shows its value for positive values of $G_t - Z_t$, and column 4 for negative values, $G_t - Z_t$ being equated to the Z in column 1. However, mainly because growth rate decreases throughout the life of a fish, a graph of year-class bulk against time tends to be convex (dome-shaped) with a "tail" to the right; whereas the exponential segments used to approximate it are all concave upward. Even the tail, which is concave upward, is less concave than the individual exponential segments (Fig. 10.1). The result is that the arithmetic mean of the initial and final values of each of these segments is a somewhat better average to use than is expression (10.4). In practice, when the fish's life is divided into intervals of suitable length, there is little difference between (10.4) and the arithmetic mean for each interval.

The formulae above rather disguise the simplicity of the procedure, which is illustrated in Example 10.2. Computations are carried out in tabular form, and are extremely flexible. Age differences in growth rate, in rate of fishing, and in natural mortality rate, different minimum size limits, and different seasonal distributions of growth, fishing, and natural mortality — all these can be examined easily and directly. Moreover, there is no need to worry about whether growth conforms to some special law, nor is there any restriction on the value of b in the weight–length relationship.

If computations are being done by hand, the number of steps can be reduced by dividing the fish's life into intervals (t) that are of different lengths. At ages where F

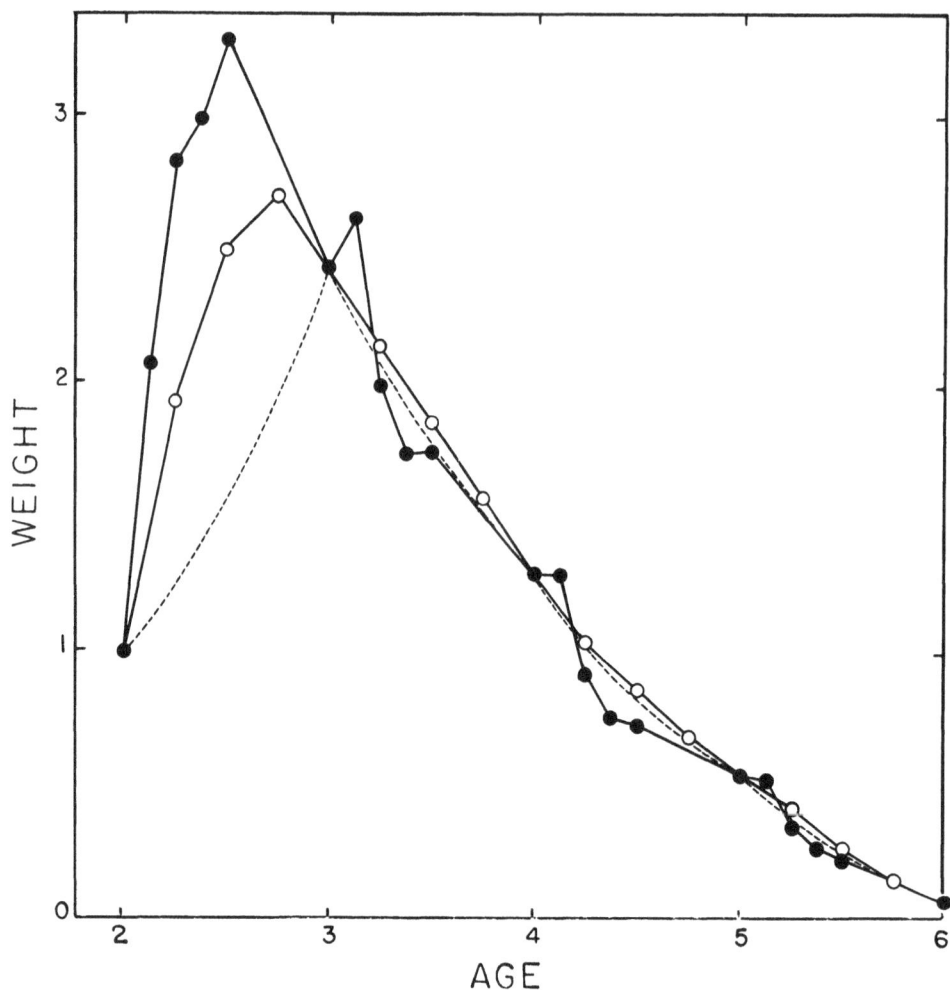

FIG. 10.1. Course of change in weight of a year-class in the populations of Table 10.3 (solid circles) and Table 10.4 (open circles). The dotted line comprises successive segments of exponential curves obtained by computation from the net instantaneous rate of increase or decrease for each full year.

or G is changing rapidly, it may be desirable to make it as short as a month or two[2], but if these parameters are relatively steady, a year or even several years may be a sufficiently fine division.

10.3.2. COMPUTER PROCEDURE. For machine computation there are Fortran programs by Paulik and Bayliff (1967) for both the arithmetic-mean procedure (using expression 10.2) or the exponential mean (expression 10.4). The best plan is to use a single standard short time interval, for example 1 month, $\frac{1}{2}$ month, or 1/10 of a year, which eliminates any appreciable difference between the sample mean and either of

---

[2] In an analogous computation of the production (sense of Ivlev) of fingerling salmon, Ricker and Foerster (1948) used half-month intervals while the young fish were very small and both the growth rate and the natural mortality rate were changing rapidly.

240

the above means at all stages of life. These programs are available at the Nanaimo Station of the Fisheries Research Board of Canada; see also L. E. Gales program FRG-708 in Abramson (1971).

10.3.3. COMPUTATION OF BEST MINIMUM SIZE. One important function of computations of this sort is to discover the *minimum size limit* which gives maximum yield from a given weight of recruits. We have defined the *critical size* as the size at which the instantaneous rates of growth and of natural mortality are equal (Section 1.6.4). At that time and size the year-class has its maximum bulk. If the brood could all be cropped at once, that would be the best time to do it. However, instantaneous cropping is possible only in piscicultural establishments, where a pond can be drawn down and all the fish removed. If cropping must be spread over a period of time, some loss in efficiency of cropping occurs. The aim should be to keep such losses to a minimum; this is done by taking some of the fish when they are less than the critical size, and some when they are greater. The smaller the fishing rate, the broader the range of sizes that should be taken (Ricker 1945c) — that is, the smaller should be the minimum size limit.

10.3.4. SEASONAL INCIDENCE OF NATURAL MORTALITY. In yield computations of all types a persistent minor worry is our (usual) lack of information concerning *when* natural mortality takes place. Often it is advisable to examine two or more possibilities and see what difference there is in the results obtained. If a fishery is restricted to a short season one might, for example, postulate that for practical purposes there was no natural mortality during the fishing season. With a longer season, the instantaneous rate of natural mortality might be divided in proportion to the length of time involved — part being combined with the rate of fishing, and the remainder acting by itself.

EXAMPLE 10.2. EQUILIBRIUM YIELD OF BLUEGILLS, PER UNIT RECRUITMENT, FOR MUSKELLUNGE LAKE, INDIANA. (From Ricker 1958a.)

Data concerning the stock of *Lepomis macrochirus* in Muskellunge Lake are shown in Table 10.3. (Growth and natural mortality were determined from samples taken and experiments made in 1941–42; however, the level of fishing shown is that believed characteristic of 1939–40, before the war decline in fishing.) Fish growth was read from scales. Computed mean lengths were interpolated on a smooth curve at quarter-year intervals, and were converted to weight using the "year-class-*b*" (*b'* of expression 9.4). The "year" in this case is the growth year, which is considered to last 6 months, from about May 1, when new circuli begin to appear on the scales, to the end of November when the lake is well cooled. Thus the quarter-years of the growth curve are really 1/8 year long on the calendar. Lengths and weights are indicated in columns 3 and 4 of Table 10.3, on this basis, May 1 being considered as the start of the year. Column

241

TABLE 10.3. Instantaneous rates of growth (G), natural mortality (M), and fishing mortality (F) for bluegills of Muskellunge Lake, distributed according to their observed or (for M) hypothetical seasonal incidence; and the computation of equilibrium yield, in successive fishing seasons, from 1000 kg of recruits at age 2.

1	2	3	4	5	6	7	8	9	10	11	12	13
Date	Age (yr)	Fork length (mm)	Wt (g)	$\log_e$(wt)	G	M	F	G-F-M	Wt change factor	Wt of stock (kg)	Avg wt (kg)	Yield
May 1	2	95	13	2.56	0.81	0.075	0	+0.735	2.086	1000		
June 16	$2\frac{1}{8}$	109	29	3.37	0.41	0.075	0.04	+0.295	1.343	2086	2444	98
Aug. 1	$2\frac{1}{4}$	122	44	3.78	0.28	0.075	0.14	+0.065	1.067	2801	2894	405
Sept. 15	$2\frac{3}{8}$	135	58	4.06	0.17	0.075	0	+0.095	1.100	2988		
Nov. 1	$2\frac{1}{2}$	145	69	4.23	0	0.300	0	-0.300	0.741	3287		
May 1	3	145	69	4.23	0.15	0.075	0	+0.075	1.078	2435		
June 16	$3\frac{1}{8}$	153	80	4.38	0.13	0.075	0.33	-0.275	0.760	2625	2310	762
Aug. 1	$3\frac{1}{4}$	160	91	4.51	0.11	0.075	0.17	-0.135	0.874	1995	1870	318
Sept. 16	$3\frac{3}{8}$	165	101	4.62	0.08	0.075	0	+0.005	1.005	1744		
Nov. 1	$3\frac{1}{2}$	170	110	4.70	0	0.300	0	-0.300	0.741	1752		
May 1	4	170	110	4.70	0.07	0.075	0	-0.005	0.995	1297		
June 16	$4\frac{1}{8}$	175	118	4.77	0.07	0.075	0.33	-0.335	0.715	1291	1107	365
Aug. 1	$4\frac{1}{4}$	178	125	4.84	0.05	0.075	0.17	-0.195	0.823	923	841	143

Sept. 16	4⅜	182	131	4.89	0.04	0.075	0	−0.035	0.966	759		144
Nov. 1	4½	185	137	4.93	0	0.300	0	−0.300	0.741	734		52
May 1	5	185	137	4.93	0.04	0.075	0	−0.035	0.966	544		
June 16	5⅛	188	143	4.97	0.03	0.105	0.33	−0.405	0.667	525	438	
Aug. 1	5¼	191	148	5.00	0.04	0.140	0.17	−0.270	0.763	350	308	
Sept. 16	5⅜	193	153	5.04	0.03	0.200	0	−0.170	0.844	267		
Nov. 1	5½	195	158	5.07	0	1.200	0	−1.200	0.844	226		
May 1	6	195	158	5.07	0	1.200	0		0.301	68		
Totals					2.51	3.52	1.68	−2.690		2287		

5 is the natural logarithm of weight, and the difference between the natural logarithms of two adjacent values is the instantaneous growth rate for the interval concerned (column 6).

Fishing in Muskellunge Lake occurred almost wholly during the period June 16–September 15; records kept in 1941 showed that 66% of the total pole-hours were in June 16–July 31, and 34% were later. (May 1 to June 15 was closed to fishing at that time.) Accordingly, of a total fishing rate of 0.5, 0.33 is assigned to the second eighth of the year, and 0.17 to the third (column 8). In the year of recruitment not many age 2 individuals would be large enough to be caught during the second eighth, but nearly all would be vulnerable by the end of the third eighth, and the F-values are adjusted accordingly.

Natural mortality is estimated as equal to about 0.6, from tagging and age-composition studies (Ricker 1945a). There is some evidence that it is at least fairly well distributed throughout the year; and here it is divided up equally: 0.075 is assigned to each of the four summer eighths, and 0.3 to the winter half (column 7). At age 5 the natural mortality is made to increase progressively, because older fish are relatively scarcer.

Column 9 is the resultant of growth and all mortality: i.e. (G–M–F) or (G–Z). From this a "change factor" is obtained, equal to e^{G-Z}, and obtainable from any exponential table or from Appendix I (column 12 when G–Z is positive; column 3 when it is negative). In column 11 of Table 10.3 successive population weights are computed, starting with an arbitrary 1000 kg (Fig. 10.1). Column 12 is the arithmetic average of adjacent stock sizes, and in column 13 these are multiplied by F to give the yield obtained during each interval. Columns 12 and 13 need be computed only for the intervals when there was a fishery.

A convenient check is provided by summing the instantaneous rates for each year, or for the whole series, and comparing with the appropriate figure in column 11. For example, the grand total of G–Z is –2.690, and $1000e^{-2.69} = 68$, as in column 11.

The sum of column 13 indicates that 2.29 kg of bluegills are caught from the lake for every kilogram of age 2 recruits. (Fish age 6 and older would not add to this appreciably.)

Of the numerous variations of the Table 10.3 conditions that can be examined, we will mention here only the possibility of opening the period May 1 to June 15 to fishing. What would its effect be on yield? In the absence of any increase in total amount of fishing, a likely distribution of fishing rates under the new conditions would be 0.15, 0.20 and 0.15, respectively, in the first three eighths of the biological year for fully-recruited fish. (The actual distribution of course would depend on the fishermen themselves.) Used in a table like 10.3, these rates indicate practically no change in yield per unit recruitment. In practice, however, opening the spring season would likely increase overall fishing effort for the year, and this results in some increase in computed equilibrium yield (cf. Example 10.3).

EXAMPLE 10.3. EQUILIBRIUM YIELD WHEN FISHING IS CONSIDERED TO ACT THROUGHOUT THE YEAR. EFFECTS OF VARYING MINIMUM SIZE AND OVERALL RATE OF FISHING. (From Ricker 1958a.)

In an earlier treatment of the data of Example 10.2, fishing and natural mortality were considered as acting at a uniform instantaneous rate throughout the year (Ricker 1945c); growth was not considered uniform, but the decreasing instantaneous rate was divided among the four *quarters* of the statistical year (instead of the first four eighths). Table 10.4 is a computation made on this basis.[3] The same total instantaneous rates of growth, fishing, and natural mortality are used in each year, but the computed yield is less: 1.98 kg per kilogram of recruits, instead of 2.29 kg. Figure 10.1 shows the reason for this difference: Table 10.3 permits the large excess of growth over mortality which exists in spring (May 1–June 15) to increase the stock to a high level, and the fishery acts on it at that high level; also, $\frac{5}{8}$ of the natural mortality occurs after the fishing is over for the year. In Table 10.4, by contrast, fishing and natural mortality are brought into play with full force from the beginning of the year (for fully-recruited ages).

For some purposes, however, failure to use a true seasonal distribution of these various factors is not important. The absolute level of yield obtained, per unit recruitment, may then be somewhat fictitious; but *changes* in that level will be accurate enough, relatively, and can provide most of the information sought. In particular, a computation like Table 10.4 is completely suitable for examining effects of an overall increase or decrease in rate of fishing, and reasonably suitable for examining changes in minimum size — though the more realistic Table 10.3 is just as easy to construct.

To determine yields from a variety of size limits, it is not necessary to repeat the whole computation in Table 10.3 or 10.4 each time.[4] Suppose, for example, we were examining in Table 10.3 the effect of a limit which would protect all age-2 fish. Then the F entries between age $2\frac{1}{8}$ and $2\frac{3}{8}$ become zero, and there are corresponding changes in columns 9–11. The yield at that age is of course zero. The new regime permits the survival to age $2\frac{3}{8}$ of 3574 weight units of stock ($= 1000e^{1.275}$), instead of the 2988 shown in the table. However, from that age onward these fish are subject to the same conditions as before; so the new yield will be 3574/2988 or 1.196 times the old yield of fish age 3 and older, namely $1.196 \times 1784 = 2134$ weight units. Thus the proposed change would decrease by 7% (from 2287 to 2134) the yield per unit weight of recruits.

Changes in rate of fishing and in minimum size for Muskellunge bluegills were examined (Ricker 1945c) in a computation similar to Table 10.4. In addition to F = 50% (applicable to 1939–40), there was used F = 30% which was close to the

---

[3] There are minor differences in the earlier computation, notably that growth was estimated by taking tangents at the even years, halves and quarters. Consequently, Table 10.4 here is not directly comparable with table 1 of Ricker (1945c).

[4] For a worked-out example of this type of computation, see columns 9–11 of table 8 of Chatwin (1959.)

245

TABLE 10.4. Instantaneous rates of growth (G), natural mortality (M), and fishing mortality (F), for bluegills of Muskellunge Lake. In contrast to Table 10.3, fishing as well as natural mortality is divided evenly through the year, while growth is distributed through the whole year but a separate rate is used for each quarter; however, the sums of G, M, and F are the same in both tables. Recruitment occurs mainly during the age 2.5–2.75 interval, to which a reduced value of F is assigned. (122 mm fork length was the legal limit of size, but the fish did not at once become fully acceptable to fishermen.)

1	2	3	4	5	6	7	8	9	10	11
Age	Mean length (mm)	Mean wt (g)	G	M	F	G-F-M	Wt change factor	Wt of stock (kg)	Avg wt (kg)	Yield (kg)
2	95	13						1000		
			0.81	.15	0	+.660	1.935			
2¼	109	29						1935		
			0.41	.15	0	+.260	1.297			
2½	122	44						2510		
			0.28	.15	.055	+.075	1.078		2608	143
2¾	135	58						2705		
			0.17	.15	.125	−.105	0.901		2572	321
3	145	69						2438		
			0.15	.15	.125	−.125	0.883		2294	287
3¼	153	80						2152		
			0.13	.15	.125	−.145	0.865		2007	251
3½	160	91						1862		
			0.11	.15	.125	−.165	0.848		1720	215
3¾	165	101						1579		
			0.08	.15	.125	−.195	0.823		1439	180
4	170	110						1299		
			0.07	.15	.125	−.205	0.815		1179	147
4¼	175	118						1059		
			0.07	.15	.125	−.205	0.815		961	120
4½	178	125						863		
			0.05	.15	.125	−.225	0.798		776	97
4¾	182	131						689		
			0.04	.15	.125	−.235	0.790		616	77
5	185	137						544		
			0.04	.18	.125	−.265	0.767		480	60
5¼	188	143						417		
			0.03	.34	.125	−.435	0.647		344	43
5½	191	148						270		
			0.04	.50	.125	−.585	0.557		210	26
5¾	193	153						150		
			0.03	.70	.125	−.795	0.452		109	14
6	195	158						68		
Total			2.51	3.52	1.68	−2.690				1981

1942 rate of fishing in this lake, and also the rather large value of F = 100%. This last figure constitutes a rather extreme extrapolation from the observed data, but was included for purpose of illustration. The relative yields for these three different rates of fishing, and for six different minimum sizes, were as follows:

Minimum fork length mm	Rate of fishing		
	0.3	0.5	1.0
102	76	96	110
116	77	99	120
122	76	100	125
128	75	99	128
140	71	95	125
149	65	88	119

The yields shown are relative to 1939–40 conditions (F = 0.5), these being taken as 100. As it turned out, the optimum or "eumetric"ᴬ size limit for getting greatest yield from recruits at the 1939–40 rate of fishing was approximately the legal minimum actually in use. For the reduced fishing of the war years (F = 0.3) the best limit would have been somewhat less, and for any rate of fishing substantially greater than 0.5 the best minimum would be somewhat greater than 122 mm (5 inches total length).

However, what is of most interest is the rather close *agreement* among calculated yields at each rate of fishing. For example, with minima anywhere from 102 to 140 mm, for F = 0.5, yield is never less than 95% of ʰhe maximum. This same stability has appeared in parallel computations (by this or other methods), for most other fisheries examined to date, and it has a number of implications. *First*, there is considerable leeway allowed for errors in the data from which the computation of minimum size is made. *Secondly*, it is evidently not important to determine the *exact* optimum minimum size for maximum yield. *Third*, if it were known that a certain minimum size is best from the point of view of regulating the size of the stock so as to obtain optimum recruitment, then a considerable adjustment of the minimum could be made to meet this requirement without sacrificing any significant part of the yield from whatever recruits actually appear. *Fourth*, if either the individual size of the fish caught, or the catch per unit of effort, are important considerations in respect to the fishery, either of these can be favored by the regulations to a considerable degree without significant loss of yield. *Fifth*, if the minimum size has to be specified as what a given mesh of net will catch rather than as a fixed limit based on measurement of individual fish, this will usually be almost as effective as a sharp cut-off size (though the fate of the rejected fish needs to be considered: whether they survive or die). *Finally*, if it is desirable to have a uniform minimum standard apply to a number of bodies of water, or even to different kinds of fish, for which the optimum minima are different, this will be possible without any great sacrifice of yield, provided the optima are not *too* diverse.

10.4. Estimation of Equilibrium Yield – Baranov's Method

10.4.1. THEORY AND COMPUTATIONS. Baranov (1918, p. 92) developed expressions for yield that are applicable to stocks of fish in which average increase in length is the same in successive years among commercial-sized fish, and weight is proportional to the cube of length. To facilitate the combination of growth and mortality into one expression, the instantaneous total mortality rate is expressed in terms of the unit of time in which the fish grows a unit of length; so that, in effect, length can be used as a measure of time. The following symbols will be used:

l fish length, in centimeters for example

d annual increase in length of a fish, in the same unit as l

Z/d instantaneous rate of decrease in numbers of a year-class, referred to the interval of time in which it grows one unit of length (Z/d = Baranov's K)

L length of a fish at recruitment

a a constant such that the mean weight (w) of fish, of a given length l, is equal to al^3 (a = Baranov's w)

R number of fish recruited annually at length L, recruitment being at a constant absolute rate throughout the year

N'_0 a constant; described by Baranov as the hypothetical number of fish that would have existed at the time when $l = 0$ if the mortality rate Z were constant back to that time. (N'_0 is not used in actual calculations)

From the definitions above, the number of recruits of length L in each interval $1/d$ of a year long is:

$$\frac{R}{d} = N'_0 e^{-LZ/d} \tag{10.5}$$

Baranov shows, by an argument similar to those of Section 1.5.6, that the *number* of fish of commercial size in a steady-state population is:

$$\bar{N} = \int_{l=L}^{l=\infty} N'_0 e^{-lZ/d} dl = \frac{N'_0 e^{-LZ/d}}{Z/d} = \frac{R}{Z} \tag{10.6}$$

This is the same as expression (1.32).

The *weight* of the commercial population is:

$$\bar{B} = \int_{l=L}^{l=\infty} N'_0 al^3 e^{-lZ/d} dl$$

$$= \frac{aL^3 N'_0 e^{-LZ/d}}{Z/d} \left(1 + \frac{3}{LZ/d} + \frac{6}{(LZ/d)^2} + \frac{6}{(LZ/d)^3}\right) \tag{10.7}$$

248

The integration factor which appears in brackets can be designated by the letter Q for convenience (Baranov's q):

$$Q = 1 + \frac{3}{LZ/d} + \frac{6}{(LZ/d)^2} + \frac{6}{(LZ/d)^3}$$

(10.8)

By comparing with (10.6), another form of (10.7) is:

$$\overline{B} = aL^3\overline{N}Q$$

(10.9)

Rearranging (10.9), (10.8) may be written in the form:

$$Q = \frac{\overline{B}/\overline{N}}{aL^3}$$

(10.10)

The numerator of the RHS of (10.10) is the average weight of a fish in the population, while the denominator is the weight of a recruit. Thus Q is a factor that reflects the gain in weight made by an average fish from time of recruitment to time of death.

The mean weight of the commercial stock, (10.7), may also be written in terms of recruitment, R:

$$\overline{B} = \frac{RaL^3Q}{Z}$$

(10.11)

Having found the mean population on hand in numbers and in weight (expressions 10.6 and 10.11) a year's catch is obtained by multiplying these by the rate of fishing, F:

$$\text{Catch in numbers} = C = F\overline{N} = \frac{FR}{Z}$$

(10.12)

$$\text{Yield in weight} = Y = F\overline{B} = \frac{FRaL^3Q}{Z} = CaL^3Q$$

(10.13)

Expression (10.13) is well adapted to examining the effects of a change in fishing rate (F) or in size at recruitment (L). Baranov illustrates the former in his figure 10, and the latter in figure 11.

10.4.2. FAILURE OF CONDITIONS. The assumptions underlying Baranov's method are rather restricting, compared with those of Section 10.3. Some of the difficulties which arise are as follows:

1. In some populations length increments do not remain even approximately constant over the main range of commercial sizes.

2. With some fishes the exponent in the length-weight relationship deviates 'considerably from 3.

249

3. There is no flexibility in respect to mortality: all recruited broods must be considered subject to equally severe attack, whether by man or by natural woes.

4. Usually fish do not suddenly become catchable at some specific size; rather, their vulnerability increases over a range of sizes and ages, which in some instances occupies a number of years. This difficulty is minimized by making L the middle of the range of increasing vulnerability; but often this middle value is not easy to decide, and in any event there may be a need to estimate rate of fishing for each year of recruitment individually.

In spite of these drawbacks, the Baranov method is easy and quick, and it can be of real value, particularly when effects of only small deviations from existing conditions are being examined. For that matter, calculations involving large deviations will usually be of doubtful applicability, no matter what method is used.

EXAMPLE 10.4. POPULATION AND CATCH OF NORTH SEA PLAICE AT VARIOUS RATES OF FISHING AND NATURAL MORTALITY, BY BARANOV'S METHOD. (Modified from Ricker 1958a.)

Baranov's application of his method was to North Sea plaice (*Pleuronectes platessa*) about 1906, but here the data are for more recent conditions, given by Beverton (1954, pp. 97, 158a–c). Growth in length of plaice is not in fact linear, but it is not far from it at ages 5 to 10, which make up the bulk of the catch: the increase averages 3.0 cm per year over that range. Let us examine first the actual situation where natural mortality, M, is 0.163, and fishing mortality, F, is 0.665 (Beverton's estimates), thus $Z = 0.828$. We have:

$$d = 3.0 \text{ cm/yr}$$
$$Z/d = 0.828/3.0 = 0.276$$
$$L = 25.2 \text{ cm (mean length at recruitment)}$$
$$LZ/d = 6.95$$
$$aL^3 = 143.4 \text{ g (mean weight at recruitment)}$$
$$a = 0.00892$$

From expression (10.8) we calculate:

$$Q = 1 + \frac{3}{6.95} + \frac{6}{6.95^2} + \frac{6}{6.95^3} = 1.574$$

Thus an average fish has a chance to increase in weight by 57%, after recruitment, before it is caught or dies. If R is the annual *number* of recruits, the mean *weight* of stock on hand is, from (10.11):

$$\bar{B} = R \times \frac{143.4}{0.828} \times 1.574 = 273R \text{ grams}$$

or 273 times the yearly number of recruits of 25.2 cm. The yield is, from (10.13):

$$Y = 0.665 \times 273R = 182R \text{ grams}$$

that is, 182 times the yearly number of recruits.

To examine the effect upon yield of having other average recruitment sizes (obtained by using other sizes of mesh in the trawls), appropriate changes are made in L, aL^3, and LZ/d. Still using $F = 0.665$ and $Z = 0.828$, a schedule can be calculated as follows:

1. Mean length at recruitment, in cm (L)	15	20	25.2	30	40
2. Mean weight at recruitment, in grams (aL^3)	30.1	71.4	143.4	241	571
3. LZ/d	4.14	5.52	6.95	8.28	11.04
4. Yield per recruit of length L, in grams	53	103	182	283	605
5. Survival from $l = 15$ cm to the recruitment length	1.0	0.762	0.575	0.443	0.250
6. Yield per fish reaching 15 cm, in grams	53	78	105	125	151

Lines 5 and 6 above are necessary to put the yields on a comparable basis, because in the pre-recruitment phase the fish are decreasing from natural mortality. The latter is $M = 0.163$, or 0.0543 on a centimeter-of-growth basis. The factors in row 5 are therefore calculated from $e^{-0.0543(L-15)}$, where L is the recruitment size under consideration.

It appears that *increasing* the mesh size would tend to increase yield under these circumstances; the same conclusion comes from Beverton and Holt's method, described below. Quantitatively, by using Baranov's method the estimated yields for small L are somewhat too small, and those for large L too great, because this computation does not take into account that absolute yearly increase in length actually decreases as age increases.

10.5. ESTIMATION OF EQUILIBRIUM YIELD — METHOD OF BEVERTON AND HOLT

This method is available in publications of Graham (1952), Beverton (1953), Parrish and Jones (1953), Beverton and Holt (1956, 1957), and the lecture notes of Beverton (1954). It resembles Baranov's, but uses the more widely applicable Brody–Bertalanffy age-length relationship described in Section 9.6. Applicability of this relationship to any population can be tested by plotting a Walford graph. In cases where it adequately describes the growth in length of commercial-sized stock, this procedure removes the first difficulty mentioned in Section 10.4.2, though the others remain.

The following symbols are used:

t age in years; it can be measured from any convenient origin: oviposition, hatching, or the start of the calendar year in which these occur

t_0 the (hypothetical) age at which the fish would have been zero length if it had always grown according to the Brody–Bertalanffy relationship

t_R age of recruitment to the fishery (*average* age at which fish become vulnerable to the gear under consideration) ($t_{p'}$ of Beverton and Holt)

251

$r = t_R - t_0$

N_0 hypothetical number of individuals that reach the hypothetical age t_0 annually

R   yearly number of recruits which enter the fishery at age t_R ($= R'$ of Beverton and Holt)

t_λ "the end of the life-span", or maximum age attained

λ $= t_\lambda - t_R$

F   instantaneous rate of fishing — considered constant over the life-span after recruitment

M   instantaneous rate of natural mortality — considered constant after age t_0

Z   instantaneous total mortality rate — considered constant after age t_R; $Z = F + M$

C   catch, or yield in numbers (Y_N of Beverton and Holt)

Y   yield in weight units (Y_W of Beverton and Holt)

L_∞ average asymptotic length of a fish, as determined by fitting expression (9.9) — see Sections 9.6.9 and 9.6.11.

W_∞ the average asymptotic weight of a fish (i.e. the weight corresponding to the average asymptotic length L_∞). This is estimated from expression (9.4), or from (9.3) using Pienaar and Ricker's (1968) adjustment.

K   Brody growth coefficient (Section 9.6.1), determined by fitting expression (9.9) — see Sections 9.6.9 and 9.6.11.

Over the period of time before recruitment, the initial number N_0 of fish decreases by natural mortality only, so that the number at recruitment is:

$$R = N_0 e^{-Mr} \qquad (10.14)$$

After recruitment, catch in numbers is equal to the rate of fishing times the average population:

$$C = F \int_{t=t_R}^{t=t_\lambda} R e^{-Z(t-t_R)} \, dt \qquad (10.15)$$

and yield in weight is therefore:

$$Y = F \int_{t=t_R}^{t=t_\lambda} R w_t e^{-Z(t-t_R)} dt \qquad (10.16)$$

Omitting F, the integral (10.16) would be the sum of the yearly average bulk of all fish in a year-class, for all the years that it contributes to the fishery. If recruitment is invariable from year to year, this is equal to the weight of commercial stock on hand.

Expression (9.9) of Section 9.6.2 describes the mean length of a fish at age t, when growth is of the Brody–Bertalanffy type. Provided this type of growth prevails over the fishable life span (it can be assumed to apply to the conventional pre-recruitment period), and when growth of the *brood* is isometric (year-class $b' = 3$), we may cube each side of (9.9) and multiply by a' of (9.4), obtaining:

$$w_t = W_\infty(1 - e^{-K(t-t_0)})^3 \tag{10.17}$$

Expanding (10.17) gives:

$$\overline{w}_t = W_\infty(1 - 3e^{-K(t-t_0)} + 3e^{-2K(t-t_0)} - e^{-3K(t-t_0)}) \tag{10.18}$$

Substituting (10.18) for w_t in (10.16), and integrating (Beverton, 1954, p. 45), gives

$$Y = RFW_\infty \left(\frac{1 - e^{-Z\lambda}}{Z} - \frac{3e^{-Kr}(1 - e^{-(Z+K)\lambda})}{Z + K} \right.$$

$$\left. + \frac{3e^{-2Kr}(1 - e^{-(Z+2K)\lambda})}{Z + 2K} - \frac{e^{-3Kr}(1 - e^{-(Z+3K)\lambda})}{Z + 3K} \right) \tag{10.19}$$

When examining different recruitment ages it is convenient to combine (10.14) with (10.19):

$$Y = FN_0e^{-Mr}W_\infty \left(\frac{1 - e^{-Z\lambda}}{Z} - \frac{3e^{-Kr}(1 - e^{-(Z+K)\lambda})}{Z + K} \right.$$

$$\left. + \frac{3e^{-2Kr}(1 - e^{-(Z+2K)\lambda})}{Z + 2K} - \frac{e^{-3Kr}(1 - e^{-(Z+3K)\lambda})}{Z + 3K} \right) \tag{10.20}$$

This is essentially an expanded form of Beverton's (1953) expression (4), except that the conventional starting point is N_0 fish at age t_0, rather than those in existence at a conventional mean age of entry to the fishing grounds (age of recruitment in the Beverton–Holt sense).

For many purposes (10.20) is more complex than is necessary, or even desirable. Selection of the quantity t_λ, the greatest age considered, is always somewhat arbitrary and the terms containing $\lambda = t_\lambda - t_R$ are all close to unity except when Z and λ are both small. The expression can be simplified by omitting such terms, that is, by making $t_\lambda = \infty$. In that manner we obtain:

$$Y = FN_0e^{-Mr}W_\infty \left(\frac{1}{Z} - \frac{3e^{-Kr}}{Z + K} + \frac{3e^{-2Kr}}{Z + 2K} - \frac{e^{-3Kr}}{Z + 3K} \right) \tag{10.21}$$

To see how yield varies with rate of fishing and age of recruitment we vary F and $r (= t_R - t_0)$ in expression (10.20) or (10.21). A typical computation for a moderately long-lived fish, the North Sea haddock, is shown in Beverton and Holt's (1957) figure 17.26 (Fig. 10.2 here). The yield contours or "isopleths" indicate a ridge of high production that starts near the origin of the graph and curves upward and to the right. For any given rate of fishing, F, the maximum yield is calculated to be taken at

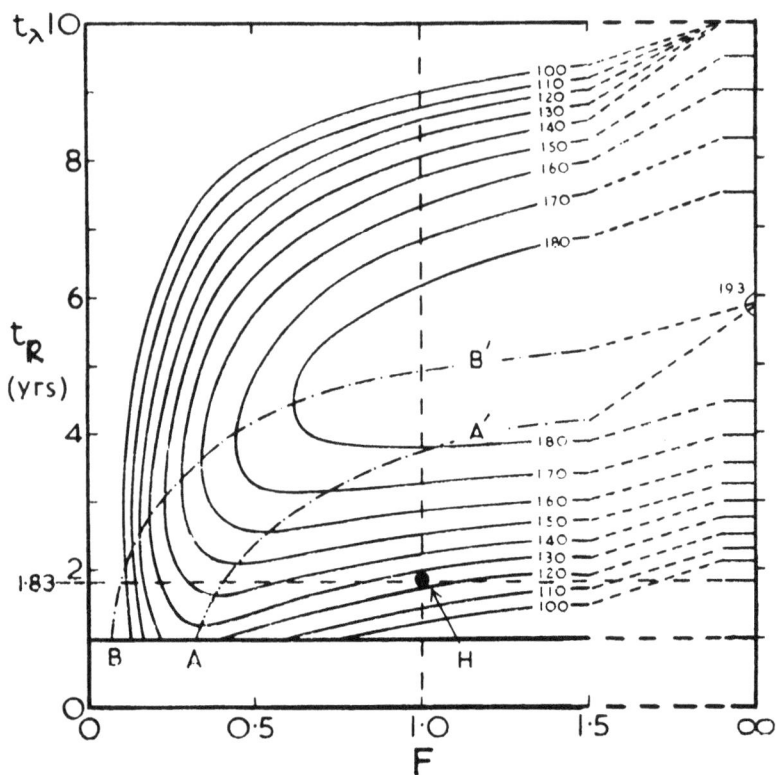

FIG. 10.2. Yield contour diagram for North Sea haddock. Yields shown are in grams per fish reaching conventional age 1, computed by the Beverton–Holt method using M = 0.20, K = 0.20, W_∞ = 1209 g, t_0 = –1.066 yr, t_λ = 10 yr. Point H represents rate of fishing and mean recruitment age in 1939. Ordinate – mean age of recruitment to the fishery (t_R here); abscissa — instantaneous rate of fishing. (Reproduced with slight modification from Beverton and Holt 1956, by permission of the Controller of Her Majesty's Stationery Office.)

the point where the perpendicular from F grazes a contour's left edge; for example, for F = 0.5 the perpendicular is tangent to the contour of 172 g, approximately, and referring this point to the vertical axis, this maximum is obtained when mean age of recruitment is a little less than 4.2 years. The line B–B' in Fig. 10.2 is the locus of all such tangents, and is called by Beverton and Holt the line of *eumetric fishing*. Equally, to find the maximum yield for any mean age of recruitment (for example 3 years) a horizontal line can be drawn to the point where it grazes the bottom of one of the contours, in this case 168 g; and the necessary rate of fishing is about 0.9, found on the abscissa below. The locus of best yields for a given recruitment age is the line A–A'.

A Fortran computer program for Beverton–Holt yield computation was written at the College of Fisheries, University of Washington, Seattle, and is also available as Program BHYLD at the Nanaimo Biological Station of the Fisheries Research Board of Canada (Pienaar and Thomson 1973).

10.6. ESTIMATION OF EQUILIBRIUM YIELD — JONES' MODIFICATION OF THE BEVERTON–HOLT METHOD

Jones (1957) proposed that the Beverton–Holt yield equation (10.16 above) be integrated in a form that would permit it to be evaluated using tables of the incomplete beta function. The resulting expression is:

$$Y = \frac{FN_0 e^{F r} W_\infty}{K} (\beta[X,P,Q] - \beta[X_1,P,Q]) \tag{10.22}$$

$X = e^{-Kr}$

$X_1 = e^{-K(t_\lambda - t_0)}$

$P = Z/K$

$Q = b' + 1$, where b' is the exponent in the population weight:length relationship (expression 9.4).

β = the symbol of the incomplete beta function

Values of $\beta[X,P,Q]$ have been tabulated by Wilimovsky and Wicklund (1963) over the ranges of X, P, and Q that are of most interest in this work. Alternatively, the function can be integrated by computer using a program by L. E. Gales (Program FRG-701 of Abramson 1971).

An advantage of (10.22) over (10.20) is that it makes it possible to deal with populations in which the weight:length exponent differs from 3. Also, the arithmetic is somewhat simpler, and although accurate interpolation in the incomplete beta table would take time, for practical purposes it is sufficient to use linear interpolation (see Example 10.5).

If t_λ is large, the expression comparable to (10.21) is:

$$Y = \frac{FN_0 e^{F r} W_\infty}{K}(\beta[X,P,Q]) \tag{10.23}$$

If for no other purpose, it is useful to make a computation using one of Jones's formulae, with $b' = 3$, to check the arithmetic of the computations using the Beverton–Holt formula — or vice versa. This applies even if the work has been done by computer; indeed, it is especially necessary then.

10.7. APPROXIMATIONS IN BEVERTON–HOLT YIELD COMPUTATIONS

10.7.1. COMPARISON OF THE LONG AND SHORT FORMS. The shortened expressions (10.21) and (10.23) are the preferable ones when Z and t_λ have values characteristic of a reasonably intensive fishery — that is, when $Z = 0.5$ or more and when t_λ represents the greatest age observed in a sample of 500 to a few thousand individuals. If, for prediction purposes, Z is given a considerably smaller value and t_λ is not changed, the full expressions (10.20) and (10.22) then describe a population in which an appreciable fraction of the fish reach age t_λ each year and then suddenly perish. This is true, for example, of the isopleths to the left of A, approximately, in Fig. 10.2.

Although the above danger is avoided by using expression (10.21) or (10.23) — which imply that the old fish continue to die off gradually and evenly at the same rate as younger fish — this may often be biased somewhat in the opposite direction. In some populations it has been shown that natural mortality rate increases among mature and older fish (cf. Ricker 1949a; Kennedy 1954b; Tester 1955), and the Beverton–Holt formulae do not allow for age variation in this statistic, other than a sudden increase to 100%. For a more exact treatment the method of Section 10.3 is available.

10.7.2. EFFECT OF USING $b' = 3$ AS AN APPROXIMATION. (1) If W_∞ is calculated from L_∞ using $W_\infty = a\, L_\infty^{b'}$, it is important that the correct exponent b' be used. For example, if $L_\infty = 10$, and $b' = 3$ is used as an approximation when true $b' = 3.25$, W_∞ and thus yield is underestimated by 44% (provided the same a' is used in both cases).

(2) Assuming that a correct W_∞ has been obtained, what is the remaining effect of using $b' = 3$ as an approximation in (10.20) or (10.21)? This can easily be discovered from Wilimovsky and Wicklund's table. For example, for a cod population similar to that of Example 10.5 below, the following figures are obtained (using X = 0.55, P = 5):

b'	Beta function $\times 10^3$
2.75	1.960
3.0	1.703
3.25	1.482
3.5	1.292

The effect of using $b' = 3$ is to underestimate yield if true b' is less than 3, and to overestimate it when $b' > 3$. The greatest difference within the range above is –24%, but usually it will be considerably less since few fish have an exponent b' as large as 3.5. Paulik and Gales (1964) have illustrated effects of such differences for a number of types of population.

In any event the absolute level of yield from a Beverton–Holt computation is not usually of any great interest, as it merely shows what is obtained from a unit number of fish of some conventional age. What *is* of interest is the difference in yield that will result from varying t_R or F, and the relative error in such differences, when using an incorrect b', tends to be much less than that in the absolute figures. This rule of course applies also to the approximations used in yield computations of other sorts.

EXAMPLE 10.5. COMPUTATION OF EQUILIBRIUM YIELDS FOR A COD FISHERY BY THE BEVERTON–HOLT METHOD. (Modified from Halliday 1972.)

Mortality and Bertalanffy growth statistics for the stock of cod (*Gadus morhua*) in ICNAF regions 4Vs and 4W were estimated as follows:

F   0.49 λ 8.8 yr $(= t_\lambda - t_R)$
M   0.20 W_∞ 11.41 kg

Z	0.69	K	0.14
t_0	+0.07 yr	b'	3.07
t_R	4.2 yr	X	0.550 (= e^{-Kr})
r	4.13 yr (= $t_R - t_0$)	P	4.93 (= Z/K)
t_λ	13 yr (see below)	Q	4.07 (= $b' + 1$)

The statistic t_λ above is defined by Halliday as the "maximum age of significant contribution to the fishery," instead of the end of the life span. This seems questionable for it implies that, at rates of fishing less than those observed, substantial numbers of cod reach age 13 and then die immediately; whereas in fact Atlantic cod are rather long-lived fish, some surviving to 25 years at least. When t_λ is put equal to a figure of this order all terms containing λ in (10.19) become negligible, so the shorter (10.21) can be used.

Calculations can most conveniently be started from a conventional round number of fish at age t_0, say 1000. With $b' = 3$ as an approximation, the yield equation (10.21) can be used; it becomes:

$$Y = F' \times 10^3 \times e^{-0.2r} \times 11.41 \left(\frac{1}{F + 0.2} - \frac{3e^{-0.14r}}{F + 0.34} + \frac{3e^{-0.28r}}{F + 0.48} - \frac{e^{-0.42r}}{F + 0.62} \right) \quad (10.24)$$

For the observed F = 0.49 and r = 4.13 the yield is 577 kg. Values for other rates of fishing and other mean recruitment ages can be found by varying F and r.

Alternatively, we could use Jones's solution (10.23), obtaining:

$$Y = \frac{F \times 10^3 \times e^{Fr} \times 11.41}{0.14} \left(\beta[e^{-0.14r}, \frac{F + 0.2}{0.14}, b' + 1] \right) \quad (10.25)$$

For F = 0.49, r = 4.13, and $b' = 3$, $\beta = 1.921 \times 10^{-3}$ and the yield is 582 kg; the 1% difference between this figure and the 577 above is because linear interpolation was used in the incomplete beta table.

However, Jones' method permits us to use the true $b' = 3.07$, so the beta function becomes 1.847×10^{-3} and the yield is 559. Thus the approximate $b' = 3$ gives a result only 4% greater than the true value, so that using the approximate exponent makes little difference in this example.

To produce a graph such as Fig. 10.2 is quite tedious, because the contour lines must be interpolated among the computed yield values. An easier presentation of the data is shown in Fig. 10.3, which shows yields for each integral age of entry and at closely-spaced series of rates of fishing.[5] (A similar graph can be plotted using age

---

[5]Halliday's yield graph (1972, fig. 7) is not directly comparable to Fig. 10.3. It shows yield in kilograms per fish *of age 1*, which is 0.93 year later than the conventional age t_0. Thus his figures are approximately $10^{-3}/e^{0.2 \times 0.93} = 1.203 \times 10^{-3}$ times those computed above on the basis of 1000 fish of age t_0, using (10.24) or (10.25). Also, Halliday used $t_\lambda = 13$ whereas Fig. 10.3 uses $t_\lambda = \infty$. From the fish's point of view the age 1 starting point used by Halliday is as arbitrary as the time t_0, as it is unlikely that natural mortality rate would remain at M = 0.2 back to age 1.

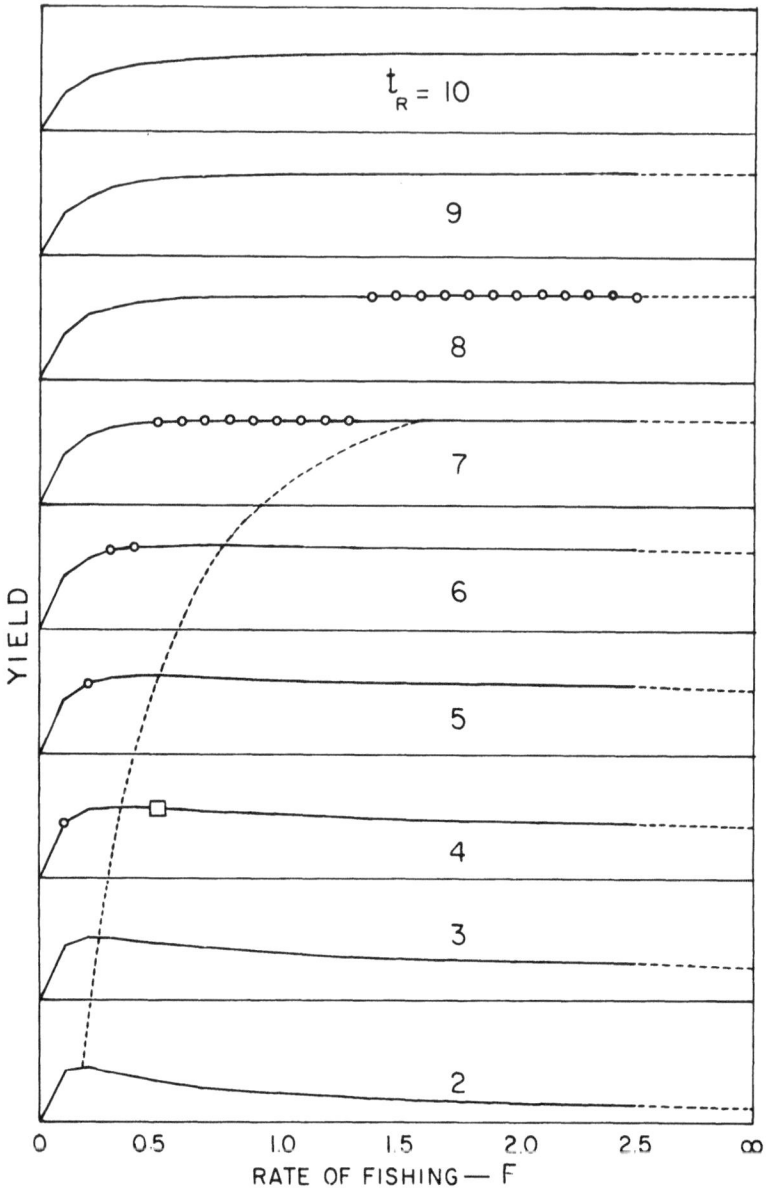

Fig. 10.3. Yield from a cod stock per unit number of fish of conventional age $t_0 = 0.07$, plotted against rate of fishing (F), for recruitment ages (t_R) from 2 to 10. The ordinate divisions represent 1 kg of yield per fish at t_0. (Data from Halliday 1972; see the text.)

of entry on the abscissa and a series of about 10 values of F in the range of interest). In either case the ordinate scale represents the yields read directly from a computer printout. There is a maximum of yield at intermediate levels of effort for recruitment ages of 7 or less; for larger t_R, yield continues to increase to an asymptote. The observed yield maxima for each age of entry are joined to form a broken line that indicates (on the abscissa) the rate of fishing needed for best yield at each t_R.

For any given fishing effort, the value of t_R that gives greatest yield is circled on Fig. 10.3. The circles form an ascending curved progression analogous to the eumetric fishing curve of Fig. 10.2. Within the range of F values shown, the best age of entry is 8 years or less, and beyond age 6 the gain is minute.

The actual situation of the fishery at the time of Halliday's analysis was close to the square point of Fig. 10.3. To obtain maximum yield at the observed F = 0.49, the mean age of entry should be increased to t_R = 5. Alternatively, to obtain maximum yield using the observed mean age of entry the rate of fishing should be reduced to about F = 0.37.

10.8. Changes in Age Structure and Biomass Resulting from Fishing

For fishes having moderate to long life-spans, even a little fishing can cause a marked change in age structure. Figure 10.4 shows the biomass contributed by successive ages of a moderately long-lived stock, based on mortality rates of a stock of lingcod (*Ophiodon elongatus*). A fishing rate of only F = 0.1 reduces the equilibrium representation of old fish in the stock very substantially. At F = 0.4 there will be only a handful of fish left that are older than age 7, although these originally made up half the biomass of the stock. Yet F = 0.4 corresponds here to a rate of exploitation

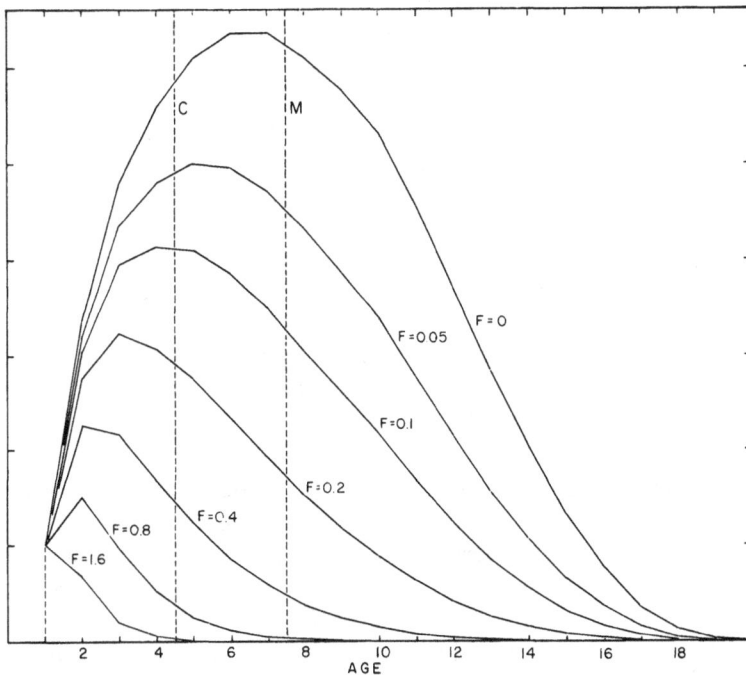

Fig. 10.4. Equilibrium weight of fish present at successive ages, for different rates of fishing (*f*) in terms of a unit weight of recruits at age 1, for a moderately long-lived stock. (After Ricker 1963, fig. 1.)

259

of about 30%, which is usually considered quite moderate. For F = 0.8 (exploitation about 44%) there are very few fish older than age 4.

If the annual recruitment and the growth rate of the fish do not change, there are corresponding decreases in biomass of the stock. F = 0.1 reduces total biomass to about half the original, while at F = 0.4 it is 16% of the original and at F = 0.8 it is only 7%. The corresponding increases in rate of exploitation are insufficient to compensate for such large decreases in biomass, so with constant recruitment the catch would decline greatly at large F. Since quite a number of long-lived stocks have supported moderate fisheries for a considerable period of time, this suggests that their absolute annual recruitments must have increased substantially when the mature population was first fished down. In some cases growth rates also increased appreciably.

10.9. TEMPORARY EFFECTS OF A CHANGE IN THE RATE OF FISHING

Previous sections of this chapter have described the *equilibrium* catches and stocks to be expected under stated conditions of growth and rate of fishing, with steady recruitment. However, the *immediate* effect of a change in fishing effort is often quite different from its long-term effect; it is important to know what will happen along the road to maximum equilibrium yield, assuming measures for achieving the latter are actually adopted. Apart from that, in most stocks rates of fishing have changed drastically during the present century, especially since 1950; thus the effects of such changes on the catches obtained need to be interpreted.

It is fairly obvious that, in any given season, increased fishing will make for greater catch *at that time*, and less fishing will mean less catch, whatever may happen later. The pattern of change from immediate yield to equilibrium yield became known when Baranov (1918)[6] and Huntsman (1918) simultaneously described the effect of a sustained change in mortality rate upon a stock's age composition and upon the catch taken from it. Huntsman showed by pyramidal diagrams that, by imposing a conditional fishing mortality rate (m) of 1/4 upon an unfished stock in which natural mortality rate (n) was 1/7, the relative number of old fish in the population decreased progressively. Baranov illustrated the same process by examining the effect of an increase in mortality, A, from 0.2 to 0.5, using the graph reproduced here as Fig. 1.1; he particularly emphasized the temporary nature of the large increase in catch which follows such an increase in rate of fishing.

Unfortunately, neither of these presentations created much impression at the time. Only during the later 1930's, after Thompson and Bell's (1934) excellent exposition and illustrations became available, did a general appreciation of these effects become evident in our fishery literature. Today the sequence of catches obtained during an expanding fishery is usually described as "the fishing-up effect" or "the removal of accumulated stock"; in Russian, *molozhenie* or "juvenation of the age structure." The reverse process could appropriately be called "replacement of stock" or restoration of an earlier age structure.

---

[6] According to Zasosov (1971), Baranov's classical paper was printed and distributed separately in 1916, although the journal containing it is dated 1918.

To illustrate these situations, schedules like Table 10.1 or Table 10.3 can be used, but each year must be treated separately during the period of transition from the old to the new rate of fishing. The contrast between equilibrium yields and temporary catch potential is illustrated in Fig. 10.5: three different rates of fishing (F = 0.3, 0.8, and 1.3) have equilibrium levels of yield that are much alike, the intermediate level being slightly the best. A yield *four times* as great, however, is taken in the first year of the change from F = 0.3 to F = 1.3. Similar short-term potentialities exist

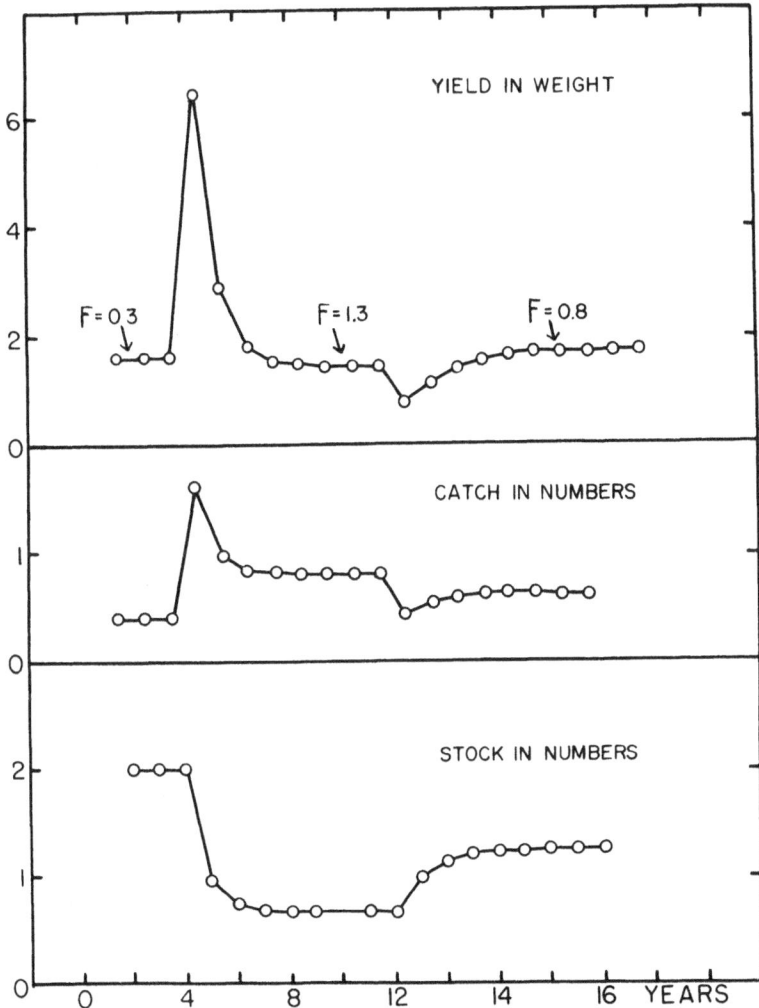

FIG. 10.5. Trends of yield in weight, catch in numbers, and stock size in numbers, for a stock in which natural mortality (M) is 0.2 throughout, and rate of fishing (F) changes from 0.3 to 1.3, and then to 0.8. The first year of each change is marked by the high peak and low trough, respectively, on the yield curve. Values were computed using a model of the type of Table 10.1, with an appropriate age–weight distribution, the same for all years.

in any new or lightly-fished stock that consists of many age-groups.[7] A stand of virgin timber affords a close analogy.

Familiar and even obvious though these relationships now appear, their discovery in 1916 represented a major feat of imaginative analysis. Furthermore, their practical value to date may have been greater than that of all the various determinations of equilibrium yield of the kinds described in Sections 10.2–10.8, for two reasons. (1) Temporary effects of changes in rate of fishing tend to be much greater than the equilibrium effects which are calculated on the basis of constant recruitment; thus it has been easier to check theory against practice and to make useful predictions. (2) Constant recruitment seems unlikely to be a suitable basis for predicting true equilibrium yield at different levels of fishing, in anything more than a minority of stocks (cf. Chapter 11); but it is a suitable basis for predicting the *immediate* effect of a change in fishing, because the increased or decreased year-classes resulting from change in stock density (caused by change in fishing) usually take some years to "grow into" the usable stock.

Moreover, these temporary changes in yield bulk very large in the view of fishermen whenever new regulations are contemplated; because of this, goals which seem desirable from the equilibrium-yield standpoint must sometimes be approached quite gradually. Conversely, knowledge of the direction and magnitude of expected *temporary* increases or decreases in yield makes it possible to avoid mistaking them for indications of long-term prospects.

EXAMPLE 10.6. COMPUTATION OF YIELDS DURING THE PERIOD OF TRANSITION FROM A SMALLER TO A LARGER RATE OF FISHING. (From Ricker 1958a.)

Table 10.5 shows the effect, upon the population of Table 10.3, of doubling the rate of fishing at all ages. Divisions of the year are condensed to the two fishing periods and the long period between. Column 2 shows the resultants of growth and natural mortality taken from Table 10.3, to which are added the new mortality rates of column 3, giving the new instantaneous rates of population change (column 4) and corresponding change factors (column 5). The latter are applied to previous equilibrium population weights at the *start* of each age shown in column 6 (from column 11 of Table 10.3). Fish of each age decrease in bulk during year 1 as shown in column 7; the average for each period was computed and multiplied by the instantaneous rate of fishing to give the yield shown in column 8. In year 2 the overwinter survivors *of each age* in year 1 are computed and their weight is entered at the start of the next greater age: for example, $2497 \times 0.878 = 2192$; $1056 \times 0.741 = 782$; etc. During

---

[7] This effect, perhaps more than any other, accounts for the fisherman's nostalgia for the "good old days" when, for a few years, catch per hour or per set was so much greater than at present. Of course other factors may also be involved. Certain types of relationship between stock density and recruitment can produce a similar effect (Section 12.2), though usually less extreme. There may also be increased wariness on the part of the fish, or bad memory on the part of the fisherman (cf. Kennedy 1956, p. 47). Finally, simply increasing the amount of gear implies a decline even in equilibrium catch per unit, which usually becomes apparent long before the level of maximum sustained yield is reached (cf. Table 10.2).

TABLE 10.5. Effect of doubling the rate of fishing at all ages, upon the population of Table 10.3. (See text for explanation.)

1	2	3	4	5	6	7	8	9	10	11	12	13	14
						Year 1		Year 2		Year 3		Year 4	
Age	G-M	F	G-M-F	Weight change factor	Previous equilibrium weight of stock	Initial weight	Yield	Initial weight	Yield	Initial weight	Yield	Initial weight	Yield
2⅛	+0.335	0.08	+0.255	1.290	2086	2086	191	2086	191	2086	191	2086	191
2¼	+0.205	0.28	−0.075	0.928	....	2691	726	2691	726	2691	726	2691	726
2⅜					....	2497		2497		2497		2497	
	−0.130	0	−0.130	0.878									
3⅛	+0.055	0.66	−0.605	0.546	2625	2625	1339	2192	1118	2192	1118	2192	1118
3¼	+0.035	0.34	−0.305	0.737	....	1433	423	1197	353	1197	353	1197	353
3⅜					....	1056		882		882		882	
	−0.300	0	−0.300	0.741									
4⅛	−0.005	0.66	−0.665	0.514	1291	1291	645	782	391	654	327	654	327
4¼	−0.025	0.34	−0.365	0.694	....	664	191	402	116	336	97	336	97
4⅜					....	461		279		233		233	
	−0.370	0	−0.370	0.691									
5⅛	−0.075	0.66	−0.735	0.480	525	525	256	319	156	193	94	161	79
5¼	−0.100	0.34	−0.440	0.644	....	252	70	153	43	93	26	77	22
5⅜					....	162		99		60		50	
Totals							3841		3094		2932		2913

263

year 2 fishing occurs and the population decreases at the same rate as in year 1, but the yield is less for age 3 and older. By year 4 the new equilibrium population structure is established, shown in column 13.

The change from the old to the new conditions is completed in four years, which is the number of vulnerable age-groups of fish present in significant numbers. In the first year of change the yield rises from 2.29 to 3.84 kg (per kilogram of age 2 recruits), then falls to 3.09, to 2.93, and finally to the new equilibrium value 2.91 kg.

10.10. ALLEN'S METHOD OF CALCULATING BEST MINIMUM SIZE

Allen (1953) suggested a method of computing the best minimum size of fish for maximum yield. It implies "knife-edge" recruitment (Section 11.1.2), but approximate adjustments can be made when this condition is only approximated. Suppose that the existing minimum weight of the fish harvested is w_R, the mean weight of fish in the catch is $\bar{w}$, and the exploitation ratio for the stock is E. Allen shows that if $w_R > E\bar{w}$ the value of w_R is less than what will provide maximum sustainable yield, whereas if $w_R < E\bar{w}$ the value of w_R is too great for MSY.

This method requires an estimate of E, which is the ratio of the number of fish caught from a year-class to the total number present when it became vulnerable to fishing. When rates of fishing (F) and of natural mortality (M) are unchanging or change proportionally throughout life, $E = F/Z$. Thus the information needed for this method is similar to what is required for one of the more complete analyses of the earlier sections of this chapter.

CHAPTER 11. — RECRUITMENT AND
STOCK-RECRUITMENT RELATIONSHIPS

11.1. TYPES OF RECRUITMENT

11.1.1. REPRODUCTION. The reproduction accomplished by a fish stock can be assessed at any stage: eggs, larvae, fry, juveniles, smolts, and so on. Of most interest in practical fishery work is the number of recruits to the usable stock. As used here, recruitment is the process of becoming catchable; for an individual fish it is the moment or interval during which it becomes in some degree vulnerable to capture by the fishing gear in use. Three types of situations can be distinguished.

11.1.2. KNIFE-EDGE RECRUITMENT. All fish of a given age become vulnerable at a particular time in a given year, and their vulnerability remains the same throughout their lives (or at least for two consecutive complete years). Few fish populations approximate this ideal.

11.1.3. RECRUITMENT BY PLATOONS. Vulnerability of a year-class increases gradually over a period of 2 or more years, but during any year (fishing season) each individual fish is either fully catchable or not catchable. Thus a year-class is divided into two different platoons: recruited and not recruited. The fish of the recruited platoon of any age are of larger average size than the unrecruited ones, but there is often a broad overlap of sizes. Platoon recruitment is typical when fishing attacks a population during a breeding migration and non-maturing fish do not mingle with the maturing ones.

Let the number of fish in the recruited platoon of any age-group be N_R, and let the total number of fish of that age be N. Then it is clear that the ratio N_R/N represents the ratio of the rate of exploitation (u) of that age-group (considered as a whole) to the rate of exploitation of fully-vulnerable ages.

11.1.4. CONTINUOUS RECRUITMENT. There is a gradual increase in vulnerability of members of a year-class over a period of two or more years, related to the increasing size of the individual fish, or a change in their behavior or distribution, or any combination of these. This is probably the commonest type of recruitment; each fish becomes more and more likely to be caught as it grows larger and older, until the limit of maximum vulnerability is reached. This situation is indicated for the halibut population of Example 5.5, where the smaller tagged fish were retaken less frequently during the first year or two after tagging than they were during later years — that is, their catchability gradually increased.

For convenience in calculation, however, continuous recruitment can often be treated as though it were platoon recruitment. In that event what appears as the ratio

N_R/N of the previous section has no objective meaning in terms of platoons, but it *is* the ratio of the rate of exploitation of the whole age-group to the rate of exploitation of fully-vulnerable ages.

Both platoon recruitment and continuous recruitment have often been approximated by knife-edge recruitment in yield computations, in which event the computed mean age at which the fish first become catchable need not be a whole number.

11.2. ESTIMATION OF RECRUITMENT — BIOSTATISTICAL METHOD

Early in this century, as soon as age determinations began to be made on a large scale, the relative abundance or "strength" of successive year-classes came to be judged from their percentage representation in the catch over a period of years. Such data were commonly plotted as ordinary columnar histograms — for example the famous series for Atlanto–Scandian herring begun by Hjort (1914) and continued by several subsequent authors (Nikolsky 1965, fig. 28).

Later the utilized stock (V) of successive year-classes, as defined in Section 8.1, came to be used as a more quantitative but still only relative estimate of recruitment. Utilized stock is strictly proportional to recruitment if rates of fishing (F) and natural mortality (M) have not changed over all the years involved, a situation that rarely occurs. However, moderate fluctuations in either F or M do not seriously interfere with the usefulness of V for this purpose, particularly when M is small: in the latter event, if a fish is not caught one year, it will probably still be available in one or more future years.

If F changes drastically, and particularly if it has experienced a sustained trend upward or downward, utilized population (V) becomes less useful as an index of recruitment. However, it is frequently possible to make adjustments for this based on known fishing efforts and a reasonable natural mortality rate (Ricker 1971b).

The principal disadvantage of the biostatistical method of estimating recruitment is that it requires a rather long series of catch statistics, with annual determinations of the age structure. It is never an absolute estimate of recruitment, since naturally-dying individuals are not included, but of itself this is no great disadvantage.

11.3. ESTIMATION OF RECRUITMENT — ALLEN'S METHOD

11.3.1. PROCEDURE. To treat recruitment quantitatively at successive ages it is best to define it as full vulnerability to the gear in use, and to assume the platoon type of recruitment described in Section 11.1.3. We need then to know what fraction of individuals of a given age in the catch joined the vulnerable platoon between the previous and the current fishing season, and what fraction represents the survivors of the vulnerable fish of earlier years.

Allen (1966b, 1968) proposed a method of estimating this, in the first instance for Type 1 fisheries in which fishing is assumed to occur before natural mortality in each biological year. The method requires knowledge of age composition of a representative sample of the catch in at least two successive years. These need not be reduced to a common sample size. The following symbols are used:

Q_1 = number of fish of all fully-recruited ages in the sample of year 1

Q_2 = number of fish of same year-classes (*not* ages) as Q_1 in the sample of year 2

p_1 = number of fish of an incompletely-recruited age t in the sample of year 1

p_2 = number of fish, of the same year-class as p_1, taken at age $t + 1$ in the sample of year 2

Year-class strengths vary and this will affect the size of Q_2 and of p_2, relative to Q_1 and p_1. However, the change will be in proportion for both; that is, the ratio:

$$B_2 = \frac{p_2/p_1}{Q_2/Q_1} \qquad (11.1)$$

is independent of absolute strengths of year-classes in the population. It is also independent of the relative size of the two samples.

Suppose that the instantaneous natural mortality rate of the recruited platoons of incompletely recruited age-groups is M_i and that of completely recruited age-groups is M_r, while F represents the rate of fishing for all vulnerable fish. Let a ratio T be defined as the survival rate of fully recruited ages divided by the survival rate of the recruited platoons of incompletely-recruited ages; that is:

$$T = \frac{e^{-(M_r+F)}}{e^{-(M_i+F)}} \qquad (11.2)$$

Allen shows that the proportion of new recruits of age $t + 1$ in year 2 is approximately equal to:

$$W_2 = \frac{B_2 - T}{B_2} \qquad (11.3)$$

T is difficult to estimate. Usually it will be unknown and the assumption that $T = 1$ must be used: that is, that natural mortality rate is the same for vulnerable fish of both incompletely recruited and fully recruited ages. Then (11.3) reduces to:

$$W_2 = \frac{B_2 - 1}{B_2} = 1 - \frac{Q_2/Q_1}{p_2/p_1} \qquad (11.4)$$

11.3.2. ILLUSTRATION. Allen's mathematical development of expression (11.3) is rather involved. Instead of reproducing it here, I will illustrate its applicability by means of numerical models. Table 11.1 shows a simple Type 1 population in which year-class abundance is the same over the period of years involved, and all members of all age-groups have the same *natural* mortality rate. The instantaneous natural mortality rate is M = 0.3; thus the natural survival rate is $e^{-0.3} = 0.7408$. The instan-

267

TABLE 11.1   Model of a stock with recruitment occurring at ages 3–6, as follows: 10% of each year-class becomes vulnerable at age 3; 50% of the previously non-vulnerable fish become vulnerable at age 4; 80% of the previously non-vulnerable fish become vulnerable at age 5; all of the previously non-vulnerable fish become vulnerable at age 6. Natural mortality rate is M = 0.3 throughout. Fishing mortality precedes natural mortality in each biological year (Type 1 fishery). The number of fish at the start of age 3 is 10,000 in each year. For vulnerable fish the rate of fishing is F = 0.4; rate of exploitation, u = 0.3297; total instantaneous mortality rate, Z = 0.7; survival rate, S = 0.4966. For non-vulnerable fish Z = M = 0.3, S = 0.7408.

1	2	3	4	5	6	7	8	9
		From previous year's nonvulnerable fish		Vulner-able carry-over	Total vulner-able fish		Catch	
Age	Total	Not vul-nerable	Vulner-able			Total	Newly vulner-able	% new
3	10,000	9,000	1000	....	1,000	330	330	100
4	6,668	3,334	3334	496	3,830	1263	1099	87.0
5	2,470	494	1976	1902	3,878	1278	651	50.9
6	366	0	366	1926	2,292	756	121	16.0
7	0	0	0	1138	1,138	375	0	0
8	0	0	0	565	565	186	0	0
9	0	0	0	281	281	93	0	0
10–18	0	0	0	275	275	91	0	0
Total	19,504	12,828	6676	6583	13,259	4372	2201	50.3
Ages 3–5						2871		
Ages 6 and up						1501		

taneous rate of fishing is F = 0.4, and fishing occurs before natural mortality each year; thus the rate of exploitation is $u = 1 - e^{-0.4} = 0.3297$, and the survival rate from fishing is 0.6703. The total survival rate of the vulnerable members of the stock is $S = e^{-(0.3+0.4)} = 0.7408 \times 0.6703 = 0.4966$.

At ages 0, 1, and 2 the fish are all too small to be captured. At age 3, 1000 out of 10,000 become vulnerable (column 4), and $0.3297 \times 1000 = 330$ are caught (column 7); of the 670 remaining, $0.7408 \times 670 = 496$ survive to be carried over into the vulnerable stock at the beginning of age 4 (column 5). At age 4 there are $0.7408 \times 9000 = 6668$ survivors of the non-vulnerable stock of the previous year (column 2). Half of these become vulnerable at age 4 (column 4), and these join the 496 survivors of the first year's vulnerable fish, for a total of 3830 vulnerable (column 6). Of these, $0.3297 \times 3830 = 1263$ are caught, 0.2592 of the 2567 survivors die naturally, and $0.4966 \times 3830 = 1902$ survive the year (column 5 at age 5). Survivors of the age-4 non-vulnerable fish number $0.7408 \times 3334 = 2470$ at age 5 (column 2), and 80% of these are vulnerable, or 1976 fish (column 4). The total vulnerable fish of age 5 number $1976 + 1902 = 3878$, of which 0.3297 or 1278 are caught. At age 6 there are 366 survivors of the 494 non-vulnerable fish of age 5, and all have become vulnerable; added to the 1926 carry-over they make 2292 vulnerable fish and the catch is 756. From age 6 onward the rate of decrease in catch from one age to the next in Table 11.1 reflects the total survival rate; ages 10 and older are condensed into a single line.

Since Table 11.1 represents a series of year-classes of identical abundance, any two successive catch entries in column 7 can also be used to represent members of the same year-class. In applying formula (11.4) we first calculate:

$$\frac{Q_2}{Q_1} = \frac{91 + 93 + 186 + 375}{91 + 93 + 186 + 375 + 756} = \frac{745}{1501} = 0.4963$$

(In this illustration Q_2/Q_1 is an estimate of survival rate, but this is true only when there is no variation in year-class strength from year to year.)

The following calculations of percentage recruits in each year's catch are independent of year-class strength.

For age 4: $p_2 = 1263$; $p_1 = 330$; so from (11.4):

$$W_2 = 1 - \frac{0.4963}{1263/330} = 87.0\%$$

For age 5: $p_2 = 1278$; $p_1 = 1263$; $W_2 = 1 - \frac{0.4963}{1278/1263} = 51.0\%$

For age 6: $p_2 = 756$; $p_1 = 1278$; $W_2 = 1 - \frac{0.4963}{756/1278} = 16.1\%$

These computed values of W agree closely with the true values in column 9 of Table 11.1. The percentage of new recruits in the catch as a whole is $2201/4372 = 50.3\%$.

11.3.3. EFFECT OF CHANGE IN RATE OF FISHING. Table 11.2 follows the population of Table 11.1 through two years in which rate of fishing increases first to 0.8 and then to 1.2. Calling these years 2 and 3:

$$\frac{Q_3}{Q_2} = \frac{129 + 132 + 265 + 533}{152 + 155 + 311 + 627 + 1262} = 0.4224$$

Again applying expression (11.4):

For age 4: $p_3 = 2562$; $p_2 = 551$; $W_3 = 1 - \frac{0.4224}{2562/551} = 90.9\%$

For age 5: $p_3 = 2272$; $p_2 = 2110$; $W_3 = 1 - \frac{0.4224}{2272/2110} = 60.7\%$

For age 6: $p_3 = 1158$; $p_2 = 2136$; $W_3 = 1 - \frac{0.4224}{1158/2136} = 22.1\%$

269

TABLE 11.2. The population of Table 11.1 in two consecutive subsequent years, characterized by rates of fishing of $F = 0.8$ and 1.2, respectively. Column numbers are as in Table 11.1, columns 2–4 being omitted because they are the same as Table 11.1 in both years. For year 2, $F = 0.8$, $u = 0.5507$, $Z = 1.1$, and $S = 0.3329$ for vulnerable fish. In year 3, $F = 1.2$, $u = 0.6988$, $Z = 1.5$, and $S = 0.2231$ for vulnerable fish.

1	5	6	7	8	9	5	6	7	8	9
			Year 2					Year 3		
				Catch					Catch	
Age	Vulnerable carry-over	Total vulnerable fish	Total	Newly vulnerable	% new	Vulnerable carry-over	Total vulnerable fish	Total	Newly vulnerable	% new
3	....	1,000	551	551	100	....	1,000	699	699	100
4	496	3,830	2110	1836	87.0	333	3,667	2562	2330	90.9
5	1902	3,878	2136	1088	50.9	1275	3,251	2272	1381	60.8
6	1926	2,292	1262	202	16.0	1291	1,657	1158	256	22.1
7	1138	1,138	627	0	0	763	763	533	0	0
8	565	565	311	0	0	379	379	265	0	0
9	281	281	155	0	0	188	188	132	0	0
10–18	275	275	152	0	0	185	185	129	0	0
Totals	6583	13,259	7304	3677	50.3	4414	11,090	7750	4666	60.3

Here again the percentage recruitment at each age agrees with figures in the model (year 3 of Table 11.2). Thus estimates of recruitment by this method are independent of changes in rate of fishing, as long as the latter affect all age-groups proportionally.

11.3.4. EFFECT OF A DIFFERENCE IN NATURAL MORTALITY RATE BETWEEN RECRUITED AND UNRECRUITED PLATOONS. If recruited and unrecruited platoons in a population differ in the size or behavior of their fish, possibly they differ also in respect to natural mortality rate (M). By constructing a table similar to Table 11.1 it can be shown that this makes no difference to the estimate of percentage recruitment obtained from expression (11.3) or (11.4).

However, when the recruited and unrecruited parts of an age-group have different natural mortality rates it is not possible to compute total abundance or total mortality rate of the fish at any incompletely-recruited age, even though the rate of fishing can be estimated for its recruited members.

11.3.5. EFFECT OF A DIFFERENCE IN NATURAL MORTALITY RATE BETWEEN INCOM-PLETELY-RECRUITED AND FULLY-RECRUITED AGES. Table 11.3 is constructed with the same parameters as Table 11.1, except $M = 0.6$ at ages 3–6 for both the vulnerable and the invulnerable platoons of each age. We estimate first:

$$\frac{Q_2}{Q_1} = \frac{39 + 40 + 80 + 161}{39 + 40 + 80 + 161 + 324} = \frac{320}{644} = 0.4969$$

which is the same as for Table 11.1 within rounding error. From (11.2):

$$T = \frac{e^{-(0.4+0.6)}}{e^{-(0.4+0.3)}} = \frac{0.3679}{0.4966} = 0.7408$$

Expressions (11.1) and (11.3) are now applied:

For age 4: $p_2 = 935$; $p_1 = 330$; $B_2 = 935/330 \times 0.4969 = 5.702$;
$\quad W_2 = (5.702 - 0.741)/5.702 = 87.0\%$

For age 5: $p_2 = 702$; $p_1 = 935$; $B_2 = 702/935 \times 0.4969 = 1.511$;
$\quad W_2 = (1.511 - 0.741)/1.511 = 51.0\%$

For age 6: $p_2 = 324$; $p_1 = 702$; $B_2 = 324/702 \times 0.4969 = 0.929$;
$\quad W_2 = (0.929 - 0.741)/0.929 = 20.2\%$

These results can be compared with values of W_2 obtained by assuming $T = 1$, as follows:

Age	$T = 0.741$	$T = 1$
4	87·0%	82·5%
5	51·0%	33·8%
6	20·2%	−7·6%

TABLE 11.3. Model of a stock similar to that of Table 11.1, but with $M = 0.6$ between ages 3 and 6, and $M = 0.3$ from age 6 onward.

1	2	3	4	5	6	7	8	9
		From previous year's nonvulnerable fish		Vulner-able carry-over	Total vulner-able fish		Catch	
Age	Total	Not vul-nerable	Vulner-able			Total	Newly vulner-able	% new
3	10,000	9,000	1000	....	1000	330	330	100
4	4,939	2,470	2469	368	2837	935	814	87.1
5	1,356	271	1085	1044	2129	702	358	51.0
6	201	0	201	783	984	324	66	20.4
7	0	0	0	489	489	161	0	0
8	0	0	0	243	243	80	0	0
9	0	0	0	121	121	40	0	0
10–15	0	0	0	119	119	39	0	0
Total	16,496	11,741	4755	3167	7922	2,611	1,568	60.1

Estimates of W_2 for $T = 0.741$ agree with the last column of Table 11.3, whereas when $T = 1$ is used they differ considerably; for age 6 an impossible negative figure is obtained.

In practice, when recruitment is of the platoon type, the recruited platoon of any age-group would probably tend to resemble the older recruited fish rather than their unrecruited siblings, in respect to natural mortality. For example, if they had become vulnerable because they had joined in the spawning migration and breeding activities, they would be exposed to both the external and the physiological hazards of mature fish rather than those of immature fish of the same age. On the other hand, if the recruitment process is actually a gradual increase in vulnerability of all fish of a year-class, it is more likely that all fish of the incompletely-recruited ages would have a natural mortality rate progressively different from that of the fully-recruited fish as you move toward younger ages, and an estimate of T becomes more important. If natural mortality rate could be estimated for each year, then a value of T could also be calculated for each year and used in expression (11.3).

11.3.6. EFFECT OF UNDERESTIMATING AGE OF FULL RECRUITMENT. In practice, year-class strengths may vary from year to year, and even if they do not there is sampling variability in the number of fish found at each age; so it is not always easy to decide what age represents the first age of full recruitment. In Table 11.1, suppose that the age of full recruitment were estimated one year too low. In that event:

$$\frac{Q_2}{Q_1} = \frac{745 + 756}{1501 + 1278} = \frac{1501}{2779} = 0.540$$

272

whereas the true value is 0.496. The result is that values of B_2 are too small, and hence the estimated proportion of new recruits (W_2) is also too small, at each age. In addition, one less age-group has been included in the "incompletely recruited" category, so total recruitment may be significantly underestimated. The comparison for Table 11.1 is as follows:

Age	Catch	New recruits in catch	
		$B_2 = 0.497$	$B_2 = 0.540$
3	330	330	330
4	1263	1099	1082
5	1278	651	467
6	756	121	0
Older	745	0	0
Total	4372	2201	1879

Thus new recruits are estimated as 43% of the total catch, instead of the true figure 50%. This suggests that, when in doubt, the older of two possible boundary ages of full recruitment should be used, unless it reduces the number of fish in the recruited ages so much that the sampling error becomes ridiculous.

11.3.7. EFFECT OF SEASON OF FISHING AND OF NATURAL MORTALITY — APPLICATION TO TYPE 2 POPULATIONS. If the fishing and natural mortalities in Table 11.1 or 11.2 operate concurrently instead of consecutively, catches are smaller and natural deaths greater, but nothing else changes. Total mortality is unaffected, and ratios of successive catches are the same as before. In Table 11.1, for example, the rate of exploitation becomes $FA/Z = 0.4 \times 0.5034/0.7 = 0.2876$, and the catches of column 7 are 288, 1102, 1115, 659, 327, 162, 81, and 79. These figures give the same values of B_2 and W_2, within the limits of rounding.

11.3.8. ESTIMATION OF ABSOLUTE RECRUITMENT. The absolute number of recruits produced by a year-class can be regarded as either: (1) the total number of fish in the year-class at the start of the year when it begins to be recruited; or (2) the sum of the fish of that year-class that first became vulnerable to fishing during each of its years of recruitment. To estimate either figure it is necessary to have information on the size of the stock (and hence the rate of exploitation) of the fully-recruited fish, which may be available from any of the methods described earlier. For (1) this is needed in only one year — the initial year of recruitment for the year-class in question; and as noted above, it requires also the assumption that recruited and unrecruited platoons of the year-class have the same natural mortality rate. For (2) the rates of exploitation must be available for all the years of recruitment.

For the platoon type of recruitment, definition (2) above might be regarded as *the* definition of the number of recruits produced by a spawning. However, if recruitment is gradual rather than by platoons (Section 11.1.3), the only definition of number of recruits that has any consistent meaning is the first one, i.e. the size of the year-class at the time it begins to become vulnerable to the fishery.

273

11.4. EFFECTS OF ENVIRONMENT UPON RECRUITMENT

11.4.1. GENERAL. The biggest difficulty in examining the effect of stock density on net reproduction is that year-to-year differences in environmental characteristics usually cause fluctuations in reproduction at least as great as those associated with variation in stock density over the range observed — sometimes much greater. Sometimes these fluctuations show significant correlation with one or more measured physical characteristics of the environment. To the extent that this is so, their effect can be removed from the total variability by some kind of regression analysis.

Detecting relationships between environment and some measure of an animal's reproduction, or abundance, has a long history; and the subject has an intrinsic interest quite apart from its use to reduce the variability of the parent-progeny relationship. The procedures most used are described in elementary statistical texts; the discussion here mainly concerns problems of interpretation. More complex methods have been proposed (e.g. by Doi 1955a, b) but are not considered here.

In general it is not too difficult to discover correlations, even quite "significant" ones, but it is necessary to be cautious in deducing causal relationships from them. It is well known, for example, that correlations between "time series" are particularly likely to be accidental (involve no causal relationship) when both quantities have a unidirectional trend over a period of years. A correlation is more likely to have meaning when the two quantities vary the direction of their trend, in parallel fashion. However, even these cases sometimes prove to be related (if at all) by way of some third factor whose mode of operation may be unknown and whose very existence is at first unsuspected.

It spite of these dangers, it would be foolish to accept the defeatism of those who argue that because a regression or correlation is based on "the theory of errors", any information it provides is bound to contain error and hence will be of little value. Actually, soundly-considered regression analysis does exactly the opposite: from an originally large variability ("error") whose causes are unknown, it separates out quantitatively the components ascribable to each of a number of factors, so that the unidentified variability or residual error is substantially reduced.

11.4.2. ADDITIVE AND MULTIPLICATIVE EFFECTS. Consider the progeny of a single spawning of a fish stock up to the time they become usable — the recruits of that year-class. The effect of a unit change in an environmental factor might be to change the number of recruits by some constant quantity, or it might change it to some constant multiple or fraction of the initial value, or it might act in some more complex manner. In practice, we should expect the effect of the physical environment normally to be multiplicative rather than additive: if conditions are favorable, all fry have a chance of benefitting; if unfavorable, a certain fraction (not a fixed number) will be lost. To make multiplicative effects amenable to linear regression analysis, the logarithm of the observed effect is used rather than its actual value. The logarithms have an additional advantage: they commonly make the variability of the number of recruits produced (Y-values) more nearly uniform over the observed range of environmental effects (X-values).

These advantages, however, are obtained only at a price; and the price is that the "expected" or "most frequent" value of Y, calculated from the logarithmic relationship for some particular X, is not the arithmetic mean of actual observed Y values at that X: rather it is their geometric mean (GM), which is always less than the

274

corresponding arithmetic mean (AM). The relationship between the two can be represented by an expression modified from formula (8) of Jones (1956, p. 35). For any series of items whose distribution is log-normal:

$$\log_{10}(AM/GM) = 1.1518s^2(N-1)/N \tag{11.5}$$

where s is the standard deviation of the normally-distributed base-10 logarithms of the items, and N is the number of items in the series.[1] If the data are in terms of natural logarithms the formula becomes:

$$\log_{10}(AM/GM) = 0.2172s^2(N-1)/N \tag{11.6}$$

where s is the standard deviation from the regression line of the normally-distributed natural logarithms of the variates.

Table 11.4 shows a selection of values computed by using $(N-1)/N = 1$ in (11.6). The formula assumes that the logarithms of the variates are normally distributed, but even if the distribution is not especially close to normal this relationship will provide an approximate adjustment. That is, average reproduction can be estimated from a computed geometric mean and the standard deviation of the logarithms.

Another, approximate, method of converting GM to AM values is by computing the expected GM values for the series and comparing their total with that of the observed values (cf. Example 11.4).

TABLE 11.4. Relation between (1) the standard deviation of the base-10 logarithms of variates whose logarithms are normally distributed, and (2) the ratio of the arithmetic mean to the geometric mean of those variates, for large values of N.

Standard deviation of logarithm	Ratio: AM/GM	Standard deviation of logarithm	Ratio: AM/GM
0.05	1.007	0.55	2.230
0.10	1.027	0.60	2.598
0.15	1.061	0.65	3.066
0.20	1.112	0.70	3.667
0.25	1.180	0.75	4.445
0.30	1.270	0.80	5.459
0.35	1.384	0.85	6.794
0.40	1.529	0.90	8.694
0.45	1.711	0.95	10.951
0.50	1.941	1.00	14.183

---

[1] In Jones's formula AM = μ; GM was put = 1, hence $\bar{x} = 0$; and $a = 0$. Before discovering this formula, a number of values of AM/GM had been worked out by calculating and averaging actual series, using Pearson's (1924) table II: there was agreement to the second decimal.

A base-10 log standard deviation of 0.5 corresponds to a 1-in-20 chance of a single observed reproduction being as small as 1/10 of the geometric mean or as great as 10 times that mean — a total spread of 100:1. When variability in reproduction is greater than this, the concept of an *average* reproduction becomes rather tenuous.

11.4.3. CURVED REGRESSIONS. For any environmental condition there is typically an intermediate most favorable range, with less favorable conditions above and below. For example, water can be either too cold or too warm for successful incubation of eggs: the optimum is intermediate. Consequently, a graph of reproduction (Y) against temperature (X) would have a maximum and would probably be dome-shaped; hence it could not be straightened by any simple transformation of either or both scales. The mathematical procedure is then to find the regression of Y (or log Y) on X *and* X^2. Even higher powers of X can be used, but data for fish stocks would rarely warrant it. A simpler procedure is to fit a curved line, or two or three straight lines, freehand to the graph — which can be justified at least for preliminary analysis (Rounsefell, 1958).

11.4.4. SECULAR TRENDS. If the data exhibit any important trend or trends extending over periods of years comparable to the total length of the series, it is usually necessary to remove this trend before examining year-to-year effects of environmental factors. Several methods can be used.

1. It is sometimes possible to fit a regression of Y against time (of linear, quadratic, or even higher order), calculate the "expected" value for each Y, and subtract this from the actual Y to obtain a series of *residuals* (as was done, for example, by Milne 1955, p. 476). These residuals can then be plotted and tested against the various environmental factors.

2. If the series is long and the trend irregular, a moving average of 5, 7, or 9 items will provide an "expected" trend line from which the residuals can be measured. Care should be taken that the averaging does not remove variability which can be related to the factors to be examined.

3. A more satisfying procedure is available when the trend in Y is related to a trend in an environmental factor (X) whose influence on Y is well established. In that event the regression of Y on X will take care of the trend, and again residuals can be computed for use with other factors. But this procedure should *not* be used with environmental factors selected only because of their correlation with Y, in the absence of independent evidence of an actual effect on Y, because of the time-series correlation danger discussed above.

11.4.5. EXPLORATORY CORRELATIONS. In general, there can be an indefinitely large number of environmental factors which could be selected for comparison with a record of reproduction or year-class abundance. For example, the temperature, rainfall, etc., in each of a series of months, and in various combinations of months, might be examined (Hile 1941; Henry 1953; Dickie 1955; Ketchen 1956; and others).

276

The usual way to assess possible relationships is to compute the coefficient of correlation for each. If more than one is tested, however, the likelihood of accidentally obtaining a "significant" correlation for one of them increases as the number of factors examined. Thus an investigator is confronted with the paradox that the more factors he tests, the more likely he is to include the effective ones in his search, but the less likely he is to be able to recognize them. If all factors tested seem equally possible *a priori*, the level of significance (P-value) for a single effect can be made more realistic by increasing it in relation to the number of factors — at least as an approximation (cf. Fisher 1937, p. 66). For example, if four factors are examined and one of them is apparently "significant" with a P-value of 0.02, the probability that this factor is really related to abundance is not 98% but about 92% ($= 1 - 4 \times 0.02$).

However, the situation is usually more complicated. There is almost always some provisional hypothesis of a possible relationship behind each correlation tested, even though some may seem far-fetched. Also, we tend to test first the relationships which seem most likely to be appropriate, or which are suggested by gross inspection of the data. The very fact that we have thought of testing a factor is some reflection of its possible significance. As a rule, then, the likelihood of one of the first-tested correlations being "real" is much greater than that of (say) the tenth one, tried on the strength of a wild idea, even though the formal statistical probability be the same for both. To help his readers assess the reality of observed correlations, an investigator should publish details of r and P values for *all* the factors he has examined, whether they seem significant or non-significant. He should also indicate his *a priori* estimate of the likelihood of each, even if only in a general way. Scrupulous attention to these matters will forestall many an embarrassing volte-face.

In general, tentative relationships deduced from an exploratory study involving several to many factors must be confirmed by additional information. This information can be more observations of the kind already used, as they accumulate in the future. With fish populations, ten years or so is usually required to obtain confirmation in this manner. To get a quicker answer, experiments or observations can sometimes be made to determine the exact causal nature of any relationship suggested by the correlation — which is desirable anyway, whenever possible. An observed correlation gains vastly in acceptability if the implied biological process can be demonstrated to occur, even if only qualitatively.

11.4.6. DIFFICULTY OF OBTAINING EVIDENCE OF SIGNIFICANT EFFECTS FROM SHORT SERIES OF OBSERVATIONS. As a rule we expect *several* environmental factors to be fairly important in determining year-class abundance. If so, no one of them can be really outstanding, and none will be apt to have a "significant" correlation when series of less than, say, 15 to 25 years are available. For example, suppose that five and only five independent and uncorrelated factors determine the variation in reproduction of a fish species, and that they are all of equal importance. Then the "coefficient of determination" (ρ^2) for each is 1/5 or 0.20, and the coefficient of correlation is $\rho = \sqrt{0.20} = 0.447$. Nineteen pairs of values are necessary to establish an estimated correlation of $r = 0.45$ as "significant" at the $P_{0.05}$ level. Hence if only 15 years' observations were available, there might be no significant effects demonstrable even though all pertinent possibilities had been examined. In practice, one or two of the five r's above would likely exceed the $P_{0.05}$ level by chance, while the others would fall well below it; and adding more years' observations would almost certainly shift the order of these r-values. In such cases the effect which initially seems most "significant" may decrease in apparent relative importance or even subside into insignificance, while some originally "nonsignificant" effect may become demonstrably important, as future years' data are added to a correlation series. Such shifts have often been observed.

11.4.7. EFFECTS OF TWO OR MORE FACTORS CONSIDERED SIMULTANEOUSLY — MULTIPLE REGRESSION. If measurements of all environmental characteristics examined are all available for the same period of years, the best method of analysis is that of multiple regression, or its close ally, partial correlation. This is particularly true if all the relationships are reasonably close to linear. For a multiple regression analysis, it is not necessary that the separate factors examined be independent. For example, the joint effects of sea temperature, salinity, and wind velocity upon survival of pelagic eggs of a fish might be examined, in a situation where these three are all somewhat correlated among themselves. The "standard regression coefficients" provide estimates of the relative value of each factor for predicting survival. They do *not* tell whether it was temperature, salinity, wind, some unmeasured factor like current speed, or some combination of these, which actually affected survival directly. The square of the adjusted multiple correlation coefficient, R_A^2, represents the fraction of the total variability in survival which is related to all the factors examined, whether or not the latter are correlated.

The superiority of multiple regression over single-factor analysis consists in the fact that it will separate the effects of two correlated factors and indicate their relative value for predictive purposes. This is especially advantageous in connexion with antagonistic effects. Suppose, for eaxmple, that fry survival is strongly favored by lower temperatures (over the range examined), and is rather weakly favored by slow currents, but that years of low temperature usually have strong currents. In that event, a simple correlation of fry survival with current speed would be positive in spite of the fact that the biological relationship is negative. When enough years' observations are at hand, multiple regression or partial correlation will uncover the true relationship and provide an estimate of the importance of each effect in the absence of the other.

Multiple regressions can be handled to many terms on a desk-size electronic computer, though their interpretation can become difficult. Without such apparatus, or for a trial run, it is desirable to limit the number of factors considered to three or four, and good judgment is required in selecting factors for examination:

1. Preference should be given to factors which are likely to affect the organism directly, as indicated by known or plausible biological relationships.

2. Of two or more closely-correlated factors, only one should be used; if it is impossible to give one of them preference on the basis above, it should be done arbitrarily. "Close" correlation, for this purpose, would be upward from $r = 0.8–0.9$, depending on the number of other factors which have to be included.

3. Factors represented by fairly accurate quantitative measures are to be preferred to those only grossly or subjectively classified (for example, as 1, 2, and 3, corresponding to light–medium–heavy).

For the use of "path coefficients" in illuminating relationships discovered, see papers by Davidson et al. (1943) and by Li (1956), and their references to Sewall Wright's contributions.

11.4.8. REGRESSION ANALYSIS BY STAGES. Whether because of the large number of factors to be examined, because of non-linearity of some relationships, or because the data are not complete for all factors, it is sometimes necessary to do an analysis in successive stages (Rounsefell 1958). One or a few factors are used each time, and the "residuals" computed from each fitting are used for the next one. In such work, the environmental factors should themselves first be tested by pairs; any which exhibit moderate correlation *and* seem likely to have independent effects on the Y value should be included in the same multiple regression, if possible. Apart from that, factors should preferably be dealt with in the order of the size of each one's correlation (whether positive or negative) with the effect in question; in this way the variability of the residuals will be reduced most quickly. Probabilities of significance can be estimated from the r or R for each regression, and an overall P-value can be obtained by transforming and combining the separate P's to a x^2 value (Fisher 1950, section 21.1).

EXAMPLE 11.1. POSSIBLE RELATION OF CHUM SALMON CATCHES IN TILLAMOOK BAY TO WATER FLOW AND OTHER FACTORS. (From Ricker 1958a, after Henry 1953.)

The method of exploratory regression was used by Henry to examine relationships between chum salmon landings and stream flows at the time the eggs which produced each brood were being spawned or were in the redds — that is, in November–April, 4 to $3\frac{1}{2}$ years previously. Of 32 kinds of flow examined, for individual months or combinations, significant or suggestive (P = 0.15 or less) correlations were found only for the maximum flow in early November and for the minimum flow in February (or some combination of months which included February). Further trials indicated that minimum flow from January 15 to March 20 produced a regression with apparent significance of P = 0.01 (this flow index is shown in column 7 of Table 11.6. The correlation coefficient was 0.63, showing that 40% of the variation in catch is associated with this index of stream flow *over the years in question*. The best prediction equation, using this variable, was:

$$Y = -493.6 + 2.059x_1 \qquad (11.7)$$

where Y is the expected catch in thousands of pounds, taken from the brood affected by the flow in question, and x_1 is the minimum flow in cubic feet per second. However, Henry emphasizes that it is unlikely that as strong a relationship, with precisely this flow index, would persist into the future, though *some* index of minimum flow during the winter might well do so. The biological relation to be postulated is that low water in winter exposes eggs to drought, frost, or suffocation.

Henry also combined two factors which exhibited suggestive relationships: maximum water flow early in November (x_2 — in cubic feet per second), and maximum air temperature in January or February (x_3 — in degrees Fahrenheit), into a multiple regression with the above, as follows:

$$Y = 346.5 + 0.9731x_1 + 0.06610x_2 - 7.782x_3 \tag{11.8}$$

However, application of this expression reduces the residual variability of catches only slightly, compared with residuals from expression (11.7).

11.5. THE RELATION BETWEEN STOCK AND RECRUITMENT

11.5.1. GENERAL. Considering that fish change their food and habits as they grow, fish of a given age may, to varying degrees, be in competition with, or be preyed upon by, other ages of the same species. Consequently, a completely adequate description of the effect of stock density on recruitment should be based on measurements of the density of each age-group in the population separately (or combinations of ecologically-equivalent ages), together with an index of the effectiveness of each. However, such an analysis requires information on a larger scale than anything yet in sight. An approximation to this approach (Ricker 1954a,b) is based on the possibility that, among the population characteristics affecting reproduction and recruitment, abundance of *mature spawners* is often sufficiently outstanding in importance (or is sufficiently well correlated with other important factors) to make it of real value for analysis and prediction. Although cannibalism of young by adults is possible in many species, it is likely that the effect of parental stock density upon recruitment is usually exerted *via* the density of the eggs or larvae they produce, survival of the latter being affected by density-dependent competition[2] for food or space, compensatory predation, etc.

11.5.2. RECRUITMENT CURVES. In principle at least, one can census a year-class at a number of stages: fertilized eggs, larvae, fingerlings, and various older ages. Of most interest in production studies is the number of recruits to the fishable stock produced by each year-class; for many populations this number is determined mainly

---

[2] There has been considerable disagreement, particularly among entomologists, concerning the role and perhaps even the reality of density dependence, in relation to animal abundance. A brief review by Solomon (1957) summarizes the controversy from a point of view similar to what underlies the argument here; see also Ricker (1955b).

during the first year, particularly during egg and larval stages (Cushing and Harris 1973, Jones 1973). During this time mortality rate can vary either moderately or severely owing to differences in environmental conditions from year to year, but it must also bear some relation to the size of the existing stock.

A graph of recruits against spawners is called a recruitment curve; reproduction curve is a more general term, applicable when the progeny are censussed at *any* life-history stage. The points on such graphs tend to be rather dispersed because of environmental effects, so attempts have been made to work out possible interactions between adults and their progeny and to deduce what kind of an average curve each would produce. For a description of biological situations that can produce one or other of several simple relationships, refer to Ricker (1954b, 1958a), Beverton and Holt (1957), and particularly to recent papers by Chapman (1973) and Cushing (1973). Unfortunately our knowledge of population regulatory mechanisms in nature is so slight that it is usually difficult to choose among different curves on this basis, so we usually fit the simple curve that looks most reasonable. However, of the two curves most used, the Ricker type is more appropriate when cannibalism of young by adults is an important regulatory mechanism, or when the effect of greater density is to increase the time needed by young fish to grow through a particularly vulnerable size range, or when there is a time lag in the response of a predator or parasite to the abundance of the young fish it consumes, with resulting overcompensation for higher *initial* densities of the prey species. The Beverton–Holt curve is likely to be appropriate when there is a ceiling of abundance imposed by available food or habitat, or when a predator can adjust its predacious activity immediately and continuously to the abundance of the prey under consideration.

Some general characteristics desirable in a curve of recruits (R) against parents (P) are as follows:

1. It should pass through the origin, so that when there is no adult stock there is no reproduction.

2. It should not fall to the abscissa at higher levels of stock, so that there is no point at which reproduction is completely eliminated at high densities. (This is not a logically necessary requirement, but it appears reasonable and accords with observations available.)

3. The *rate* of recruitment (R/P) should decrease continuously with increase in parental stock (P). In theory at least, this condition might be violated within some intermediate range of stock densities, but the only example reported has recently been given a different interpretation (Ricker and Smith 1975).

4. Recruitment must exceed parental stock over some part of the range of P values (when R and P are measured in equivalent units); otherwise the stock cannot persist.

281

11.6. RICKER RECRUITMENT CURVES

11.6.1. FIRST FORM. The family of curves proposed by Ricker (1954b, 1958a, 1971c, 1973c) can be written in various ways, of which two are in general use. One is:

$$R = \alpha P e^{-\beta P} \qquad (11.9)$$

R number of recruits
P size of parental stock (measured in numbers, weight, egg production, etc.)
α a dimensionless parameter
β a parameter with dimensions of $1/P$

Figure 11.1 is an example of such a curve. The slope (differential) of (11.9) is:

$$(1 - \beta P)\alpha e^{-\beta P} \qquad (11.10)$$

Equating this to zero, and since $\alpha e^{-\beta P}$ cannot $= 0$, the maximum level of recruitment is obtained when the spawning stock is:

$$P_m = \frac{1}{\beta} \qquad (11.11)$$

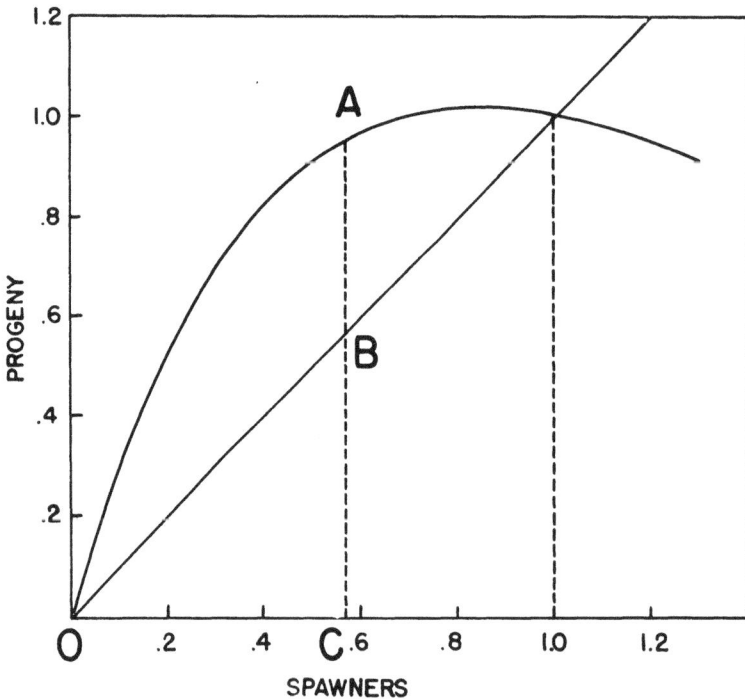

FIG. 11.1. Example of a reproduction curve of the type $R = \alpha P e^{-\beta P}$ or $R = P e^{a(1-P/P_r)}$, with $a = 1.119$. Point A is any point on the curve, the distance AB representing the surplus reproduction which must be removed by fishing if the stock is to remain in equilibrium at this level. The distance AB becomes a maximum a little farther to the left on the curve.

Substituting (11.11) in (11.9), maximum recruitment is:

$$R_m = \frac{\alpha}{\beta e} = \frac{0.3679\alpha}{\beta} \tag{11.12}$$

Expression (11.9) can be fitted to data, and P_m and R_m estimated, regardless of what units are used for R and P. For example, P might be biomass of mature stock, and R the number of age-2 fish produced by that stock. But when recruits and parents are measured in comparable units such that there is a level of *replacement abundance* at which R = P, this replacement level (P_r) can be determined by substituting R = P in (11.9):

$$P_r = \frac{\log_e\alpha}{\beta} \tag{11.13}$$

11.6.2. SECOND FORM. With R and P in the same units, expression (11.9) can be modified by introducing a parameter $a = P_r\beta = \log_e\alpha$ (from 11.13). Substituting $\beta = a/P_r$ and $\alpha = e^a$ in (11.9):

$$R = Pe^{a(1-P/P_r)} \tag{11.14}$$

In this form, replacement abundance P_r appears as an explicit parameter, which is a convenience whenever P_r can be estimated. Another advantage of (11.14) is that the single parameter a completely describes the shape of the curve (Fig. 11.2). And since $a = P_r/P_m$, the size of spawning stock needed for maximum recruitment is immediately known. For $a>1$ maximum recruitment occurs when spawners are less than the replacement level, and the curve becomes steeper and more dome-like as a increases; for $a<1$ the recruitment maximum is at a stock level greater than replacement.

11.6.3. GEOMETRIC AND ARITHMETIC MEANS. Expressions (11.9) and (11.14) provide estimates of geometric mean values of R at a given P. To convert to arithmetic mean values, expression (11.5) or (11.6) can be used, with s^2 equal to the variance of the points from the linear regression line of (11.15) or (11.16).

If the distribution of values of R at given P is log normal, the GM curve estimates the most probable value of the recruitment obtained in any year from the observed P, while the AM curve estimates the long-term arithmetic average value of recruitments obtained at that P.

11.6.4. FITTING THE CURVE. The easiest way to fit (11.9) or (11.14) to a body of data is to shift P to the left side and take logarithms, as was originally suggested by Rounsefell (1958):

$$\log_e R - \log_e P = \log_e\alpha - \beta P \tag{11.15}$$

$$\log_e R - \log_e P = a - aP/P_r \tag{11.16}$$

283

The slope of the regression of $(\log_e R - \log_e P)$ on P is an estimate of β or a/P_r, and the ordinate intercept is an estimate of $\log_e \alpha$ or a. This geometric mean line can then be converted to an AM line using expression (11.6). The work may be easier if base-10 logarithms are used, as in Example 11.3 below. Fitting the GM line can be done using computer program RICKER-1 of P. K. Tomlinson (Abramson 1971).

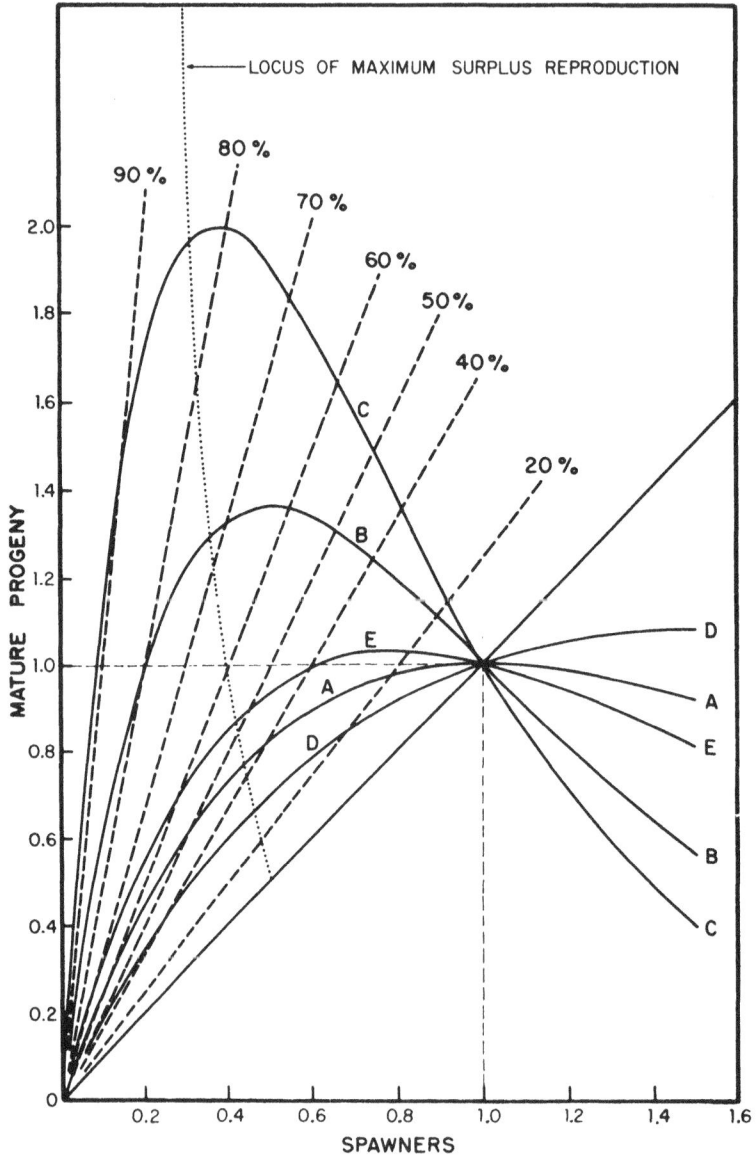

FIG. 11.2. Reproduction curves conforming to the Ricker relationship. The point where the curves cut the diagonal is the replacement level of stock and reproduction. The broken lines from the origin are loci of equilibrium reproduction for the rates of exploitation indicated. (See Table 11.5 for the parameters.)

An alternative method of fitting, used by Cushing and Harris (1973), is "simply" to minimize the sum of squares of deviations from the curve by trial and error, using a computer. (Estimates obtained from the logarithmic fitting can be used as initial trial values.) This procedure gives an AM line which can be very close to the one obtained using logarithms and (11.6); however, the logarithmic procedure will usually have the advantage of stabilizing variance of points from the line.

11.6.5. OTHER STATISTICS. Appendix III summarizes a series of statistics derived from (11.9) and (11.14). Item 19 is probably of greatest interest, showing how to find maximum sustainable yield (maximum equilibrium catch or maximum surplus reproduction). Figure 11.2 illustrates that MSY is obtained at the point where the curve is parallel to the replacement line, i.e. when its slope is 1. Spawners needed (P_s) are found by equating the slope to 1 and solving by trial; the two forms of this equation are as follows:

$$(1 - \beta P_s)\alpha e^{-\beta P_s} = 1 \qquad (11.17)$$

$$(1 - aP_s/P_r)e^{a(1-P_s/P_r)} = 1 \qquad (11.18)$$

(If (11.18) is used, it is convenient to put $P_r = 1$ until after P_s has been located.)

From the value of P_s so obtained, R_s is computed from (11.9) or (11.14), and MSY is computed as $C_s = R_s - P_s$. Alternatively, the arithmetic mean value of R_s can be computed using (11.6), and again C_s is found by difference.

Another point of interest concerns maximum recruitment (R_m — items 9 and 10 of Appendix III). R_m is equal to $\alpha/\beta e$, obtained from spawners $P_m = 1/\beta$; thus the rate of recruitment at maximum absolute recruitment is $R_m/P_m = \alpha/e$. The maximum rate of recruitment occurs when P→0, and is equal to α (item 3 of Appendix III). Thus maximum *absolute* recruitment (for this model) always occurs when *rate* of recruitment is $1/e$ or 37% of its maximum. The difference in instantaneous mortality rate between the two situations is exactly 1.

11.6.6. ILLUSTRATIONS. Figure 11.2 shows 5 reproduction curves that conform to (11.9) and (11.14). The corresponding parameters and certain characteristic quantities are shown in Table 11.5. Notice that a wide variety of shapes is possible within the framework of one simple mathematical expression.

A variety of questions can be answered from Table 11.5 and Appendix III. For example, at one time it was required that 50% escapement be allowed in Alaska salmon streams; thus we might like to know the shape of the Ricker curve for which catch and escapement are equal at maximum sustainable yield (MSY). From item (21) of Appendix III, $a = 0.5 - \log_e(1 - 0.5) = 1.193$; thus the curve lies a little below E of Fig. 11.2, passing through the point where the locus of maximum surplus reproduction cuts the line of 50% exploitation. For this curve the spawning stock required for MSY is 42% of the primitive average stock abundance.

285

TABLE 11.5. Parameters of the five reproduction curves of Figure 11.2, and one other, computed from Appendix III. The replacement level of stock is $P_r = 1000$.

Curve	D	A	E	F	B	C
Parameter a of (11.14)	0.667	1.000	1.250	1.500	2.000	2.678
Parameter α of (11.9)	1.948	2.718	3.490	4.482	7.389	14.556
Parameter β of (11.9)	0.000667	0.001	0.00125	0.0015	0.002	0.002678
Maximum recuitment (R_m)	1072	1000	1027	1102	1359	2000
Spawners needed for maximum recruitment (P_m)	1500	1000	800	667	500	373
Maximum sustainable yield — MSY (C_s)	198	330	447	587	935	1656
Spawners needed for MSY (P_s)	456	433	415	397	361	314
Recruitment at MSY (R_s)	654	763	862	984	1296	1970
Rate of exploitation at MSY (u_s)	0.304	0.433	0.519	0.596	0.722	0.841
Limiting equilibrium rate of exploitation	0.486	0.632	0.714	0.777	0.865	0.932

EXAMPLE 11.2. FITTING A RECRUITMENT CURVE TO A COD POPULATION. (Data from Garrod 1967.)

Figure 11 of Garrod (1967) illustrates success of reproduction for Arcto-Norwegian cod. An index of year-class strength is plotted against an index of stock weight, both indices being obtained by various adjustments to rather complex primary data. In Fig. 11.3 the reproduction curve of expression (11.9) is fitted to Garrod's data, using the ordinary regression of $\log_e R - \log_e P$ against P, because P is likely to be much more accurate than R. The constants are $\log_e \alpha = 1.774$, $\alpha = 5.89$, and $\beta = 0.01861$. Variance from the regression line is $s^2 = 0.3180$, and from (11.6) the multiplier 1.165 is obtained to convert the GM line to an AM line; both are shown on Fig. 11.3.

Neither line is a perfect fit to the apparent trends in the data, though they probably lie within limits of possible random error. An empirical curve would be steeper and would have a higher dome; however, Garrod (p. 179) suggests that both R and P are too small for the years 1937–43, and that R is somewhat too large in 1949 and too small in 1950. If these adjustments were made, the fit would be considerably better. The large year-to-year variability is typical of the known history of this stock.

There is no direct way to compute the replacement level of stock from Fig. 11.3. However, the primitive stock size is not likely to have been much if any less than the largest shown on the graph, and making allowance for underestimation of points for the earlier years, I have drawn a possible replacement line to pass through (P = 180, R = 50). In the natural state of this stock a rather small annual contingent of recruits (starting at age 3, say) joined an existing body of older fish that was many

286

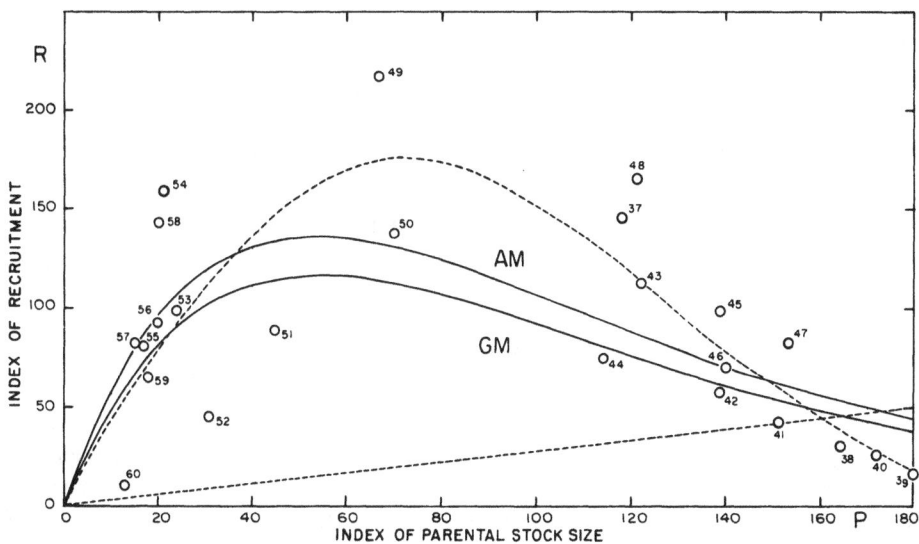

FIG. 11.3. Graph of recruitment against parental stock for Arcto-Norwegian cod. Solid curves—Ricker reproduction curves for geometric and arithmetic mean values; the broken curve is drawn freehand (see the text). (Data from Garrod 1967.)

times more numerous. Usually these recruits replaced the annual natural deaths among older fish, which may have amounted to about 10% (in Example 2.5 a total death rate of 25% was estimated for a period when a considerable fishery already existed).

Under conditions of intensive exploitation, however, the cod population assumes a new character. Stock older than age 3 is less numerous, or at least less in total bulk, and is much younger, and the average number of recruits per year is much greater, both relatively and absolutely. Figure 11.3 shows that maximum sustainable recruitment should be less than maximum recruitment, but not much less. From (11.11), the stock size which gives maximum recruitment is $P_m = 1/0.01861 = 53.7$, and from (11.12) the corresponding recruitment is $R_m = 0.3679 \times 5.89/0.01861 = 117$ from the GM curve, and $117 \times 1.165 = 136$ from the AM curve. Thus P_s should be somewhat less than 54, the AM of R_s values should be a little less than 136, and the maximum sustainable yield would be the difference, or about 80.

To estimate sustainable *yields* for different ages of recruitment and different fishing mortality rates, growth rates and natural mortality rates need to be known and combined with recruitment information. Walters (1969) estimated equilibrium yields for this stock by the method of Section 12.4.2, but the natural mortality rates he used are so impossibly high that the result is not realistic. Any simulation of the actual history of the stock would have to be done year by year, and would have to take account of the fishing-up effect as well as changes in recruitment.

A possible reason for the cod curve having this shape is found in the observation of Ponomarenko (1968) that cod in the Barents Sea eat large numbers of their own

287

young. As mentioned in Section 11.5.2, population control by cannibalism is one of the situations which can lead to the reproduction curve of expression (11.9).

EXAMPLE 11.3. FITTING A RECRUITMENT CURVE TO STATISTICS OF TILLAMOOK BAY CHUM SALMON. (Modified from Ricker 1958a, after Henry, 1953.)

Chum salmon (*Oncorhynchus keta*) of Tillamook Bay mature mostly at age 4, as described in Example 11.1; so each year's catch (C) can be considered as largely the progeny of the spawning stock of 4 years earlier. Henry says that much the same group of fishermen fished the bay over the years included in Table 11.6, so that year-to-year variation in rate of exploitation was probably not large except in 1932: in that year economic conditions greatly reduced the catch in November and December (Henry 1953; p. 11, 17; the quantities taken in October suggest there would have been a better-than-average catch if fishing had continued). To make an objective recruitment analysis of these data, it would be necessary to know also the escapement (P) in each year, so that total return (R = P + C) could be related to P four years earlier. No such data are available, but for the purpose of illustration I have assumed C = P each year; thus C serves as an estimate of P, and 2C of four years later serves as an estimate of the resulting R (Table 11.6). To obtain a curve of the form of expression (11.16), an ordinary regression line was fitted to $\log_{10}(R/P)$ against P, the unit being 1000 lb throughout. The slope was −0.0003919 and the Y-axis intercept +0.6295. These values are converted to terms of natural logarithms and applied in (11.16):

$a/P_r = 0.0003919/0.4343 = 0.0009024$

$a = 0.6295/0.4343 = 1.450$

$P_r = 1.450/0.0009024 = 1607$ thousand lb

The equation becomes:

$$R = Pe^{1.450(1-P/1607)} \tag{11.19}$$

Figure 11.4A shows the log data and the line fitted; the lower curve of Fig. 11.4B is the antilogged GM curve. Variance of the points from the fitted curve is 0.0791. To find the corresponding AM curve we have, from (11.5):

$$\log_{10}(AM/GM) = 1.1518 \times 0.0791 \times 20/21 = 0.0869$$

Therefore AM/GM = 1.222, and this factor is used to convert values of R computed from (11.19), which produces the upper curve of Fig. 11.4B. Notice that for the AM curve 1607 thousand lb is no longer the replacement level of stock, nor is this equal to 1607 × 1.222. The actual value can be calculated by iteration or estimated graphically: it is about 1830 thousand lb.

The variance of the values of log R (column 4 of Table 11.6) from their mean is 0.0932, while their variance from the fitted curve is 0.0791. Thus the variance is not greatly reduced by using the curve, but the points are so distributed that no conceivable curve could do much better.

TABLE 11.6. Catches of Tillamook Bay chum salmon, and the log data used to fit a reproduction curve of the form (11.14). (Logarithms are to the base 10.) Column 2 is the catch of salmon in thousands of pounds, which is considered also to represent the size of the spawning stock at an estimated rate of exploitation of 50% (see the text *re* 1928 and 1932). Column 4 is the logarithm of the progeny generation (catch plus escapement), which is estimated as twice the catch 4 years after the brood year shown in column 1. Column 5 is the difference between 4 and 3, and it is regressed against P in column 2. Column 6 is the "expected" log R, computed from $\log R = \log P - 0.000392P + 0.630$. Column 7 was used in Example 11.1: it represents the minimum water flow in cubic feet/second in certain spawning streams during January 15–March 20 of the year *following* spawning. Each flow therefore can affect the catch taken in the calendar year 3 years later. (Data from Henry 1953, and personal communication.)

1	2	3	4	5	6	7
Year	P	log P	log R	log(R/P)	Computed log R	Minimum flow
1923	644	2.81	3.55	0.74	3.19	
1924	854	2.93	3.75	0.82	3.23	
1925	931	2.97	3.37	0.40	3.24	
1926	244	2.39	2.67	0.28	2.92	
1927	1764	3.25	3.28	0.03	3.19	
1928	(2804)	....	....	....	....	
1929	1171	3.07	3.04	–0.03	3.24	
1930	234	2.37	2.83	0.46	2.91	
1931	947	2.98	3.06	0.08	3.24	
1932	(89)	....	....	....	....	
1933	552	2.74	2.94	0.20	3.15	795
1934	336	2.53	3.16	0.63	3.03	380
1935	572	2.76	2.93	0.17	3.17	665
1936	1189	3.08	2.94	–0.14	3.24	515
1937	438	2.64	3.54	0.90	3.10	640
1938	725	2.86	3.72	0.86	3.21	821
1939	427	2.63	2.88	0.25	3.09	945
1940	439	2.64	2.86	0.22	3.10	490
1941	1756	3.24	3.19	–0.05	3.18	348
1942	2651	3.42	2.98	–0.44	3.01	486
1943	379	2.58	2.87	0.29	3.06	344
1944	361	2.56	3.25	0.69	3.05	646
1945	777	2.89	2.94	0.05	3.22	572
1946	482	....	....	....	....	....?
1947	374	....	....	....	....	....
1948	895	....	....	....	....	....
1949	436	....	....	....	....	....
Mean[a]	828	....	3.131	0.305	....	....
Variance[a]	349651	....	0.0932	0.1313	....	....?

[a]Omitting 1928, 1932, and 1946–49.

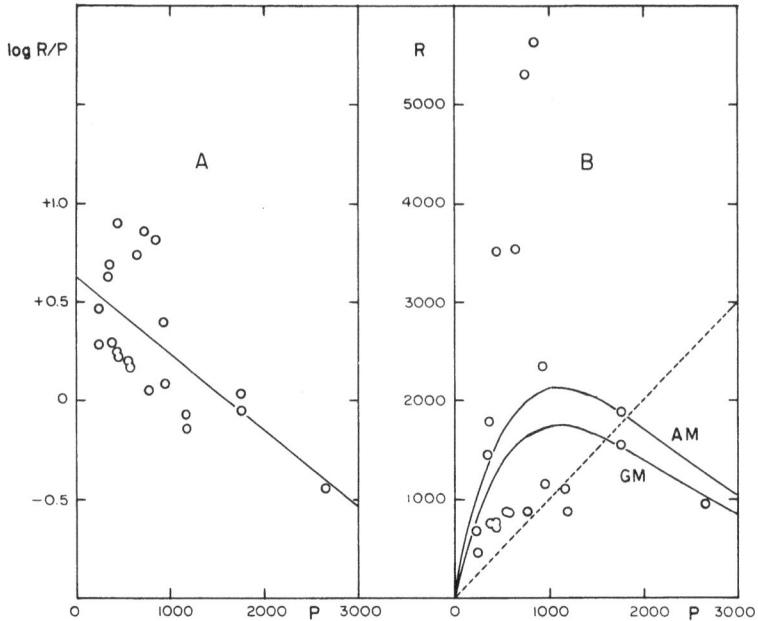

FIG. 11.4. Approximate recruitment curve for Tillamook Bay chum salmon. (A) Base-10 logarithms of R/P plotted against P and fitted with an ordinary regression line; (B) Recruitment plotted against parental stock, with the Ricker recruitment curve (GM) and its arithmetic mean transformation (AM). The scale is 1000's of lbs. of fish. (Data from Table 11.6.)

To estimate maximum equilibrium catch (MSY) we first estimate the necessary number of spawners, P_s, by equating the slope of (11.19) to 1 (line 2 of Appendix III):

$$(1 - 1.450P_s/P_r)e^{1.450(1-P_s/P_r)} = 1$$

Solving this by trial gives $P_s/P_r = 0.400$; $P_r = 642$ thousand lb, $P_s = 257$ thousand lb, from which the GM value of R_s is calculated from (11.19) as 1533 thousand lb. The AM value of optimum recruitment is 1.222 times this, or 1872, and the estimate of MSY is $1872 - 642 = 1230$ thousand lb, taken at a constant rate of exploitation of $1230/1872 = 66\%$. If the stock could be forecast fairly accurately each year, a somewhat larger average take would be possible by fishing large recruitments more heavily and smaller ones less heavily (Ricker 1958b, Larkin and Ricker 1964).

Another statistic of interest is the limiting or density-independent rate of reproduction (item 3 of Appendix III). This is equal to $\alpha = e^a = e^{1.45} = 4.26$. According to this, if the stock were reduced to a very low ebb each spawner would produce, on the average, a little more than 4 recruits.

What relation has this analysis to the relationship between reproduction and environmental factors which was develop in Example 11.1? Because in that example the relationship was fitted to the catches themselves and not to their logarithms, and a slightly different series of years was used, a repetition of Henry's simpler fitting

290

(expression 11.7) was made using the variables of this example (log R and x_1). The numerical relationship of stream flow to catch is about the same, though somewhat less "significant" than that of (11.7) above ($r = 0.57$ for 12 degrees of freedom).

To test the factors parental abundance and minimum flow together (still assuming a rate of exploitation of 50%), residual deviations of the log catches (i.e. column 4 less column 6 in Table 11.6) were related to the minimum flows three years earlier (column 7 of Table 11.5). However, the resulting coefficient of correlation becomes smaller ($r = 0.45$) rather than larger. Thus the two factors mentioned are to some extent "competing" for the same variability in the size of the catch. Only additional information would decide which has more influence on reproduction, and whether one of them could be ignored, for practical purposes.

11.7. Beverton–Holt Recruitment Curves

11.7.1. Formulae. The curves of another family, proposed by Beverton and Holt (1957), are hyperbolic in shape and have the formula:

$$R = \frac{1}{\alpha + \beta/P} \qquad (11.20)$$

R and P are as above and α and β are new parameters. When R and P are in the same units and $R = P$ at replacement, expression (11.20) too can be put into a form involving P_r:

$$R = \frac{P}{1 - A(1 - P/P_r)} \qquad (11.21)$$

In these circumstances the parameters of (11.20) are related to those of (11.21) as follows:

$$\beta = 1 - A; \qquad \alpha = A/P_r$$

Parameter A can have values from 0 to 1, and it completely describes the shape of the curve (Fig. 11.5).

The slope and other statistics from (11.20) and (11.21) are derived by Ricker (1973c) and are shown in Appendix III here.

11.7.2. Fitting the curve. The easiest way to fit (11.20) to a body of data would appear to be by linear regression of $1/R$ on $1/P$, since (11.20) can be transformed algebraically to:

$$\frac{1}{R} = \alpha + \frac{\beta}{P} \qquad (11.22)$$

Alternatively, Paulik (1973) recommends using:

$$\frac{P}{R} = \beta + \alpha P \qquad (11.23)$$

291

and regressing P/R on P. Either way it is 1/R rather than R which is estimated, so that the R-line obtained by inverting computed values of 1/R represents the harmonic mean of expected recruitments at each P. This tends to be considerably below an arithmetic mean line fitted to observed R. The line of least squares fit of R to P using (11.20) can be obtained by successive trials using a computer. If a computer is not available, this line can be approximated by applying an adjustment to one of the above lines, proportional to the mean difference between observed and computed values of R (Example 11.4).

11.7.3. ILLUSTRATIONS. Figure 11.5 shows a selection of curves that conform to expressions (11.20) and (11.21). They differ from those of Fig. 11.2 in that (1) recruitment approaches its maximum asymptotically as the number of spawners increases, so that there is no "dome"; (2) at stock densities less than replacement, recruitment never exceeds the replacement level; (3) for values of A near unity recruitment can be almost the same over nearly the whole range of spawner abundance, diving steeply to the origin at very small numbers of spawners; (4) the locus of maximum sustainable yields is a straight line rather than a curve. The parameters and certain other characteristics of these curves are given in Table 11.7.

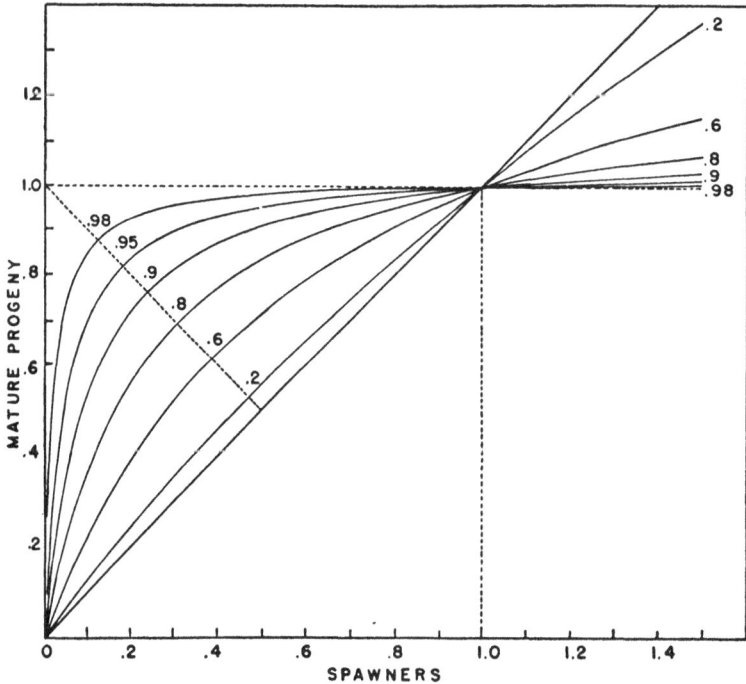

FIG. 11.5. Beverton–Holt reproduction curves. Numbers by each curve are the value of A in expression (11.21); see Table 11.7 for other parameters. The diagonal from 1.0 on the ordinate is the locus of maximum surplus reproduction.

292

TABLE 11.7. Parameters of the reproduction curves of Fig. 11.5, and characteristics computed from Appendix III. The replacement level of stock is 1000.

Curve	1	2	3	4	5	6
Parameter A in (11.21)	0.2	0.6	0.8	0.9	0.95	0.98
Parameter α in (11.20)	0.0002	0.0006	0.0008	0.0009	0.00095	0.00098
Parameter β in (11.20)	0.8	0.4	0.2	0.1	0.05	0.02
Maximum recruitment (R_m)	5000	1667	1250	1111	1053	1021
Maximum sustainable yield (C_s)	56	225	383	519	635	752
Spawners needed for MSY (P_s)	472	387	309	241	183	124
Recruitment at MSY (R_s)	528	612	692	760	818	876
Rate of exploitation at MSY (u_s)	0.106	0.367	0.553	0.683	0.777	0.858
Limiting equilibrium rate of exploitation (when P → 0)	0.2	0.6	0.8	0.9	0.95	0.98

11.7.4. SAMPLE PROBLEM. The problem of what curve provides maximum sustainable catch when $u_s = 0.5$ can also be solved for the Beverton–Holt model. From item 21 of Appendix III, the necessary value of A is $1 - (1 - u_s)^2 = 0.75$. The curve thus lies a little below that marked 0.8 in Fig. 11.5, and requires a spawning stock that is 33% of the primitive abundance (from item 16 of Appendix III).

EXAMPLE 11.4. FITTING A RECRUITMENT CURVE TO A PLAICE POPULATION. (Data from Beverton 1962.)

Table 11.8 and Fig. 11.6 show estimates of recruitment plotted against adult population for North Sea plaice. The outer (right-hand) points suggest that recruitment may approach an asymptote which is also the maximum, so a Beverton–Holt recruitment curve should be appropriate. Expression (11.22) cannot be used because it gives a negative value of β and thus a line that does not pass through the origin. However, expression (11.23) yields $\beta = 0.1606$ and $\alpha = 0.06312$ from an ordinary regression. This determines the lower line of Fig. 11.6, which is a line of harmonic mean recruitments. Computed values of R for the observed P add up to 355.9, whereas the observed values total 437 (columns 4 and 2 of Table 11.8). Thus the computed values are multiplied by $437/355.9 = 1.228$ in the last column of Table 11.8 to get a line that approximates the line of arithmetic mean recruitments (upper line of Fig. 11.6).

11.8. OTHER RECRUITMENT CURVES

11.8.1. MODIFICATIONS OF THE RICKER CURVE. If expression (11.9) be regarded as a convenient basic form of recruitment curve, it can be modified in a number of ways. Modifications are most likely to take the form of (a) flatter domes, (b) steeper right limbs, (c) a reduced rate of reproduction at very low stock densities.

TABLE 11.8. Biomass of adult stock (P) and number of progeny that survive to recruitment (R) for North Sea plaice, both in arbitrary units. (From fig. 3 of Beverton 1962.)

1	2	3	4	5
			Calculated	Calculated
P	R	P/R	R	R × 1.228
9	13	0.69	12.3	15.1
9	20	0.45	12.3	15.1
9	45	0.20	12.3	15.1
10	13	0.77	12.6	15.5
10	13	0.77	12.6	15.5
10	20	0.50	12.6	15.5
10	21	0.48	12.6	15.5
10	26	0.38	12.6	15.5
11	11	1.00	12.8	15.7
11	12	0.92	12.8	15.7
12	8	1.50	13.1	16.1
12	15	0.80	13.1	16.1
18	7	2.57	13.9	17.1
18	10	1.80	13.9	17.1
19	17	1.12	14.0	17.2
20	16	1.25	14.1	17.3
21	11	1.91	14.1	17.3
21	15	1.40	14.1	17.3
26	16	1.62	14.4	17.7
32	33	0.97	14.7	18.1
35	10	3.50	14.8	18.2
45	23	1.96	15.0	18.4
54	13	4.15	15.1	18.5
70	13	5.38	15.3	18.8
82	12	6.83	15.4	18.9
88	24	3.67	15.4	18.9
Total 672	437	46.59	355.9	437.2

A *flatter dome* and more gently inclined limbs would be the usual consequences of division of the population into partially-distinct units, so that effects of density are not uniformly felt throughout the stock. Flattening of the dome is also expected when there is an upper limit of space available to a population which exhibits territorial behavior, or a limited number of safe habitat niches: in extreme cases the right limb could be horizontal, as for the Beverton–Holt curve. Flattening of the dome also occurs when there is an upper limit of food supply for which there is a "contest" in the sense of Nicholson (1954) — each successful individual gets enough to complete growth, and others die without reducing the food available to the successful ones.

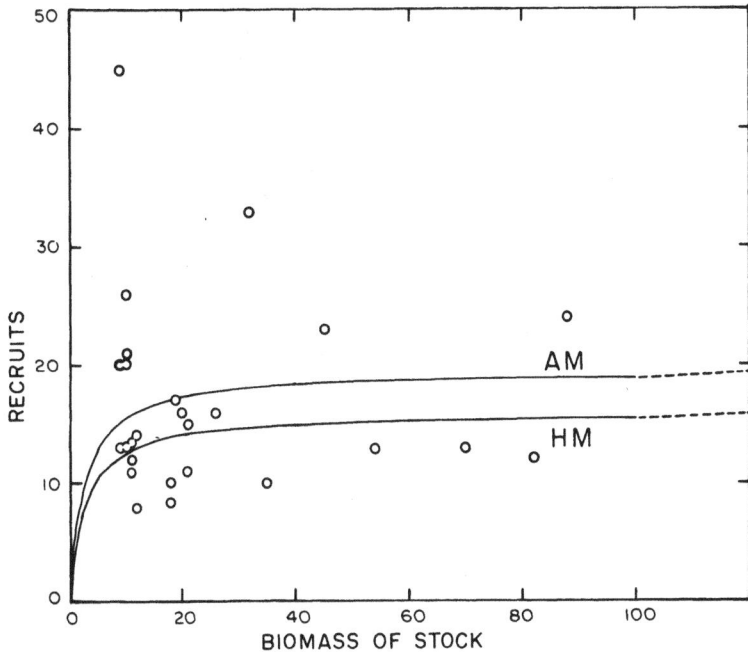

FiG. 11.6. Number of recruits of North Sea plaice plotted against the biomass of the adult stock, both in arbitrary units. Lower line—a Beverton–Holt curve of harmonic mean recruitments fitted to the data; Upper line—a corresponding arithmetic-mean curve. (From data in Table 11.8.)

A *steeper right limb* results if there is a "scramble" (Nicholson) for limited food or other requisites, so that all members of a brood get some but many do not get enough to complete development; in the limiting case none get enough and all die. For example, a large deposition of salmon eggs in a spawning area might have an oxygen demand in excess of the supply, so that no eggs would survive.

A *reduced rate of reproduction at low stock densities* might result from a need for group activity in the breeding cycle, difficulty of finding mates in a scattered population, relatively higher losses from predation when the stock is at a low level, etc. Neave (1954) has discussed this possibility and has observed it in pink salmon reproduction. This effect and the preceding one, working together, would tend to produce a *narrow dome*.

11.8.2. NON-PARAMETRIC RELATIONSHIPS. If expectation or observation suggest some irregular type of reproduction curve, a freehand line fitted to the stock-reproduction data may be preferred. Chapman (1973) suggested a convenient technique. Data are grouped into several (preferably equal) ranges on the basis of P, and mean values of P and R are found for each. These means are then joined by straight lines or curves, from which the point of maximum recruitment can be found graphically, as well as the maximum sustained yield (if the replacement value of R is known). Chapman gives examples for sockeye and fur-seal populations, and Ricker (1968) used it for sockeye. Generally this method does not supply any unique solution, since the length ranges used must be decided more or less arbitrarily, as well as the form of the final line fitted to the points. However, this arbitrariness may sometimes be less than what is involved in choosing a particular form of curve.

For *any* line or curve, empirical relationships similar to those of Appendix III can be computed graphically. In particular, when R and P are in equivalent units the maximum equilibrium catch always occurs at the point where the curve has a slope of 45° and a tangent of +1, and where its absolute vertical distance from the 45° diagonal is greatest.

11.8.3. COMPLEX RELATIONSHIPS. Paulik (1973) described a number of curves that can be fitted by using one more parameter than the Ricker or Beverton–Holt types. He also described the use of Leslie matrices, and applied this technique to the computation of recruitment and yield of chinook salmon (*Oncorhynchus tshawytscha*) using computer program GRSIM of L. E. Gales (Fishery Analysis Center, University of Washington).

The complexity of such schemes need be limited only by the availability of data. Indeed, speculative projections are frequently made far beyond the limits of what is known. These can be illuminating, but it requires continual vigilance to distinguish fact from fiction in the results.

11.9. COMPENSATORY AND DENSITY-INDEPENDENT MORTALITY RATES

As noted in Section 11.6.5, for the Ricker model the maximum rate of increase of a stock is equal to α, when P→0; hence this is its potential rate of increase when density-dependent (compensatory) effects are zero. The corresponding instantaneous rate of increase is $\log_e \alpha$. The remainder of expression (11.9), $e^{-\beta P}$, is a fraction that reduces the potential rate of increase because of interaction between members of the stock; in other words it is the compensatory mortality rate. The corresponding instantaneous compensatory mortality rate is βP.

If on the average Y eggs are produced by each adult in the stock (including both sexes), they should produce one adult at replacement density; thus at replacement the survival rate from egg to recruitment is $1/Y$ and the total instantaneous mortality rate (egg to recruit) is $-\log_e(1/Y) = \log_e Y$. Since the potential instantaneous rate of increase of the stock is $\log_e \alpha$, and at replacement its actual rate is 0, the instantaneous rate of compensatory mortality at replacement must also be $\log_e \alpha$. By difference the instantaneous rate of density-independent mortality becomes $\log_e Y - \log_e \alpha$, which by definition is the same at all levels of stock.

Corresponding statistics have been developed for the one-parameter Ricker equation (11.14), and for the Beverton–Holt equations (11.20) and (11.21); they were summarized by Ricker (1973c), and are shown here as items 26 and 27 of Appendix III. Similar statistics can readily be obtained for any other stock-recruitment relationship.

CHAPTER 12. — RECRUITMENT AND THE FISHERY

12.1. RELATION OF EQUILIBRIUM YIELD TO RATE OF FISHING, FOR DIFFERENT RECRUITMENT CURVES (EQUILIBRIUM SITUATIONS)

The practical importance of the shape of a reproduction curve is best realized by comparing the yields that different ones provide. Figure 12.1 shows the equilibrium catch taken from stocks characterized by four of the curves discussed earlier in relation to rate of exploitation. The great difference in yield potential is obvious: Curve C can provide about five times as great a catch as A, although the two stocks were the same size before exploitation. The rate of exploitation (u) necessary to achieve maximum yield increases from A to C (i.e. with increasing values of the coefficient $a = P_r/P_m$); for Type 1 fisheries the rate of fishing would be the natural logarithm of $1 - u$ (with sign changed), and the necessary fishing effort would be approximately proportional to the latter. However, catch increases even more rapidly than rate of fishing, so that the catch per unit effort at maximum sustainable yield would be about 55% greater for C than for A.

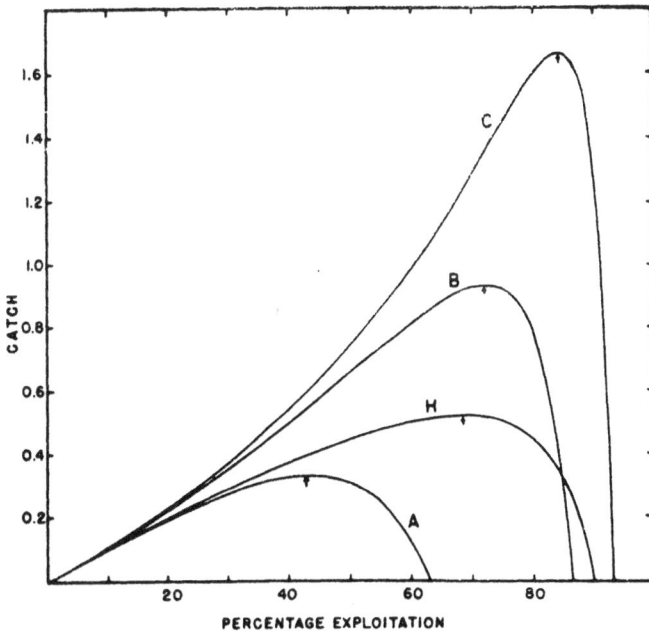

FIG. 12.1. Sustained catches obtained at various rates of exploitation from stocks having Ricker recruitment curves A, B, and C of Table 11.5, and Beverton–Holt curve No. 4 (H above) of Table 11.7. Catches are in terms of the replacement level of stock, and arrows indicate the position of maximum equilibrium catch. (From Ricker 1958b.)

The yield curves of Fig. 12.1 fall off much more rapidly on the right side than on the left. With curves of this sort, consequently, obtaining the maximum yield will always be tricky — with a more drastic penalty for too much fishing than for too little.

Comparable graphs can be constructed for any other recruitment curve (mathematical or empirical), most readily by direct measurement of yield from the curve to the replacement diagonal.

12.2. INTERACTION OF A RECRUITMENT CURVE WITH A CHANGING RATE OF FISHING

12.2.1. INCREASING EXPLOITATION. Curves like those of Fig. 12.1 show equilibrium yields, but in many fisheries equilibrium catch is known for very few, if any, levels of exploitation. Consequently it is desirable to examine the effects of changing exploitation rates upon these population parameters, for the situation where the stock available to the fishery is determined solely by the reproduction curve.

Figures 12.2 and 12.3 show the catch history of a new fishery attacking a species which has a single age of maturity and is vulnerable only near maturity. Three model populations are shown, characterized by the reproduction curves A, B, and C of Fig. 11.2 and Table 11.5. Each starts at the replacement level of stock and is exploited 10% in the first generation, 20% in the second, and so on. The solid points indicate the course of events when exploitation increases to 90%. Stabilization of fishing at 90% would soon exterminate stocks A and B, but the C curve can support removals somewhat greater than 90%.

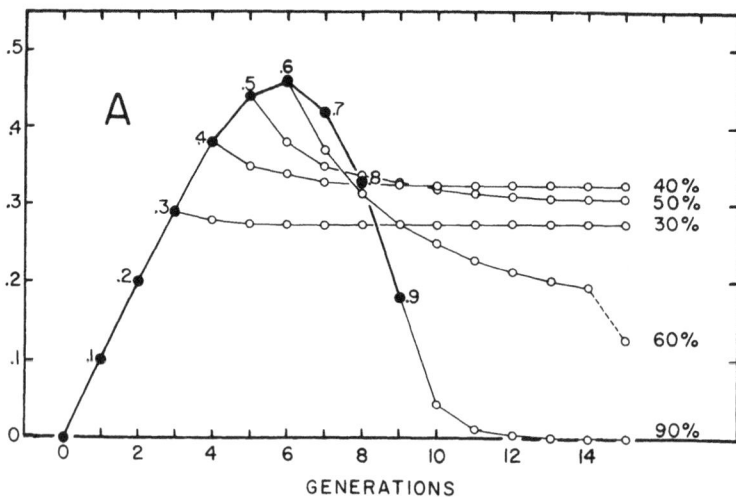

FIG. 12.2. Catches obtained from a stock whose reproduction is described by Curve A of Fig. 11.2. Solid dots — catches obtained when the rate of exploitation has successive absolute increases of 10% per generation, starting with a stock at the replacement level; open circles — catches obtained when rate of exploitation remains constant at the levels shown, starting from the position described by the solid dot to which it is joined. The maximum sustained yield is obtained at 43%, and is very slightly higher than the equilibrium level shown for 40%.

298

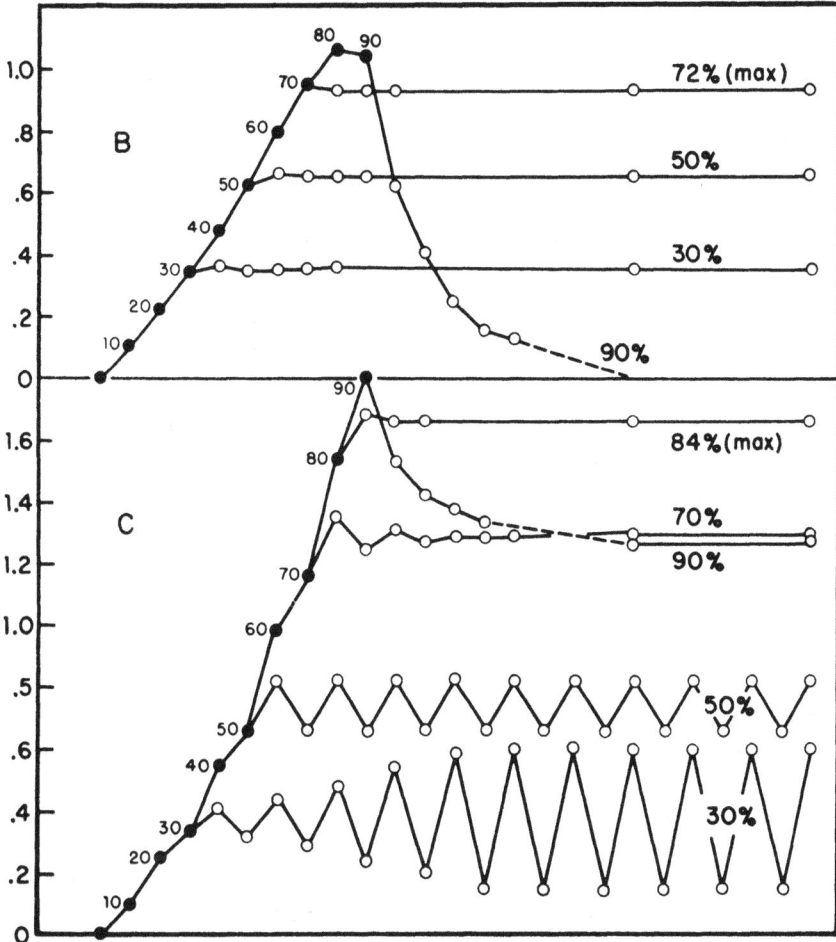

FIG. 12.3. Catches obtained from stocks whose reproduction is described by Curves B and C of Fig. 11.2. Symbols as in Fig. 12.2.

In practice fishing is more likely to level off at some intensity less than 90%. This level will be determined either by the increasing cost of fishing effort per unit catch, or by regulations resulting from the alarm provoked by a reduced absolute yield. Various stabilized levels are indicated by open circles in Fig. 12.2 and 12.3, while the equilibrium yield corresponding to *any* rate of exploitation can be calculated from items 13 and 18 of Appendix III.

For population A (Fig. 12.2), whereas the maximum equilibrium catch is achieved at 43% exploitation and 50% gives almost as much, a further increase to only 60% eventually cuts yield to less than half the maximum, while 63% gradually reduces the stock to zero.

For population B (Fig. 12.3), stabilizing the fishery at any level up to 70% exploitation means that catch becomes stabilized practically at the level already achieved, with very minor fluctuations. Maximum equilibrium catch is 0.935 of the replacement reproduction, and is achieved at 72% exploitation.

Population C (Fig. 12.3) is inherently oscillatory in the absence of fishing.[1] Even if the fishery began when the stock was at replacement level, it would develop somewhat jerkily, quite apart from any fluctuations due to environmental causes. If exploitation leveled off at 30%, catches of alternate generations would become more and more different until they came to a stable alternation between 0.15 and 0.60 (of replacement). At 50% exploitation a smaller oscillation would be established at once. At 70% a small oscillation would quickly be damped to a steady catch of 1.29. At 90% the catch would fall asymptotically to the stable level 1.27. Maximum sustained yield is 1.66, obtained with 84% exploitation.

In the above sequences, if rate of exploitation during the developmental period increases steadily to a value greater than that required for maximum sustained yield, the equilibrium catch will be less than the peak catch. What is less expected is that for Curve A there will be a maximum of catch during the developmental period even if the optimum rate of exploitation is not exceeded. In the sequence shown (Fig. 12.2) the first year's catch at 40% exploitation is 15% greater than the maximum equilibrium yield. However, the magnitude of this historical peak in landings will depend on the speed with which optimum exploitation is approached. In the extreme situation, if the A population is subjected all at once to the optimum rate of exploitation (43%), its first year catch is 30% higher than the sustainable yield ($= [0.43 - 0.33]/0.33$). Even higher peaks are possible, without exceeding the optimum level of fishing, with stocks in which the parameter α is less than 1.

12.2.2. COMPARISON OF INCREASING, DECREASING, AND EQUILIBRIUM EXPLOITATION SITUATIONS. In Fig. 12.4 the upper curve is a reproduction curve of type (11.14) with $a = 1.25$ and $P_r = 1000$. A fishery is started at equilibrium stock size and its rate of exploitation increases from generation 1 to generation 14, and then decreases to generation 17; the sequence of rates of exploitation is: 0.02, 0.06, 0.15, 0.25, 0.30, 0.35, 0.48, 0.56, 0.59, 0.62, 0.64, 0.65, 0.67, 0.61, 0.53, 0.50, and 0.46. Stocks available for exploitation each year are shown by numbered points on the recruitment curve. The curve labelled "catch" in Fig. 12.4 shows the catch in numbers taken in each generation, as well as the point of maximum equilibrium yield (MSY). The interesting thing is that any given size of stock produces a larger catch when the stock is decreasing than when it is at equilibrium or (still more) when it is increasing. Similarly, Fig. 12.5 shows that any given rate of exploitation produces a larger catch when exploitation is increasing than when it is at equilibrium or (still more) when it is decreasing. Ricker (1973b) called these effects "Mechanism 1."

---

[1] This is because of the effect discovered by Moran (1950): when the right limb of a reproduction curve crosses the replacement line at a steeper slope than 45°, there is no stable equilibrium level of abundance (see Ricker 1954b, fig. 7).

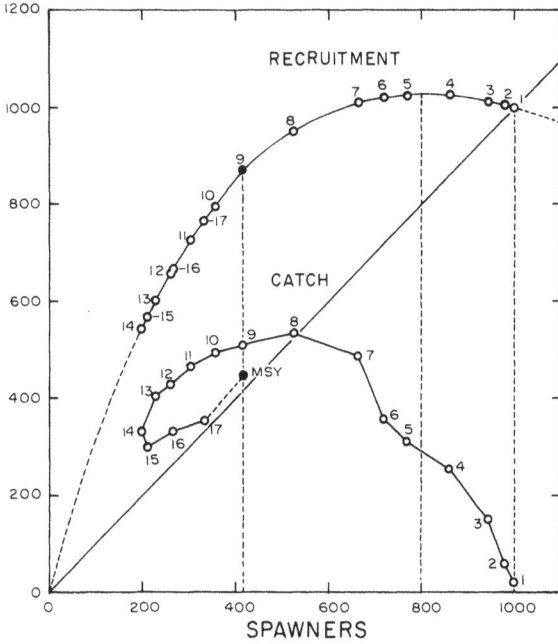

FIG. 12.4. Progress of catch and recruitment for a single stock, when subjected to the succession of exploitation rates given in Section 12.2.2. Recruitment and catch are plotted against the spawning stock that produced them, and generations are numbered in sequence. Upper curve — recruitment curve for $a = 1.25$ in expression (11.14). (From Ricker 1973b.)

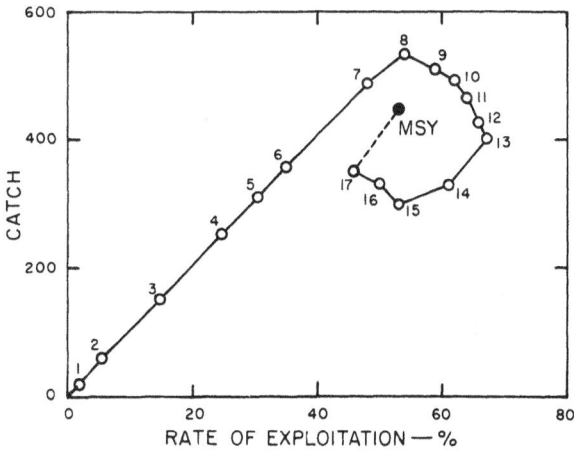

FIG. 12.5. Catch taken in successive generations from the single stock of Fig. 12.4, plotted against the rate of exploitation that prevailed during the catch year. (From Ricker 1973b.)

12.2.3. MANAGEMENT PROBLEMS. For the fishery manager, different kinds of reproduction curves present different problems:

1. For the steep Curve C the possible equilibrium *catch* is 66% greater than the original average *population*. If stabilization occurs at moderate values of exploitation, the catch from year to year will have an inherent tendency to fluctuate — though the stock as a whole will be more stable than before fishing started. These fluctuations will be superimposed on those caused by environmental variability, and will prove confusing until the slope of the reproduction curve is finally determined.

2. For Curve B stocks, maximum sustained yield is a little less than the original abundance. Stabilization of fishing leads immediately to stabilization of catch, up to quite high values of rate of exploitation. Management should be easiest in this situation, although, if the fishery initially develops beyond the point of maximum yield, some of the problems below will be encountered.

3. A population close to Curve A presents more puzzling management problems. A steadily increasing fishery will take an increasing catch up to a point considerably beyond the point of maximum sustained yield. For example, suppose exploitation increases by 10% per generation in a fish which has a 5-year generation, and the unfished equilibrium population is 1000 tons. After 25 years the average catch will have risen to 440 tons per year, taken at 50% exploitation. This is considerably greater than the possible permanent yield of 330 tons per year, but there is no way for anyone to know it. Catch has been increasing continuously. It is true that catch per unit effort will have decreased by a third or a little more, but that will scarcely be noticeable because of the fortuitous variability that would exist in any real situation. Even if noticed, it would likely be disregarded, because few expect fishing in a developed fishery to be as good as when fishermen were scarce. Thus there will seem to be no harm in continuing to fish at least at the 50% rate, or perhaps even more.

Actually there is no way to avoid a decrease in catch once exploitation has reached 50%. If by good luck fishing is held steady at that level the decrease in catch will be gradual, and the final level will be only about 30% less than the highest level achieved, or about 10% less than the possible maximum equilibrium value. But it is a critical time. If exploitation were to rise from 50% to only 60%, the catch again rises slightly, but in succeeding years it falls. Decreasing rapidly at first, then more slowly, it moves toward equilibrium at a catch about 27% of the maximum achieved (60% line in Fig. 11.7). Concurrently catch per unit effort will fall to as low as 14% of what it was originally, or 28% of what it was at maximum catch. This really will be noticeable, and a search for causes and remedies will get underway. In a salmon fishery where there is information on abundance of spawners on the redds, there will be an additional indication of scarcity: the spawners will be only 8% as numerous as in the unfished stock, or 27% as numerous as when catch was greatest. All these signs will point toward "depletion" by rule of thumb reasoning, and hence restrictions on fishing will likely be imposed. Restrictions mean temporary sacrifices, and they may or may not be made effective enough to get back to the level of maximum equilibrium catch — which, of course, has yet to be determined. However, no amount of restriction will ever get a sustained catch as large as the maximum of the developmental period, even if the optimum rate of exploitation has not been exceeded.

12.3. FISHERIES ATTACKING MIXTURES OF STOCKS HAVING DIFFERENT RECRUITMENT POTENTIALS

12.3.1. EQUILIBRIUM CONDITIONS. Ricker (1958b) examined some problems that arise when stocks having different recruitment curves are harvested together. Figure 12.6 shows the equilibrium catch taken from combinations of stocks characterized by recruitment curves A, B, and C of Table 11.5. Considering combinations of two stocks at a time, either of two situations may occur: (1) maximum yield may be available at an intermediate level of exploitation, so that both components persist (e.g. Curve A + B); or (2) maximum yield may be achieved only at a level of exploitation that exterminates one stock, leaving only one in the fishery (Curve A + C).

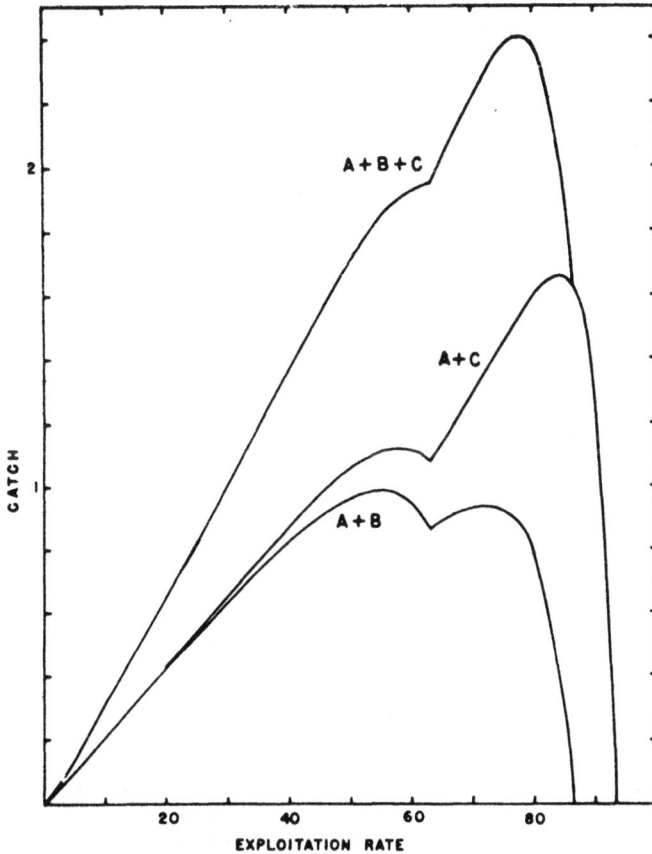

FIG. 12.6. Sustained yields when three combinations of populations A, B, and C of Fig. 11.2 are exploited in common. (After Ricker 1958b.)

303

Also, if all three of A, B, and C must be fished together, it would pay to fish at a level that exterminates A (Curve A + B + C). Observe that the combined graphs are not smooth curves, but have a re-entrant near the rate of exploitation at which any stock can no longer persist. Paulik et al. (1967) have prepared a computer program for such computations and have developed a general mathematical notation.

12.3.2. PROGRESS TOWARD EQUILIBRIUM. Since it may take many generations for an equilibrium situation such as Fig. 12.6 to become established, it is desirable to know what happens in the interim. Such histories are considered in detail by Ricker (1973b). Here we show in Fig. 12.7 the catch sequence when three unfished stocks, each in equilibrium at 1000 spawners (and recruits), are fished by the following sequence of exploitation rates in successive generations: 0.2, 0.4, 0.6, 0.75, 0.8, 0.8, 0.8, 0.8, 0.75, 0.7, 0.65, 0.6, and 0.55 from then onward. The three stocks are character-

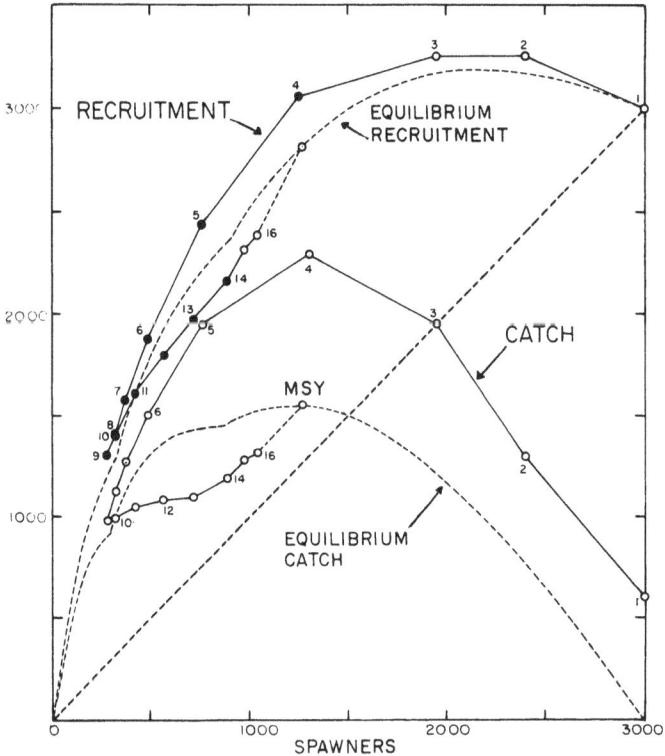

FIG. 12.7. Solid lines — catches and recruitments in successive generations plotted against the spawning population that produced them,
for a combination of three stocks, starting from a state of no fishing.
Stock parameters and rates of exploitation in successive generations
are given in section 12.3.2; the final point for each series is the equilibrium state for maximum sustainable yield at 55% exploitation. Broken
curved lines — equilibrium catches and recruitments for the spawning
populations shown on the abscissa. (The ordinate scale is the same as
the abscissal.)

304

ized by Ricker reproduction curves with $a = 1$, 1.5, and 2, respectively (Table 11.5), and thus are not as dissimilar as the three used in Fig. 12.6. The final rate of exploitation, 55%, provides approximately the level of maximum equilibrium catch (MSY) for the combined stocks, which is 84% of what could be achieved if they were to be fished independently, each at its optimum rate.

Comparing Fig. 12.7 with the similar plot for a single stock (Fig. 12.4), there is one interesting difference: *from any given total abundance of spawners a smaller recruitment is obtained during decreasing exploitation* than when exploitation is stable or is increasing. This is demonstrated by the gap between the two arms of the recruitment line in Fig. 12.7. This gap (called "Mechanism 2" by Ricker 1973b) occurs because less productive stocks (those with a recruitment curve close to the replacement diagonal) come to equilibrium at a smaller fraction of their initial abundance than do more productive ones.

In Fig. 12.7 the combined stocks reach 90% of their maximum equilibrium catch only after as many as 12 generations have elapsed from the time that the optimum (55%) rate of exploitation was established. This is a rather extreme situation, but it illustrates the time span that can be involved — 48 years for a fish that matures at the end of its fourth year. To reach practically complete equilibrium requires about 10 generations more. The maximum equilibrium yield is 68% of the maximum yield taken during the period of increasing exploitation.

Neither the upper nor the lower solid recruitment line in Fig. 12.7 represents the equilibrium level of recruitment corresponding to the spawning stocks shown on the abscissa. The equilibrium lines for recruitment and catch can be located as follows: (1) use item 16 of Appendix III to estimate the equilibrium number of spawners in each of the three stocks at each of a series of rates of exploitation; (2) compute the corresponding recruits from (11.16); (3) find the catch by difference; (4) at each rate of exploitation add together the spawners of the three stocks, and similarly for the recruitments and catches. Equilibrium catch and recruitment lines are shown in Fig. 12.7. Each has two re-entrants, which reflect the successive disappearance of the two less productive stocks at the point where the percentage of recruits removed exceeds the maximum tolerable rate of exploitation shown in the last line of Table 11.5.

Although each stock in Fig. 12.7 conforms to a Ricker curve, the combined equilibrium curve does not; nor do the curves computed from the observed catches during the period of either increasing or decreasing rates of exploitation. In practice, what is likely to be available is a series of spawning stocks and resulting recruitments that begin some years later than the fishery began. If, for example, data were available for generations 4 through 14 (the solid dots in Fig. 12.7), a fitted Ricker curve underestimates the original unexploited stock by 30%, while the estimate of optimum rate of exploitation is somewhat too large (about 65%). However, the latter figure will provide 92% of the maximum sustainable yield — which is not far off the mark (for another example see Ricker 1973b). In practice catch quotas are usually set in relation to the quantity of fish estimated to be on hand each year, but estimates of mean optimum rate of exploitation such as these are useful as a base line.

305

12.3.3. ILLUSTRATION. An application of the principles of Section 12.3 to the Skeena River sockeye fishery is given by Ricker and Smith (1975). In addition to Mechanisms 1 and 2 described above, the history of that stock is apparently complicated by interaction between different year-classes in the 4- or 5-year life cycle.

12.3.4. MANAGEMENT PROBLEMS. If two or more stocks are fished together, the management problems discussed in Section 12.2.3 become more acute (Ricker 1973b). The maximum of catch during development tends to be more pronounced when mixtures of stocks are involved (compare the catch lines of Fig. 12.4 and 12.7), so that early indications tend to encourage a higher level of overfishing and thus greater disappointments later. Even if the optimum rate of fishing is never exceeded, there is a maximum of catch that lies considerably above the level of sustainable yield.

These considerations apply not only to salmon fisheries where the catch is taken in the last year of life, but also to multiple-age fisheries. In the latter any catch maximum derived from the recruitment curve is confounded with that caused by the fishing-up effect of Section 10.9, so it can scarcely be identified or estimated from catch statistics alone. Annual sampling for age composition and computation of successive recruitments may help to separate the two.

12.3.5. STRATEGY FOR ENHANCEMENT. Finally, what is the best policy for increasing yields, when mixtures of stocks having different productivities (which implies different recruitment curves) must be fished in common? The most productive stocks are those with large R/P ratios — the most recruits per spawner. If their R/P ratio is increased, for example by hatchery rearing or improved spawning facilities, it will require an increased rate of exploitation to harvest them adequately. However, this will depress the less productive stocks still further, so that in the long run total yield may not be increased. The danger is that increased recruitments become available almost immediately, whereas the reduction of the unimproved stocks due to heavier fishing begins a generation later and is fully effective only after a number of generations.

The opposite improvement policy is to work with unproductive stocks and bring their R/P ratio up to a level somewhat above average. This will not harm naturally productive stocks. The best net yield from the system will occur if all important stocks are brought to approximately the same level of productivity, i.e. with the same mean number of recruits produced per spawner.

12.4. COMBINATIONS OF RECRUITMENT AND YIELD-PER-RECRUIT ANALYSES

12.4.1. BEVERTON–HOLT MODEL. Beverton and Holt (1957, p. 64) combined their yield equation (expression 10.20 here) with their recruitment curve (11.20) to produce a combined or "self-regenerating" equilibrium yield equation. This is their expression (6.23), in which the ratio γ (number of eggs produced per recruit) is equal to the number of recruits (R) divided into one of their expressions (6.18), (6.19), or

(6.21). Presumably other recruitment curves could be similarly incorporated with their yield equation, but in all cases the combined expression will be one of considerable complexity, and Beverton and Holt give no examples of its application.

12.4.2. WALTERS MODEL. Walters (1969) formulated the problem of combining recruitment with yield-per-recruit in analytical terms and developed computer programs that facilitate its solution. Successive ages are treated individually as regards their fecundity, natural mortality rate, rate of fishing, and mean individual weight. Any type of recruitment curve can be used. If the various parameters are not too complex, a computer search for an optimum point can be carried out in a reasonable time. The growth functions Walters uses do not allow for selective mortality within an age-group, but this could easily be incorporated. The programs are available in Fortran IV from the Institute of Annual Resource Ecology, University of British Columbia, Vancouver, B.C. (also given as Program POPS/M in Abramson 1971).

Walters' scheme can also model the transition from one state to another, year by year — for example, the yields obtained following an increase or decrease in rate of exploitation. Such a transition period will be much longer than the "fishing-up" period described in Section 10.9 (associated with constant recruitment), because here the stock has to reach equilibrium in recruits produced as well as in age composition. Indeed, for long-lived species, attaining any such equilibrium may lie so far in the future as to be of little practical interest, even supposing that the rate of fishing can be stabilized. The interim period then assumes major importance.

12.4.3. SILLIMAN ANALOG COMPUTER TECHNIQUE. The use of analog computers in fishery research was pioneered by Doi (1962), who used one to solve the catch equation and to investigate two-species predation and competition on the basis of Volterra's equations. Silliman (1966, 1967, 1969) used an analog computer to combine recruitment and yield-per-recruit analyses, by repeated trial and error. He fitted weight-at-age data with a Gompertz function, which is convenient for analog computation and is as useful as the Brody–Bertalanffy curve for describing observed growth data. Known fishing efforts were combined with observed or trial values of catchability and natural mortality rate, and these were combined with various recruitment curves. The resulting computed catches were compared with the observed course of the yield from the fishery over a period of decades, and the trial parameters were varied until good agreement was obtained.

Silliman obtained good fits to data for tunas, cod, and other species, and computed corresponding maximum sustainable yields. Naturally the fewer the arbitrary parameters or relationships that have to be postulated the more convincing are the results.

In principle, the same technique can be used with a digital computer, particularly one with graphical printout. However, routine displays on the analog screen make it possible to discover instantly the effect of modifications of parameters, so that absurd results can be rejected immediately.

CHAPTER 13. — DIRECT ESTIMATION OF THE RELATION OF EQUILIBRIUM YIELD TO SIZE OF STOCK AND RATE OF FISHING

13.1. GENERAL ASPECTS OF ADDITIONS-AND-REMOVALS METHODS

A diagram of inputs and outputs for a fish population was shown in Fig. 1.2. When data are not available to make a detailed analysis of growth, mortality, and recruitment, it is sometimes possible to relate yield directly to stock abundance or fishing effort. The methods used all involve the reasonable postulate that a fish stock produces its greatest harvestable surplus when it is at an intermediate level of abundance, not at maximum abundance. The reasons for lessened surplus production at higher stock densities are three:

1. Near maximum stock density, efficiency of reproduction is reduced, and often the actual number of recruits is less than at smaller densities. In the latter event, reducing the stock will increase recruitment.

2. When food supply is limited, food is less efficiently converted to fish flesh by a large stock than by a smaller one. Each fish of the larger stock gets less food individually; hence a larger fraction is used merely to maintain life, and a smaller fraction for growth.

3. An unfished stock tends to contain more older individuals, relatively, than a fished stock. This makes for decreased production, in at least two ways. (a) Larger fish tend to eat larger foods, so an extra step may be inserted in the food pyramid, with consequent loss of efficiency of utilization of the basic food production. (b) Older fish convert a smaller fraction of the food they eat into new flesh — partly, at least, because mature fish annually divert much substance to maturing eggs and milt.

Under reasonably stable natural conditions the net increase of an unfished stock is zero, at least on the average: its growth is balanced by natural deaths. Introducing a fishery increases production per unit of stock by one or more of the methods above, and so creates a surplus which can be harvested. In these ways "a fishery, by thinning out a fish population, itself creates the production by which it is maintained" (Baranov 1927). Notice that effects 1 and 3 above may often increase *total* production of fish flesh by the population — it is not merely a question of diverting some of the existing production to the fishery, though that also occurs.

The question of the interaction of a fish stock and its food supply occupied Petersen (1922) and others, but apparently the first comprehensive attempt at a numerical computation and catch predictions based on these relationships was by Baranov (1926). He applied them to two fisheries: North Sea plaice and Caspian vobla. Following Petersen, he assumed a constant production of fish food in the environment, all of which was consumed at all possible densities of stock—to be used partly for maintenance of stock (including production of sexual products), partly for growth in excess of natural mortality. In both his examples the reaction of the stock to the reduced fishery of World War 1, and

309

to the subsequent increase in fishing, provided the basis for the numerical computations. His argument was that in a primitive state the biomass of fish in a body of water consumes the annual food production and simply maintains itself. If a fishery starts and part of the fish are removed, the remainder eat the surplus food and so grow faster. From these premises an equation applicable to a fishery in equilibrium is readily derived:

$$B_1 + yb_1 = B$$

B original stock under conditions of no fishing
B_1 existing "basic stock"
b_1 annual catch
y ratio of the food needed to *produce* 1 kg of fish to the food needed to *support* 1 kg of fish for a year

If data for two periods of stability are available, characterized by different rates of exploitation, it is possible to calculate y and B from the observed levels of catch and rate of exploitation, the latter being put equal to $b_1/(B_1 + b_1)$. And knowing these, it is possible to look ahead and forecast future catches, given the amount of fishing effort that will be exerted.

Although the above expression is logically correct, it has not proved very useful because the assumptions underlying it are not sufficiently realistic. The assumption of a food production that is independent of the size of the fish stock is questionable, as is the assumption that a reduced fish stock will continue to consume as much in toto as did the original virgin stock. Two implied postulates we now know to be wrong: that a given amount of food will either maintain, or produce, the same fish biomass independently of the average age of the fish in question. Older fish require far more food to maintain a unit of body weight because (in spite of their lower metabolism) they have to release a large mass of eggs or sperm each year. Older fish also require more food to produce a unit of new body weight (Brett 1970, fig. 5).

Thus although Baranov emphasized that his theory was only approximate and would require adjustments, later work suggests that it was in fact too approximate to be useful. The paper was criticized by Edser (1926), and also by a series of Russian biologists of whom we need mention only Monastyrsky (1940). A synopsis of this exchange of views is available in volume 3 of Baranov's Collected Works (Zasosov 1971). But although this paper is now only of historical interest, it exhibits the author's customary originality and ingenuity, and it represents "Mark I" of the now widely used additions-and-removals method.

13.2. Parabolic Surplus Production Curve and Logistic Growth Curve — Graham's Method

13.2.1. Surplus-production parabola and derivations. It remained for Graham (1935) to introduce a simple and consistent numerical model based on reasoning like the above. He postulated that under equilibrium conditions the instantaneous rate of surplus production of a stock (= recruitment plus growth less natural mortality) is directly proportional to its biomass and also to the difference between the actual biomass and the maximum biomass the area will support

$$\frac{dB}{dt} = \frac{kB(B_\infty - B)}{B_\infty} \tag{13.1}$$

B   stock size (biomass)
B_∞ maximum stock size
k instantaneous rate of increase of stock at densities approaching zero ($k =$ V of Graham)
t time, conventionally in years

Further, when fishing removes the surplus production of the stock at the same rate as it is produced, it becomes the annual yield from a stock held in equilibrium:

$$Y_E = F_E B_E = k B_E \left(\frac{B_\infty - B_E}{B_\infty} \right) \tag{13.2}$$

$$= k B_E - \left(\frac{k}{B_\infty} \right) B_E^2 \tag{13.3}$$

B_E mass of a stock when it is in an equilibrium condition
F_E rate of fishing which maintains the stock in equilibrium at mass B_E
Y_E yield when the stock is in equilibrium

Expression (13.3) shows that on this basis the relation between equilibrium yield and equilibrium biomass is a parabola; an example is shown in Fig. 13.1B.

The differential of (13.3) with respect to B_E is:

$$k - \frac{2k B_E}{B_\infty} \tag{13.4}$$

Equating to zero gives the value of B_E for which Y_E is a maximum (called the optimum stock size, B_s):

$$B_s = \frac{B_\infty}{2} \tag{13.5}$$

Substituting (13.5) for B_E in (13.3), the maximum sustainable yield becomes:

$$Y_s = \frac{k B_\infty}{4} \tag{13.6}$$

Thus maximum equilibrium yield is obtained when stock size is exactly half the maximum equilibrium biomass; it is equal to one-quarter of the maximum biomass multiplied by the instantaneous rate of increase at very small biomass.

Also, substituting $F_s B_s$ for Y_s in (13.6) and dividing both sides by (13.5), we obtain the rate of fishing at maximum sustainable yield:

$$F_s = \frac{k}{2} \tag{13.7}$$

Since $F = qf$, the optimum level of fishing effort is:

$$f_s = \frac{k}{2q} \tag{13.8}$$

13.2.2. LOGISTIC CURVE OF POPULATION GROWTH. If a stock in equilibrium at any mass B_E is released from fishing pressure it will start growing at the rate established by expression (13.1). Integrating (13.1), the growth curve is the S-shaped "logistic" curve of Verhulst:

$$B = \frac{B_\infty}{1 + e^{-k(t-t_0)}} \tag{13.9}$$

t_0 is a constant that adjusts the time scale to an origin at the inflection point of the curve: i.e. $t - t_0 = 0$ when $B = B_\infty /2$. Sometimes it is useful to transform (13.9) to a straightline relationship in t:

$$\log_e\left(\frac{B_\infty}{B} - 1\right) = kt_0 - kt \tag{13.10}$$

13.2.3. FITTING A GRAHAM CURVE. Fitting expression (13.2) to statistics of a fishery can be done from several combinations of data. Item No. 1 below is always required; it is to be combined with any one of 2, 3, and 4.

1. Absolute size of the stock, B_E, and rate of fishing, F_E, at a stable level of abundance (i.e. when the stock is in equilibrium with the fishery); this is obtained by one of the methods described in earlier chapters.

2. Easiest to combine with the above is the level of stock, B_∞, characteristic of no fishing. Under suitable conditions this can be found by relating B_∞ to the stable B_E, proportionately to the catch per unit effort for each. (Strictly speaking, B_∞ can exist only when there is no catch, but an early stage of a fishery can be considered as corresponding approximately to natural equilibrium.) However, the "suitable conditions" just postulated may not often occur, for catch per unit effort at the start of exploitation may overestimate the true abundance of the stock. Example 13.1 illustrates the use of these two pieces of information.

3. In place of the value B_∞, a second equilibrium level of exploitation and stock may be combined with (1). Either the new B_E and F_E may be estimated independently; or, with greater risk, the new B_E may be related to that determined in (1) by using the catch per unit effort for the two situations, and F_E may be considered proportional to gear in use in each case. Example 13.2 illustrates this procedure.

4. Finally, stock parameters may be estimated from rate of growth of a stock during a period of no fishing, as shown in Example 13.3.

EXAMPLE 13.1. FITTING A GRAHAM SURPLUS-PRODUCTION CURVE, GIVEN B_∞ AND ONE EQUILIBRIUM LEVEL OF FISHING. (From Ricker 1958a.)

This illustration and the next two are freely adapted from Graham's (1935) data and computations concerning North Sea demersal fish stocks. However, the absolute level of stock indicated is fictitious.

A fishery many years in a state of steady effort and yield is characterized by a yearly catch (Y_E) of 40,000 tons, of which 30,000 tons are fish which were already vulnerable at the start of the year. The rate of exploitation of these fully-vulnerable individuals is found, by tagging, to be 30%. The vulnerable stock present at the start of the year is therefore $30,000/0.3 = 100,000$ tons, and as the stock is in equilibrium with the fishery, this represents also the vulnerable stock continuously on hand (B_E). The rate of fishing, F_E, is therefore $40,000/100,000 = 0.4$, and this must also be the instantaneous rate of surplus production (rate of recruitment plus rate of growth less rate of natural mortality).

Catch per unit effort is currently 10 tons/boat-day. However, some years earlier, immediately following a long fishing respite, catch was 22 tons/day. Considering Y/f proportional to stock, the stock characteristic of no fishing was therefore $B_\infty = 100,000 \times 22/10 = 220,000$ tons.

Applying the above information in (13.2):

$$40,000 = k \times 100,000 \left(\frac{220,000 - 100,000}{220,000} \right)$$

from which $k = 0.733$. The relation of equilibrium yield to size of stock is therefore, from (13.3):

$$Y_E = 0.733\ B_E - \frac{0.733\ B_E^2}{220,000} \tag{13.11}$$

EXAMPLE 13.2. FITTING A GRAHAM SURPLUS-PRODUCTION CURVE GIVEN TWO EQUILIBRIUM LEVELS OF FISHING. (From Ricker 1958a.)

Suppose that B_∞ is not known, but that a second level of equilibrium yield is available along with the one described in Example 13.1. The data for the two are as follows:

	First level	Second level
	$B_E = 100,000$ tons	$B_E = 60,000$ tons
	$Y_E = 40,000$ tons	$Y_E = 32,000$ tons
	$F_E = 0.40$	$F_E = 0.53$

For a trial $B_\infty = 200,000$, the value of k is estimated for each level, using (13.2):

First level: $k = 0.40 \times 200,000/100,000 = 0.80$
Second level: $k = 0.53 \times 200,000/140,000 = 0.76$

Further trials show that $B_\infty = 220,000$ makes the two estimates of k equal, its value being 0.733. The best descriptive equation can then be found as in Example 13.1: it will be (13.11). The two levels of stock and yield are indicated in Fig. 13.1B.

From (13.7) we compute that the rate of fishing at maximum equilibrium yield is $F_s = 0.733/2 = 0.366$, while the yield taken is, from (13.6), $Y_s = 0.733 \times 220,000/4 = 40,300$ tons. Thus at both of the stable levels of stock postulated in this example, extra fishing effort is being devoted to reducing the size of the annual catch. Even at the "first level," where the catch (40,000 tons) is *almost* the maximum, this same amount could be obtained from a higher level of stock at a rate of fishing of 0.333, with a saving of effort of $(0.4 - 0.333)/0.4 = 17\%$. From such considerations Graham concluded that the North Sea was being overfished, both economically and biologically.

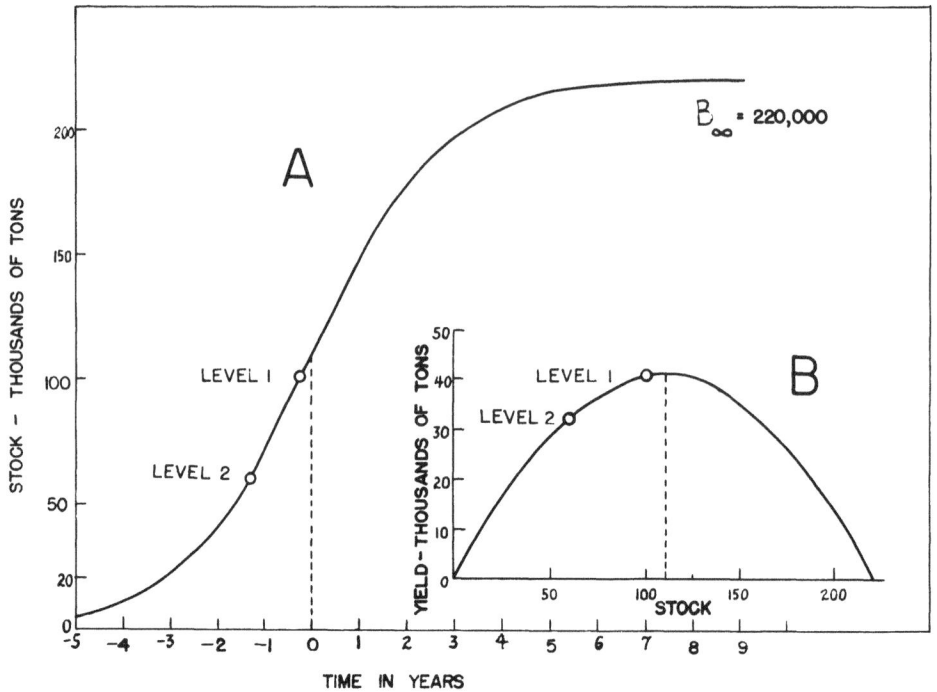

FIG. 13.1. (A) Logistic curve of increase for the population of Examples 13.1–13.3; (B) concomitant parabolic relationship of yield to stock density. The abscissal scale of A indicates the rate at which the stock, in the absence of fishing, would move toward the asymptotic level B_∞.

EXAMPLE 13.3. FITTING A GRAHAM SURPLUS-PRODUCTION CURVE AND LOGISTIC POPULATION GROWTH CURVE, GIVEN ONE EQUILIBRIUM LEVEL OF STOCK AND ITS INCREASE DURING A PERIOD OF NO FISHING. (From Ricker 1958a.)

Consider again the equilibrium state of Example 13.1, characterized by $Y_E = 40,000$ tons, $u_E = 0.30$, $B_E = 100,000$ tons, and $F_E = 0.40$. After some years, fishing suddenly ceased: let this be time $t = 0$. During 2.7 subsequent years of no fishing, yield per unit effort had increased from 10 tons/boat-day to 19 tons/boat-day. Considering this as proportional to stock, the final stock is estimated as 190,000 tons. There is no direct evidence whether stock was yet at its maximum equilibrium size, but it would seem unlikely to have occurred in so short a time.

The procedure is to use a trial value of B_∞, and compute a trial k from $F = 0.40$, using (13.2). For trial $B_\infty = 250,000$, $k = 0.40 \times 250,000/150,000 = 0.667$. Substituting in (13.10), at time $t = 0$:

$$\log_e\left(\frac{250,000}{100,000}\right) - 1 = 0.667 t_0$$

Hence the trial t_0 is 0.61, which determines a trial relationship of the type (13.9):

$$B = \frac{250,000}{1 + e^{-0.667(t-0.61)}}$$

314

Substituting $t = 2.7$, we obtain B = 200,000 tons, which is higher than the observed 190,000 tons. Further trials show that B_∞ = 220,000 tons, k = 0.733, and t_0 = 0.25 give the best fit to the data; the corresponding logistic equation of stock growth is:

$$B = \frac{220,000}{1 + e^{-0.733(t-0.25)}} \qquad (13.12)$$

The data of Example 13.1 or 13.2 are also adequate to compute (13.12), subject to the reservations already given. The value $t_0 = 0.25$ indicates that the equilibrium state of Example 13.1 is $\frac{1}{4}$ year to the left of or "previous to" the inflection point of the logistic curve, the latter occurring when B_E = 110,000 tons. Having obtained this t_0, it is convenient to use the inflection point as the origin of the logistic graph, and let B_E = 100,000 correspond to $t = -0.25$. This is the abscissal scale indicated in Fig. 13.1A. The corresponding equation is:

$$B_E = \frac{220,000}{1 + e^{-0.733t}} \qquad (13.13)$$

13.3. RELATION OF EQUILIBRIUM YIELD TO FISHING EFFORT

In the above examples surplus production has been related to size of stock under equilibrium conditions. Simply by rearranging the terms of (13.2) it can be shown that (still under equilibrium conditions) there is a similar relationship between surplus production and rate of fishing, F, hence also with effective fishing effort, f. From (13.2) it is clear that:

$$B_E = B_\infty - \frac{F_E B_\infty}{k} \qquad (13.14)$$

Substituting this back for B_E in (13.3):

$$Y_E = F_E B_E = B_\infty F_E - \left(\frac{B_\infty}{k}\right) F_E^2 \qquad (13.15)$$

Substituting $q f_E$ for F_E:

$$Y_E = F_E B_E = a f_E - b f_E^2 \qquad (13.16)$$

where $a = q B_\infty$ and $b = q^2 B_\infty /k$.

Thus under equilibrium conditions surplus production is a parabolic function of rate of fishing (F) and of fishing effort (f), as well as of stock size (B). This provides still another means of fitting the relationship. Dividing (13.16) by f_E:

$$\frac{Y_E}{f_E} = a - b f_E \qquad (13.17)$$

Hence values of yield per unit effort (Y_E/f_E) and effort (f_E) for two equilibrium levels can be substituted in (13.17) and values obtained for a and b.

315

To estimate optimum level of fishing effort (f_s), (13.16) is differentiated and equated to zero, giving:

$$f_s = a/2b \tag{13.18}$$

This is the same result as in (13.8). Knowing f_s, the maximum sustainable yield is estimated by substituting it for f_E in (13.16):

$$Y_s = \frac{a^2}{4b} \tag{13.19}$$

Thus maximum sustainable yield (Y_s) *and optimum rate of fishing* (f_s) *can be estimated directly from the relation of equilibrium yield to equilibrium effort,* without knowing the catchability (q) of the fish. Of course, if q is known from other sources it can be combined with Y_s and f_s to calculate stock size and rate of fishing under equilibrium conditions. The appropriate expressions are (13.5)–(13.8).

EXAMPLE 13.4. ESTIMATING MAXIMUM SUSTAINABLE YIELD AND OPTIMUM RATE OF FISHING FROM TWO EQUILIBRIUM LEVELS OF CATCH AND FISHING EFFORT.

From the information given in Examples 13.1 and 13.2, the two equilibrium yields of Example 13.2 were taken by the fishing efforts shown below:

First level	Second level
$Y_E = 40{,}000$ tons	$Y_E = 32{,}000$ tons
$f_E = 4000$ days	$f_E = 5300$ days
$Y_E/f_E = 10$ tons/day	$Y_E/f_E = 6.04$ tons/day

Substituting these in (13.17) we obtain $a = 22.2$ and $b = 0.00305$. From (13.19) and (13.18):

Maximum sustainable yield $Y_s = 22.2^2/4 \times 0.00305 = 40{,}300$ tons

Optimum fishing effort $f_s = 22.2/2 \times 0.00305 = 3640$ days

The figure for MSY given above is the same as was found in earlier examples. But here it has been estimated without knowing absolute stock size or the catchability of the fish. For the latter, additional information is necessary (for example, the rate of exploitation given in Example 13.1)

13.4. RELATION OF SURPLUS PRODUCTION TO SIZE OF STOCK, USING THE YEARLY INCREASE OR DECREASE OF STOCK — METHOD OF SCHAEFER (1954)

13.4.1. COMPUTATION OF ANNUAL PRODUCTION. Schaefer (1954) introduced a method of estimating surplus production for each year individually.[1] This provides

---

[1] In an interesting paper, Thompson (1950) applied the concept of a "normal catch" to the Pacific halibut fisheries. The catch selected as normal was the average yield for a period of time after the original fishing-up of accumulated stock was completed—1926-33 on the southern halibut grounds and 1926-36 on the western grounds. From the definition of normal, removals in excess of the norm

a value of Y_E that can be related to stock size or fishing effort without waiting for the stock to come to equilibrium, and thus broadens the applicability of surplus-production methods. The procedure divides the weight of each year's landings, Y, by its rate of fishing, F, to obtain an estimate of mean stock, $\bar{B}$, present during the year.

The level of stock at the turn of a year is approximately the average of the mean stocks of the year just completed and the year ahead. The difference between two initial stocks is the increase for the year in question, ΔB; that is, the increase in year 2 of any series of 3 is approximately:

$$\Delta B_2 = \frac{\bar{B}_3 + \bar{B}_2}{2} - \frac{\bar{B}_2 + \bar{B}_1}{2} = \frac{\bar{B}_3 - \bar{B}_1}{2} \qquad (13.20)$$

The surplus production or equilibrium yield, Y_E, for year 2 above is the actual yield Y_2 plus the change in stock size, positive or negative:

$$Y_E = Y_2 + \Delta B_2 = Y_2 + \frac{\bar{B}_3 - \bar{B}_1}{2} \qquad (13.21)$$

Values of F needed to compute $\bar{B}$ estimates are most directly obtained by estimating F (for at least one year) by tagging or by one of the other methods described in earlier chapters. From it the catchability, q, is estimated, and the other F's are estimated as proportional to effective fishing effort (F $= qf$ for each year).

Determined in this way, surplus production can be plotted against stock density, independently of any hypothesis relating the two. If a well-defined curve for a reasonably wide range of stock densities is obtained, it can be used empirically to define a position of maximum yield for the stock, relative to the current year's abundance. If the graph does not cover the whole of the range that is of interest, some kind of curve must be fitted to it to permit extrapolation. In his 1954 paper Schaefer adopted Graham's parabola for this purpose, although with the reservation that a curve skewed to the left seemed to have some support from observations then available.

13.4.2. FITTING A PARABOLIC RELATIONSHIP TO ANNUAL PRODUCTIONS. Because surplus production, Y_E, is being determined for each year separately, expression (13.3) can be written in the form:[2]

$$Y_E = F\bar{B} = k\bar{B} - \left(\frac{k}{B_\infty}\right)\bar{B}^2 \qquad (13.22)$$

---

were accompanied by a decrease in stock density (as shown by decrease in catch per unit effort), while catches less than the norm were accompanied by increase in catch per unit effort, during those years. There is a superficial similarity between this treatment and Schaefer's, because the latter also proceeds from a definition that catches in excess of surplus production (his "equilibrium catch") in any year must result in a decrease in stock, and vice versa. However, the difference between the two approaches is more important than the resemblance: whereas Thompson is impressed by the apparent constancy of the "normal" catch over the indicated intervals of time, Schaefer joins with Baranov, Hjort and Graham in emphasizing that equilibrium catch must change as size of stock changes. Actually, of course, the normal levels used by Thomspson did not remain normal in later years, and in addition some of the population statistics computed from them are seriously at odds with those derived from the age structure of the population or from the results of tagging experiments.

  [2] See Section 13.5 for a comparison of Schaefer's symbols with those used here.

where $\bar{B}$ and F are the mean stock and rate of fishing for each year individually. To fit (13.22) to a body of data it is convenient to divide through by $\bar{B}$, giving:

$$\frac{Y_E}{\bar{B}} = k - \frac{k\bar{B}}{B_\infty} \tag{13.23}$$

Consequently $Y_E/\bar{B}$ can be plotted against $\bar{B}$, and a functional straight line fitted (Appendix IV). Its Y-axis intercept represents k and the slope of the line is k/B_∞, so both k and B_∞ are estimated.

By arguments similar to those of Section 13.3 it can be shown that $\bar{B}$ can be substituted for B_E in expressions (13.15) and (13.16), so that a regression of Y_E/f on f for each individual year can also be used to compute the statistics of the situation.

13.4.3. MAXIMUM SUSTAINABLE YIELD IN RELATION TO STOCK SIZE AND FISHING EFFORT OF THE SAME YEAR. From the parabola fitted to population or effort, a maximum equilibrium yield can be calculated using (13.6) or (13.19). However, such maxima must be interpreted with caution. Consider the possible paths by which stock size or fishing effort can affect surplus production. Surplus production is the sum of individual growth increments plus recruitment less natural mortality. *Growth* is a function of stock size: as long as food is ample, more stock means more growth; but at higher stock densities food may become scarce and growth per unit biomass may decrease because more of the food consumed is used for maintenance and sexual products, and less for additional body tissue. Growth also depends on the average age of the stock: younger fish make larger percentage weight increases from a given food consumption than do older ones. *Stock size*, in turn, is a function of fishing effort in the current year and in earlier years, and of recruitment. In addition to usually large environmental effects, *recruitment* is a function of stock size some years previously, and this relationship may be direct or inverse, depending on the then size of the stock, particularly of the mature stock. This in turn depends partly on the fishing of earlier years. Finally, the biomass lost by *natural mortality* among the catchable stock will increase at least in proportion to stock size, but whether at a greater rate with increasing density is usually unknown.

In considering the relation of Y_E/B to B in the same year, the principal potential disturbing factors will evidently be: (a) differences in age of the stock, and (b) differences in recruitment. Neither of these is related directly to stock size in the current year. In dealing with a many-aged stock having a history of generally increasing fishing effort, the ratio Y_E/B will be increasing because the stock is becoming smaller and younger, and possibly also because recruitment has been increasing owing to the decline in big old fish (though this is not certain, and will eventually shift to a decrease if the stock is reduced far enough). If at some point fishing begins to decrease, stock size begins immediately to increase (or to decrease less rapidly), but any trend in recruitment will not change until the young fish produced by the resulting larger stocks grow into the fishery. The mean age of the fish might therefore continue to decrease for a while, despite the fact that the smaller rate of fishing is permitting greater survival of fish already recruited.

Our general conclusion is that any observed relation between Y_E and B of the same year will not necessarily reflect a balanced situation. In particular, the level of B that is computed to correspond to maximum sustainable yield will not necessarily be a good estimate of what is needed for MSY. Rather, MSY may require either a larger or a smaller stock, depending on the age structure and antecedent circumstances in the stock and fishery. Equally, any relationship between Y_E/f and f of the same year will be subject to various interpretations, in resolving which knowledge of the biology and history of the stock is essential.

Why should there be any relationship at all under these circumstances? The usual reason is that fishing effort and hence population size tend to change slowly, having sustained periods of increase or decrease. During such a trend the effort of any year will be closely correlated with the effort of several earlier years, so that the influence of those earlier years will be what determines the observed relationship.

EXAMPLE 13.5. COMPUTATION OF SURPLUS PRODUCTION FOR THE PACIFIC HALIBUT OF AREA 2, AND FITTING A GRAHAM CURVE FROM THE RELATION OF SURPLUS PRODUCTION TO BIOMASS. (Modified from Ricker 1958a, after Schaefer.)

Table 13.1 illustrates the method used by Schaefer (1954) to calculate surplus production for Pacific halibut in the southern area. It is based on his table 1, but is arranged to correspond with the usages of this Bulletin, and a longer series of years is used — from 1910 to 1957 (Anon. 1962, table 7). The yield (column 2) and the fishing effort (column 3) for each year are known. These are divided to obtain the yield per unit effort (column 4), considered to be proportional to the mean biomass present each year. The rate of fishing for 1926 was estimated in Example 5.5 above as F = 0.57 for fully-vulnerable individuals (mean age 9 and older), and progressively less for smaller ones. A mean F for all ages present in 1926 is estimated by weighting the F for each length class (Table 5.7) jointly as its abundance in the sample (third column of Table 5.3) and as the mean individual biomass of the individuals comprising it, the latter being taken as proportional to the cube of the mid-class length. The mean rate of fishing so obtained is F = 0.434,[3] which applies to 1926, when effort was f = 478,000 skates of gear. Thus catchability was q = 0.434/478,000 = 0.907 × 10⁻⁶ of the stock taken by 1 skate. The 1926 rate of fishing can be divided into the catch to obtain a mean stock estimate of $\bar{B}$ = 24.7/0.434 = 57.0 million lb in that year. Values of $\bar{B}$ for other years are then calculated in proportion to the known estimates of Y/f in column 4. (Values of F for other years can be calculated as proportional to effort, f, and are shown in column 5, but this step is not essential.) Differences between stocks 2 years apart are then divided by 2 to give an estimate of the increase or decrease in the intervening year (expression 13.20), and are shown in column 7. Following

---

[3] Schaefer (1954) used F = 0.635 for 1926, corresponding to the conditional fishing mortality rate of m = 0.47 computed by Thompson and Bell (1934); Ricker (1958a) adjusted this to F = 0.615 because a change had been made in the estimate of fishing effort in the base year 1926. The present figure is a more realistic mean value as of 1926, taking weight into account. However, the mean F corresponding to a given f must change somewhat with change in the size composition of the stock; no attempt is made here to adjust for this because the necessary data are not available.

TABLE 13.1. Yield in millions of pounds, fishing effort in thousands of skates, and derived statistics for Pacific halibut of Area 2. (From data of the International Pacific Halibut Commission, Anon. 1962.)

1	2	3	4	5	6	7	8	9	10	11	12
Year	Y	f	Y/f	F	B	ΔB	Y_E	Y_E/B	Y_E/f	5-year mean of f	3-year mean of B
	10^6 lb	10^3 skates	lb/skate	—	10^6 lb	—	10^6 lb	—	lb/skate	10^3 skates	10^6 lb
1910	51.0	189	271	0.172	298	.....	...	...	...	...	...
1911	56.1	237	237	0.215	261	-52.0	4.1	0.02	17	244	251
1912	59.6	340	176	0.309	194	-60.0	-0.4	0.00	0	312	199
1913	55.4	432	128	0.392	141	-28.5	26.9	0.19	62	349	157
1914	44.5	360	124	0.327	137	-5.5	39.0	0.28	108	354	136
1915	44.0	375	118	0.340	130	-5.5	38.7	0.30	103	362	131
1916	30.3	265	114	0.241	126	-20.5	9.8	0.08	37	336	115
1917	30.8	379	81	0.344	89	-15.0	15.8	0.18	42	329	104
1918	26.3	302	87	0.274	96	+0.5	26.8	0.28	89	332	92
1919	26.6	325	82	0.295	90	-2.0	24.6	0.27	76	374	93
1920	32.4	387	84	0.351	92	-3.0	29.4	0.32	76	396	89
1921	36.6	479	76	0.435	84	-12.0	24.6	0.29	51	435	81
1922	30.5	488	62	0.443	68	-10.5	20.0	0.29	41	464	72
1923	28.0	494	57	0.448	63	-4.5	23.5	0.37	48	475	64
1924	26.2	473	55	0.429	61	-3.5	22.7	0.37	48	475	60
1925	22.6	441	51	0.400	56	-2.0	20.6	0.37	47	471	58
1926	24.7	478	51.7	0.434	57	-1.0	23.7	0.42	50	480	56
1927	22.9	469	49	0.425	54	-2.5	20.4	0.38	44	508	54
1928	25.4	537	47	0.487	52	-5.0	20.4	0.39	38	543	50
1929	24.6	617	40	0.560	44	-7.0	17.6	0.40	28	555	45
1930	21.4	615	35	0.559	38	+0.5	21.9	0.58	35	550	42

Year											
1931	21.6	534	41	0.484	45	+8.0	29.6	0.66	55	530	46
1932	22.0	445	49	0.404	54	+6.0	28.0	0.52	63	489	52
1933	22.5	438	52	0.397	57	+3.5	26.0	0.46	59	439	57
1934	22.6	411	55	0.373	61	+5.5	28.1	0.46	68	424	62
1935	22.8	366	62	0.332	68	-0.5	22.3	0.33	61	421	63
1936	24.9	459	54	0.416	60	-1.0	23.9	0.40	52	406	65
1937	26.0	431	60	0.391	66	+8.0	34.0	0.51	79	414	67
1938	25.0	363	69	0.392	76	+0.5	25.5	0.34	70	429	70
1939	27.4	452	61	0.410	67	-3.5	23.9	0.36	53	422	71
1940	27.6	440	63	0.399	69	0.0	27.6	0.40	63	412	68
1941	26.0	426	61	0.387	67	+1.0	27.0	0.40	63	408	69
1942	24.3	378	64	0.343	71	+6.5	30.8	0.43	82	381	73
1943	25.3	346	73	0.314	80	+11.0	36.3	0.45	105	353	81
1944	26.5	314	84	0.285	93	+4.5	31.0	0.33	99	338	87
1945	24.4	303	81	0.275	89	+0.5	24.9	0.28	82	330	92
1946	29.7	351	85	0.319	94	+3.0	32.7	0.35	93	323	93
1947	28.7	334	86	0.303	95	+3.0	31.7	0.33	95	320	96
1948	28.4	312	91	0.283	100	+2.0	30.4	0.30	97	316	98
1949	26.9	299	90	0.271	99	+3.0	29.9	0.30	100	310	102
1950	27.0	282	96	0.266	106	+3.5	30.5	0.29	108	293	104
1951	30.6	321	96	0.291	106	15.0	45.6	0.43	142	277	116
1952	30.8	252	123	0.229	136	+27.0	57.8	0.42	229	272	134
1953	33.0	229	145	0.208	160	+6.0	39.0	0.24	170	262	148
1954	36.7	274	134	0.249	148	-12.0	24.7	0.17	90	252	148
1955	28.7	234	123	0.212	136	-2.5	26.2	0.19	112	262	142
1956	35.4	272	130	0.247	143	-11.5	23.9	0.17	88	..	131
1957	31.3	302	103	0.274	113	....	...	...		..	..

expression (13.21), surplus production is shown in column 8, being equal to yield plus increase in stock (or less the decrease).[4] In Fig. 13.2A the surpluses of column 8 are plotted against the mean stock of column 6. The points suggest that a maximum of surplus production exists somewhere near 30 million lb. A functional Graham curve was fitted to the graph of $Y_E/\bar{B}$ against $\bar{B}$, using two procedures:

GM regression (Fig. 13.2A) $\qquad$ $Y_E/\bar{B} = 0.612 - 3.00 \times 10^{-9}\bar{B}$

Nair–Bartlett regression $\qquad$ $Y_E/\bar{B} = 0.635 - 3.25 \times 10^{-9}\bar{B}$

The derived statistics are:

Expression	Symbol	GM	Nair
(13.23)	k	0.612	0.635
(13.23)	k/B_∞	3.00×10^{-9}	3.25×10^{-9}
	B_∞	204×10^6	195×10^6
(13.5)	B_s	102×10^6	98×10^6
(13.6)	Y_s	31.2×10^6	31.0×10^6
(13.7)	F_s	0.306	0.318
(13.8)	f_s	337,000	350,000

EXAMPLE 13.6. FITTING A GRAHAM CURVE TO PACIFIC HALIBUT USING THE RELATION BETWEEN SURPLUS PRODUCTION AND FISHING EFFORT.

Expression (13.17) indicates that another way to fit a curve to the halibut data is by regressing Y_E/f_E on f_E. The method of computing Y_E used in Example 13.5 implies that the fishing effort for each year will be that which keeps the stock in equilibrium; thus the effort values in column 3 of Table 13.1 can be equated to f_E. Two estimates of the functional regression of Y_E/f (column 10) on f are as follows:

GM regression: $\qquad$ $Y_E/f = 231.8 - 0.408 \times 10^{-3}f$

Nair–Bartlett: $\qquad$ $Y_E/f = 216.3 - 0.368 \times 10^{-3}f$

The GM regression line is shown in Fig. 13.2B. The derived statistics are obtained (for the GM line) by putting $a = 231.8$, $b = 0.408$, and substituting in (13.18) to obtain f_s. Given $q = 0.907 \times 10^{-6}$ as described in Example 13.5, F_s is equal to $f_s q = 0.258$, and the value of k is twice that, from (13.7). The other statistics are then available using the expressions indicated below:

Expression	Symbol	GM regression	Nair
(13.7)	k	0.515	0.533
(13.6)	B_∞	256×10^6	239×10^6
(13.5)	B_s	128×10^6	119×10^6
(13.19)	Y_s	33.0×10^6	31.8×10^6
	F_s	0.258	0.267
(13.18)	f_s	284,000	294,000

These are listed in the same order as in Example 13.5, although the order of computation is different.

---

[4] Schaefer called this the "equilibrium catch," as it would have to be removed to maintain stock at its current level. However, the concept applies whether or not any catching actually takes place; in fact the equilibrium "catch" can at times be negative.

13.5. Computation of a Parabolic Yield Curve when Catchability (q) Is Not Known Independently — Method of Schaefer (1957).

In dealing with a fishery for yellowfin tuna (*Thunnus albacares*) Schaefer (1957) had no direct estimate of catchability, q, so proceeded to extract one from the data. The method will not be described in detail here, but a concordance of Schaefer's symbols with those used in this handbook is as follows:

Schaefer	Here	Schaefer	Here
$\overline{f}P$	Y_E	U	Y/f or U
k_1	k/B_∞	ΔU	$q\Delta B$
k_2	q	M	qB_∞
P	B	a	k/q^2B_∞
L	B_∞	C	Y
L_u	qB_∞	$C/\overline{U}$	f

Schaefer's basic equation is, in his symbols:

$$\frac{1}{k_2}\sum_{i=1}^{i=n}\frac{\Delta U_i}{\overline{U}} = naM - a\sum_{i=1}^{i=n}\overline{U}_i - \sum_{i=1}^{i=n}\frac{C_i}{\overline{U}_i} \qquad (13.24)$$

This he solves for k_2, a, and M by dividing the data series into two parts — not necessarily equal, but having as much contrast as possible — and using these partial sums as well as the sum of the whole series to obtain three equations of the type (13.24). In general the precision of such an estimate is low, and there is no unique solution because the series can be divided in different ways. However, when the data include values of effort and yield near the sustainable maximum of the latter, quite large differences in q make little difference to the estimate of maximum sustainable yield.

13.6. Generalized Production Model — Method of Pella and Tomlinson

Expression (8) of Pella and Tomlinson (1969) is a more general form of expression (13.3), in which the exponent 2 is replaced by a variable quantity m. In our notation this is:

$$Y_E = F\overline{B} = k\overline{B} - \left(\frac{k}{B_\infty}\right)\overline{B}^m \qquad (13.25)$$

(C of Pella and Tomlinson = Y_E here; $P = \overline{B}$; $H = -k/B_\infty$; and $K = -k$). For a steady-state situation, expression (13.25) can be recast into a relation between Y_E and F or f, analogous to (13.15) and (13.16). This is Pella and Tomlinson's expression (9); in our symbols it is:

$$Y_E = F(B_\infty - B_\infty F/k)^{\frac{1}{m-1}} = qf(B_\infty - B_\infty qf/k)^{\frac{1}{m-1}} \qquad (13.26)$$

Like (13.3), the curve described by expression (13.25) begins at the origin ($\overline{B} = 0$, $Y_E = 0$), rises to a maximum, and falls again to the abscissa. The size of the stock when surplus production falls again to zero is $B_\infty^{1/(m-1)}$, found by substituting $Y_E = 0$ in (13.25). Similarly (13.26) rises to a maximum from the origin and then declines to zero when $F = qf = k$.

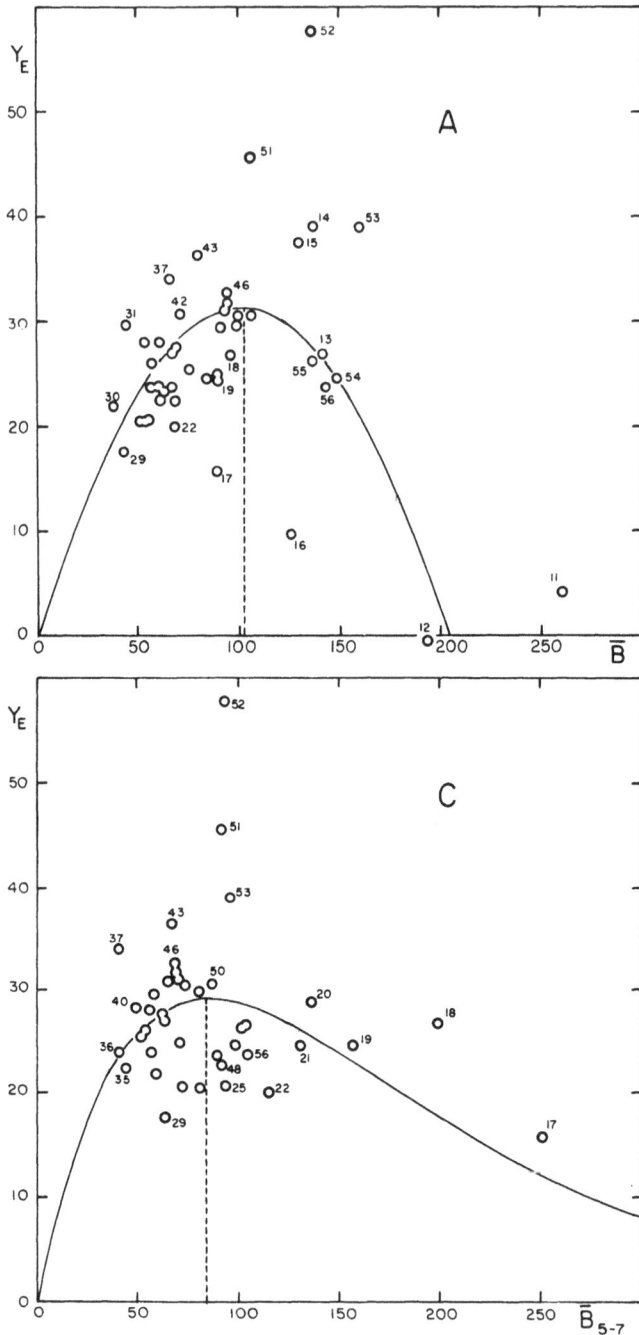

FIG. 13.2. Four approaches to estimating maximum equilibrium yield of Area 2 Pacific halibut. (A) Plot of surplus production (Y_E) against mean population biomass ($\bar{B}$) of the same year (Graham–Schaefer or logistic model, first variant); (B) Plot of Y_E against fishing effort (f) during the same year (Graham–Schaefer model, second variant); (C) Plot of Y_E against $\bar{B}_{5-7}$, which is the mean population biomass 5 to

324

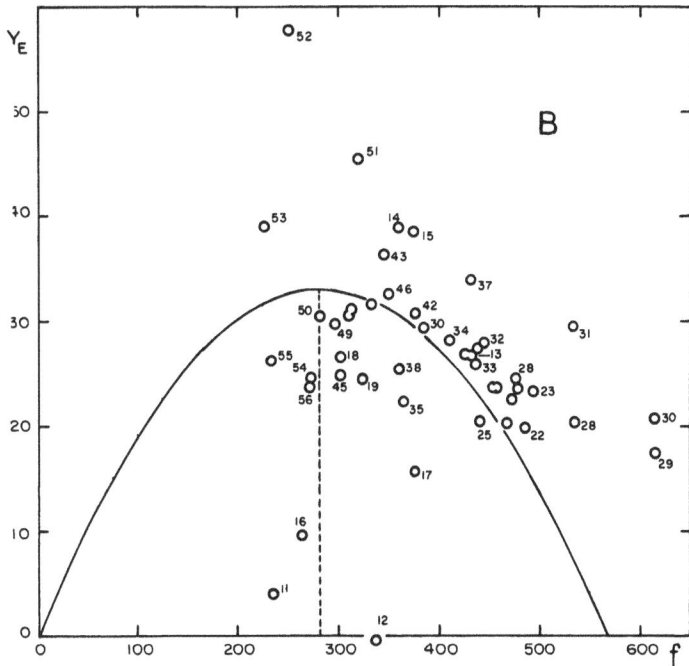

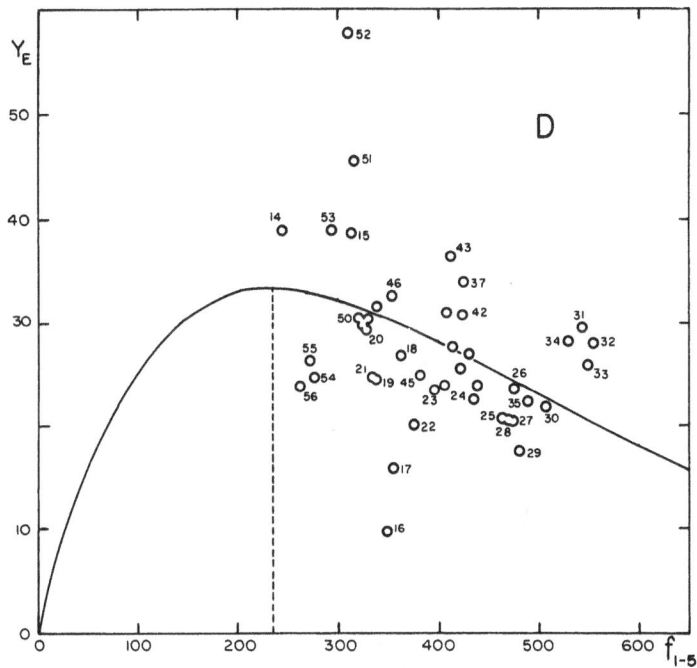

7 years earlier (Ricker or recruitment model); (D) Plot of Y_E in the current year against the mean fishing effort ($\bar{f}_{1-5}$) of the previous 5 years (Gulland–Fox or exponential model). Numbers by certain of the points identify the year in which the Y_E (ordinate value) was observed. (Data from Table 13.1.)

Pella and Tomlinson show curves of Y_E plotted against $\bar{B}$ (their fig. 1), and of Y_E against f (fig. 2). The left limb of (13.25) is steeper than the right limb when $m < 2$, and is less steep when $m > 2$. As increasing $\bar{B}$ corresponds to decreasing f, with f on the abscissa the above skewness is reversed in expression (13.26). Expression (13.25) can be fitted to surplus production estimates made by the method of Section 13.4. It is necessary to use a computer as there are four parameters to be estimated (k, B_∞, q, and m), and numerous iterations are necessary. Program GENPROD is given by Pella and Tomlinson (1969); later they modified it (GENPROD-2) in Abramson (1971). Although good fits to observed data can be obtained, unless constraints are put on the value of m the other parameters are frequently unreasonable. Consequently this model has not as yet proven very useful in practice. In interpreting the results of a GENPROD fit, the same considerations must be faced as were discussed in Section 13.4.3.

13.7. RELATION OF SURPLUS PRODUCTION TO STOCK SIZE OF THE YEARS THAT PRODUCED THE CURRENT RECRUITMENT — RICKER'S METHOD

Ricker (1958a) pointed out that recruitment is a component of surplus production (Y_E) and that it may sometimes be the dominant factor causing changes in Y_E over a period of years. This suggests relating Y_E in a given year to the size of adult stock ($\bar{B}$) that produced the major year-class being recruited that year (or to the mean size of the stocks that produced two or more important recruit classes). A graph of Y_E against this $\bar{B}$ will indicate whether there is a possible relationship, and one of the mathematical recruitment curves can be fitted if it seems warranted.

Figure 13.3 shows the surplus reproduction (adult recruits less spawners) corresponding to the five recruitment curves of Fig. 11.2. These would be directly applicable only to stocks that have a single age of maturity and are harvested when mature, but they indicate the asymmetrical shapes that production curves can attain if they are importantly influenced by recruitment.

To fit a reproduction-type curve to data of this sort the procedures described in Section 11.6 can be used. For example, in expression (11.15) R will be the value of Y_E for the current year, and P will be the mean value of $\bar{B}$ for the years that contributed importantly to current recruitment. The equation thus becomes:

$$\log_e Y_E - \log_e \bar{B} = \alpha + \beta \bar{B} \qquad (13.27)$$

Fitting a regression line provides estimates of α and β, hence also the following:

$Y_s = \alpha/\beta e$ — maximum value of Y_E, which is the maximum level of surplus production and thus the maximum sustainable yield

$B_s = 1/\beta$ — optimum stock size, i.e. that which permits maximum surplus production

$F_s = Y_s/B_s$ — instantaneous rate of fishing at maximum sustainable yield

326

EXAMPLE 13.7. RELATION OF SURPLUS PRODUCTION OF AREA 2 HALIBUT TO STOCK SIZE 5–7 YEARS EARLIER.

From Fig. 2.12 and other data it appears that halibut were being recruited to the fishery from age 3 or 4 to 8 or 9, most of them being added at ages 5 to 7. On this basis each figure for surplus production (Y_E) in Table 13.1 should be related

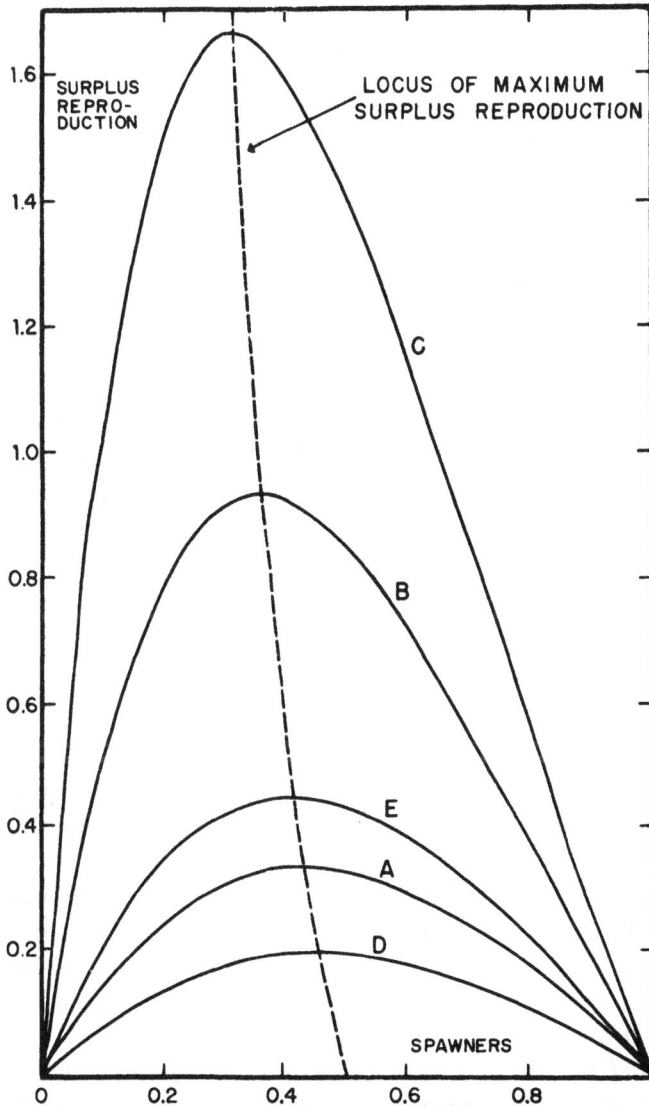

FIG. 13.3. Curves of surplus reproduction (equilibrium reproduction less adult stock, or $R_E - P_E$) plotted against adult stock density (P_E), for the five curves of Fig. 11.2. The unit of both axes is the replacement level of reproduction.

327

to the mean of the stock size 5 to 7 years earlier ($\overline{B}_{5-7}$). Column 12 of Table 13.1 gives the 3-year means of B centered on the middle year, and these are related to surplus production 6 years later in Fig. 13.2C.

The curve (13.27) was fitted; with $\overline{B}_{5-7}$ and Y_E in millions of pounds, the GM regression line has the form:

$$\log_e Y_E - \log_e \overline{B}_{5-7} = -0.06325 - 0.011867 \overline{B}_{5-7}$$

The curve is shown in Fig. 13.2C.

From this expression $\log_e \alpha = -0.06325$, $\alpha = 0.9387$, and $\beta = -0.011867$. From the relations listed in Section 13.7:

Stock size at MSY = $\overline{B}_s$ = $1/0.011867$ = 84.3 million lb

MSY = Y_s = $0.9387/0.011867 \times 2.7183$ = 29.1 million lb

Optimum rate of fishing = F_s = $29.1/84.3$ = 0.345

Using the value $q = 0.907 \times 10^{-6}$ for catchability from Example 13.5, the fishing effort needed to obtain maximum sustained yield is $f_s = F_s/q = 380{,}000$ skates.

The yield values estimated above are geometric means. Variance from the regression line is 0.05593. Applying this in (11.6), the ratio of arithmetic mean values of Y_E to the computed geometric mean values is 1.028. Thus the AM estimate of Y_s becomes 29.9×10^6 million lb, and on this basis F_s is $29.9/84.3 = 0.355$ and $f_s = 391{,}000$ skates.

13.8. RELATION OF CATCH PER UNIT EFFORT TO THE FISHING EFFORT OF YEARS IMMEDIATELY PAST — GULLAND'S METHOD

13.8.1. GENERAL POSTULATES. Gulland (1961) suggested another way of examining the relationship between the present condition of a stock and past events. He pointed out that abundance of an exploited year-class at any time depends partly on the rates of fishing and hence the levels of fishing effort that have prevailed during the years it has been in the fishery. Thus some relationship should exist between abundance and past effort, and this will be the better if recruitment and natural mortality have been reasonably steady. He compared catch per unit effort (Y/f) of a stock consisting of a number of year-classes with the mean of fishing efforts over several immediately preceding years ($\overline{f}$). The ideal number of years to be used depends partly on prevailing survival rates, which in turn are affected by rate of fishing. Gulland used the mean of fishing efforts of the 3 years immediately prior to the current one for Icelandic cod (*Gadus morhua*) and plaice (*Pleuronectes platessa*), based on the fact that about 7 year-classes contribute appreciably to the fishery; for haddock (*Melanogrammus aeglefinus*) a 2-year mean was used. Actually it is usually not crucial to obtain the exact best interval, because fishing efforts in adjacent years tend to be similar.

The relationship obtained between Y/f and the earlier $\bar{f}$ was sometimes approximately straight, sometimes curved. Whether straight or curved, Gulland (p. 9) argues that "this line, obtained from data of varying fishing effort, will be very close to the true relation between catch per unit effort (abundance) and effort in a steady state. Suppose for instance the effort has been increasing in the past few years and the average effort over the past 3 years is $\bar{f}$, then the youngest age-group will have undergone more than $\bar{f}$, and will be less abundant than in the steady state with effort f; the fish which have been in the fishery 3 years will be equally as abundant as in the steady state, while the older fish will be more abundant, having had a mean effort during their life of less than f. The overall abundance . . . will be close to that from a steady effort f."

The relationship between Y/f and $\bar{f}$ for earlier years can be differentiated to give estimates of maximum sustainable yield and optimum rate of fishing, as illustrated below. As in Section 13.3, it is not necessary to know the catchability of the population.

13.8.2. LINEAR RELATIONSHIP. If the graph of Y/f against $\bar{f}$ is linear, and the above argument applies, we may write:

$$\frac{Y_E}{f_E} = a - b\bar{f}_E \qquad (13.28)$$

For example, Gulland (1961) found that the plot of Y/f against $\bar{f}$ for Icelandic cod over many years did not indicate curvature, although there is a suggestion of it for the period 1947–58.

A functional regression line can be fitted to (13.28) and a and b estimated. Since expressions (13.28) and (13.17) are the same, estimates of optimum fishing effort and maximum sustainable yield are as shown in expressions (13.18) and (13.19).

13.8.3. CURVILINEAR RELATIONSHIP (GULLAND–FOX EXPONENTIAL MODEL). Considering the biological processes involved, the relation between catch in numbers per unit effort (C/f) and $\bar{f}$ of earlier years would be expected to be negative and concave upward. The reason is that any increase in the $\bar{f}$ affecting a year-class means a proportional increase in mean rate of fishing $\bar{F}$, and thus (if natural mortality is negligible) a proportional increase in $\bar{Z}$. The abundance of a year-class of a given initial size in any subsequent year is a function of the product ΠS_i of its survival rates S_i during the years it has been in the fishery, whereas the sum of the fishing efforts Σf_i in those years is more or less proportional to the sum of the instantaneous mortality rates ΣZ_i; and $\Sigma Z_i = -\log_e(\Pi S_i)$. Insofar as abundance is represented by catch per unit effort, we might therefore expect the logarithm of C/f in a given year to approximate a negative linear function of the earlier mortalities ΣZ_i and hence efforts Σf_i. This implies the concave relationship between C/f and $\bar{f}$ mentioned above.

The existence of appreciable natural mortality will tend to make the line *less* curved than expected from this argument. However, if yield is in terms of weight (Y) instead of numbers, the larger the mean effort $\bar{f}$, the smaller will be the average size of the fish in the catch, which tends to make the line of $Y/\bar{f}$ against $\bar{f}$ *more* curved than that of C/f against $\bar{f}$. The combination of this latter effect with the effect of natural mortality can result in a line that agrees closely with the simple relationship:

$$\log_e(Y/f) = a + b\bar{f} \tag{13.29}$$

Garrod (1969) suggested fitting such a negative exponential relation to curves obtained by Gulland's method, and Fox (1970) developed the computations in detail. In what follows, it is convenient to use Schaefer's symbol $U_E = Y_E/f_E$ for an equilibrium catch per unit effort, and U_∞ for the value of U_E when the population is at its maximum equilibrium size. Referring (13.29) to equilibrium situations on the basis of Gulland's argument quoted in Section 13.8.1, it becomes:

$$\log_e(Y_E/f_E) = \log_e U_E = a - bf_E \tag{13.30}$$

When $f_E \rightarrow 0$, $a \rightarrow \log_e U_\infty$, which means $a = \log_e U_\infty$. Taking antilogarithms and multiplying by f_E, (13.30) becomes:

$$Y_E = f_E U_\infty e^{-bf_E} \tag{13.31}$$

A functional straight line fitted to (13.30) will estimate the values of b and a, hence U_∞.

Expression (13.31) has the same form as a Ricker reproduction curve (expression 11.9), and similar statistics can be derived from it. The differential of (13.31) with respect to f_E is:

$$(1 - bf_E)U_\infty e^{-bf_E} \tag{13.32}$$

Equating to zero, and since $U_\infty e^{-bf_E}$ cannot $= 0$, (13.31) is a maximum when $bf_E = 1$; thus the effort corresponding to maximum sustainable yield is $f_s = 1/b$. Substituting this in (13.31) gives the maximum sustainable yield:

$$Y_s = \frac{U_\infty}{be} = \frac{e^{a-1}}{b} \tag{13.33}$$

Although it is usually not essential to know stock size and rate of fishing, these can be estimated if an estimate of catchability (q) is available. The best source is from independent information, from tagging for example; but an estimate (usually a rather poor one) can be obtained in a manner similar to Schaefer's procedure described in Section 13.5. Fox (1970, p. 83) gives the necessary formula.

The steady-state statistics that provide maximum yield for this Gulland–Fox exponential model can be summarized as follows:

Optimum fishing effort: $\qquad\qquad f_s = \dfrac{1}{b}$ $\qquad\qquad$ (13.34)

330

Maximum sustainable yield: $Y_s = \dfrac{U_\infty}{be} = \dfrac{e^{a-1}}{b}$ (13.35)

Catch per unit effort at MSY: $U_s = \dfrac{Y_s}{f_s} = e^{a-1}$ (13.36)

If an estimate of catchability (q) is at hand, then also:

Optimum rate of fishing: $F_s = qf_s = \dfrac{q}{b}$ (13.37)

Stock required for MSY: $B_s = \dfrac{Y_s}{F_s} = \dfrac{U_\infty}{qe} = \dfrac{e^{a-1}}{q}$ (13.38)

The following are also available:

Maximum catch per unit effort: $U_\infty = e^a$ (13.39)

Maximum stock size: $B_\infty = \dfrac{B_s U_\infty}{U_s} = \dfrac{e^a}{q}$ (13.40)

A reservation must be made concerning the yields above, similar to that made in Section 11.6.3. In fitting a line to (13.30), the logarithms of catch per unit effort are used, and the position of the line is determined on this basis. Therefore any estimated magnitude of Y_E will tend to be less than if a fit were made on a non-logarithmic basis. Computed values of Y_E can be adjusted upward using expression (11.5) or (11.6), with s equal to the standard deviation of the points from the line of expression (13.30).

This Gulland–Fox model has as rigid a form as the Graham model. The size of stock that produces maximum sustainable yield is always $1/e = 37\%$ of the maximum size (expressions 13.38 and 13.40). Greater flexibility can be achieved by fitting an empirical curve to the observed data and computing the points graphically, as Gulland (1961) did for plaice and haddock.

EXAMPLE 13.8. RELATION OF CATCH PER UNIT EFFORT OF HALIBUT TO THE FISHING EFFORT OF THE 5 YEARS IMMEDIATELY PAST.

Column 11 of Table 13.1 shows the 5-year means of fishing effort. The natural logarithm of Y/f (column 4) is regressed against the mean effort for the previous 5 years, using a functional regression,[5] and expression (13.30) takes the form:

$$\log_e(Y/f) = 5.959 - 4.258 \times 10^{-6} \bar{f}_{1-5}$$

---

[5] The figures shown were obtained from the GM regression. The symmetrical Nair–Bartlett procedure gives a very similar result: $b = 4.273 \times 10^{-6}$ (Appendix IV).

331

Values of Y computed from this expression give the curve of Fig. 13.2D. The points plotted in this figure are, however, Y_E rather than Y, so the curve is not a best fit to the points. A plot of Y against $\tilde{f}_{1-5}$ has much less scatter from the fitted line, but much of this agreement arises from the fact that f in a given year is correlated with $\tilde{f}_{1-5}$.

From (13.34)–(13.36) and (13.39):

Optimum fishing effort: $\quad f_s = 1/4.258 \times 10^{-6} = 235{,}000$ skates

Maximum sustainable yield: $\quad Y_s = e^{4.959}/4.258 \times 10^{-6} = 33.4 \times 10^6$ lb

Catch per unit effort at MSY $\quad U_s = e^{4.959} = 142$ lb per skate

Maximum catch per unit effort $\quad U_\infty = e^{5.959} = 387$ lb per skate

Given catchability $= q = 0.907 \times 10^{-6}$ (from Example 13.5), then from (13.37), (13.38), and (13.40):

Optimum rate of fishing: $\quad F_s = 0.907 \times 10^{-6}/4.258 \times 10^{-6} = 0.213$

Optimum stock size: $\quad B_s = e^{4.959}/0.907 \times 10^{-6} = 157 \times 10^6$ lb

Maximum stock size: $\quad B_\infty = e^{5.959}/0.907 \times 10^{-6} = 426 \times 10^6$ lb

The variance of the points from the ordinary regression of $\log_e(Y/f)$ on f is 0.02426. Using this for s^2 in expression (11.6), the factor by which Y/f is to be multiplied to obtain arithmetic mean values is 1.012. Thus the arithmetic mean value for catch per unit effort at MSY becomes $U_s = 144$ lb/skate, and the absolute yield becomes 33.8×10^6 lb. The difference is evidently inconsequential in this example.

EXAMPLE 13.9. INTERPRETATION OF BIOLOGICAL STATISTICS ESTIMATED FOR AREA 2 HALIBUT.

Table 13.2 compares four different sets of estimates computed for the Pacific halibut stock. The two Graham–Schaefer estimates, A and B, are based on the same basic theory but two different regression lines. Estimate B suggests a somewhat larger maximum sustainable yield than A, which would be taken from a larger stock and with considerably greater effort. The Ricker estimate C predicts an MSY close to A, but it would be taken from a much smaller stock with much greater effort, so that Y/f falls to 77 lb/skate. The Gulland–Fox estimate D predicts a yield close to B, but taken from a larger stock and with less effort, so that Y/f is 142 lb/skate. However, estimates C and D are not completely comparable with A and B or with each other, as different numbers of years were available for the regressions.

Because halibut can live to a great age, we should expect a considerable lag in adjustment of population size and age structure to the current rate of fishing (this is an aspect of the fishing-up effect of Section 10.9). Another consideration is that fishing effort was increasing fairly regularly from before 1910 up to 1929; hence there would be a strong correlation between one year's effort and the effort in any year a fixed number of years earlier, throughout this period. Since population size was decreasing steadily, the same kind of correlations exist with it. From 1930 to 1953, on the other

332

TABLE 13.2. Comparison of four estimates of vital statistics for the Area 2 halibut population. The first Schaefer estimate is from the regression on $\bar{B}$; the second is from the regression on fishing effort (f). Estimates marked with an asterisk require an estimate of catchability (q) from an outside source.

Statistic	Symbol	Unit	Schaefer ($\bar{B}$)	Schaefer (f)	Ricker	Gulland–Fox
Curve of Fig. 13.2	....	....	A	B	C	D
Maximum sustainable yield	Y_s	lb × 10⁻⁶	31.2	33.0	29.9	33.4
Rate of fishing at MSY	F_s	....	0.306	0.258*	0.355	0.213*
Fishing effort at MSY	f_s	skates × 10⁻³	337	284	391*	235
Stock size at MSY	B_s	lb × 10⁻⁶	102	128*	84	157*
Catch per unit effort at MSY	U_s	lb/skate	93	116	77	142
Maximum stock size	B_∞	lb × 10⁻⁶	204	256*	....	426*
Maximum catch per unit effort	U_∞	lb/skate	....	....	....	387

hand, effort decreased and stock increased, so the correlations disappear for awhile, but develop again as this period advances. Thus if we had only the period of increasing (or decreasing) fishing, we would scarcely be justified in drawing any conclusion whatever concerning maximum sustainable yield using the Graham–Schaefer method (Examples 13.5 and 13.6; A and B in Table 13.2).

Table 13.1 and Fig. 13.2A show that during the phase of decreasing population (1910–29) the surplus produced at any given level of stock was usually substantially less than during the phase of increasing population (1930–52). This can be a result of an older age structure during the earlier period, as large old individuals persisted in small numbers (but appreciable weight) well into the era of intensive exploitation. Notice also that during the most recent years shown, 1953–56, mean Y_E/B had fallen close to its mean in 1914–16 (when B was about the same), which may reflect the fact that older fish had again been able to accumulate in quantity after fishing effort began its decline in 1931.

A possible effect of recruitment on surplus production is indicated in the C set of estimates in Table 13.2 and Fig. 13.2C. The earliest year shown, 1917, suggests a trend toward small surpluses from large stocks. This trend is supported by the surpluses produced in 1911–16 which average 20 million lb (column 8 of Table 13.1) and consist mainly of the progeny of the year-classes 1900–10, a time when stock size was certainly much greater than 250 million lb. However, the line fitted to Fig. 13.2C evidently falls off more rapidly at large stocks than is likely on the evidence available.

Finally, there is an undeniable effect of fishing on stock abundance. This is evaluated by the Gulland method in panel D of Fig. 13.2. Although fitted with curves of similar shape, the arrangement of the points in panels C and D is very different, and they have a different logical basis. Presumably both level of recruitment and amount of antecedent fishing can have contributed to the actual level of maximum sustainable yield, and the best estimates from Table 13.2 may be the means of the last two columns.

333

APPENDIX I. TABLE OF EXPONENTIAL FUNCTIONS AND DERIVATIVES

Functions frequently needed in the calculations of this Handbook are given in terms of instantaneous mortality rate, Z, annual mortality rate, A, and annual survival rate, S.

$$(1 - A) = S = e^{-Z}$$

Note that column 5, headed A/ZS, is also equal to $(e^Z - 1)/Z$.

In general, the accuracy of the figures tabulated does not exceed about 0.05%, so that the last figure is not always wholly reliable.

The short schedule below gives a few values of Z for even values of A and S.

S	A	Z	S	A	Z
1.00	0.00	0.0000	0.50	0.50	0.6931
0.95	0.05	0.0513	0.45	0.55	0.7985
0.90	0.10	0.1054	0.40	0.60	0.9163
0.85	0.15	0.1625	0.35	0.65	1.0498
0.80	0.20	0.2231	0.30	0.70	1.2040
0.75	0.25	0.2877	0.25	0.75	1.3863
0.70	0.30	0.3567	0.20	0.80	1.6094
0.65	0.35	0.4308	0.15	0.85	1.8971
0.60	0.40	0.5108	0.10	0.90	2.3026
0.55	0.45	0.5978	0.05	0.95	2.9957

335

1	2	3	4	5	6	7	8	9	10	11	12	13
Z	S	A	A/Z	A/ZS	A^2	Z^2	A^2/Z^2	Z–A	$A^2/(Z–A)$	$Z^2/(Z–A)$	e^Z	Z
.01	.9900	.0100	.9950	1.005	.0001	.0001	.9900	.0000	1.988	2.007	1.0101	.01
.02	.9802	.0198	.9901	1.010	.0004	.0004	.9800	.0002	1.973	2.013	1.0202	.02
.03	.9704	.0296	.9851	1.015	.0009	.0009	.9704	.0004	1.961	2.020	1.0305	.03
.04	.9608	.0392	.9803	1.020	.0015	.0016	.9609	.0008	1.948	2.027	1.0408	.04
.05	.9512	.0488	.9754	1.025	.0024	.0025	.9514	.0012	1.935	2.034	1.0513	.05
.06	.9418	.0582	.9706	1.031	.0034	.0036	.9417	.0018	1.922	2.041	1.0618	.06
.07	.9324	.0676	.9658	1.036	.0046	.0049	.9327	.0024	1.909	2.047	1.0725	.07
.08	.9231	.0769	.9610	1.041	.0059	.0064	.9234	.0031	1.896	2.053	1.0833	.08
.09	.9139	.0861	.9563	1.046	.0074	.0081	.9148	.0039	1.844	2.060	1.0942	.09
.10	.9048	.0952	.9516	1.052	.0090	.0100	.9050	.0048	1.871	2.067	1.1052	.10
.11	.8958	.1042	.9470	1.057	.0108	.0121	.8967	.0058	1.860	2.074	1.1163	.11
.12	.8869	.1131	.9423	1.062	.0128	.0144	.8880	.0069	1.848	2.081	1.1275	.12
.13	.8781	.1219	.9377	1.068	.0149	.0169	.8792	.0081	1.835	2.087	1.1388	.13
.14	.8694	.1306	.9332	1.073	.0171	.0196	.8708	.0094	1.824	2.094	1.1503	.14
.15	.8607	.1393	.9286	1.079	.0194	.0225	.8622	.0107	1.812	2.101	1.1618	.15
.16	.8521	.1479	.9241	1.084	.0219	.0256	.8540	.0121	1.800	2.108	1.1735	.16
.17	.8437	.1563	.9196	1.090	.0244	.0289	.8457	.0137	1.789	2.115	1.1853	.17
.18	.8353	.1647	.9152	1.096	.0271	.0324	.8372	.0153	1.776	2.122	1.1972	.18
.19	.8270	.1730	.9107	1.101	.0300	.0361	.8294	.0170	1.765	2.129	1.2092	.19
.20	.8187	.1813	.9063	1.107	.0329	.0400	.8215	.0187	1.754	2.136	1.2214	.20
.21	.8106	.1894	.9020	1.113	.0359	.0441	.8136	.0206	1.743	2.142	1.2337	.21
.22	.8025	.1975	.8976	1.119	.0390	.0484	.8057	.0225	1.732	2.149	1.2461	.22
.23	.7945	.2055	.8933	1.124	.0422	.0529	.7980	.0245	1.721	2.157	1.2586	.23
.24	.7866	.2134	.8890	1.130	.0455	.0576	.7904	.0266	1.710	2.163	1.2712	.24
.25	.7788	.2212	.8848	1.136	.0489	.0625	.7829	.0288	1.699	2.170	1.2840	.25
.26	.7711	.2289	.8806	1.142	.0524	.0676	.7754	.0310	1.688	2.177	1.2969	.26
.27	.7634	.2366	.8764	1.148	.0560	.0729	.7680	.0334	1.677	2.184	1.3100	.27
.28	.7558	.2442	.8722	1.154	.0596	.0784	.7607	.0358	1.667	2.191	1.3231	.28
.29	.7483	.2517	.8679	1.160	.0633	.0841	.7535	.0383	1.656	2.198	1.3364	.29
.30	.7408	.2592	.8639	1.166	.0672	.0900	.7464	.0408	1.646	2.205	1.3499	.30
.31	.7334	.2666	.8598	1.172	.0710	.0961	.7393	.0434	1.635	2.212	1.3634	.31
.32	.7261	.2739	.8558	1.179	.0750	.1024	.7324	.0462	1.625	2.219	1.3771	.32
.33	.7189	.2811	.8517	1.185	.0790	.1089	.7255	.0489	1.614	2.226	1.3910	.33
.34	.7118	.2882	.8477	1.191	.0831	.1156	.7187	.0518	1.604	2.233	1.4049	.34
.35	.7047	.2953	.8437	1.197	.0872	.1225	.7119	.0547	1.595	2.240	1.4191	.35
.36	.6977	.3023	.8398	1.204	.0914	.1296	.7052	.0577	1.585	2.247	1.4333	.36
.37	.6907	.3093	.8359	1.210	.0956	.1369	.6986	.0607	1.575	2.254	1.4477	.37
.38	.6839	.3161	.8319	1.216	.0999	.1444	.6921	.0639	1.565	2.261	1.4623	.38
.39	.6771	.3229	.8281	1.223	.1043	.1521	.6857	.0671	1.555	2.268	1.4770	.39
.40	.6703	.3297	.8242	1.230	.1087	.1600	.6793	.0703	1.546	2.275	1.4918	.40
.41	.6637	.3364	.8204	1.236	.1131	.1681	.6730	.0736	1.536	2.282	1.5068	.41
.42	.6570	.3430	.8166	1.243	.1176	.1764	.6667	.0770	1.526	2.289	1.5220	.42
.43	.6505	.3495	.8128	1.250	.1221	.1849	.6606	.0805	1.517	2.297	1.5373	.43
.44	.6440	.3560	.8090	1.256	.1267	.1936	.6545	.0840	1.508	2.305	1.5527	.44
.45	.6376	.3623	.8053	1.263	.1313	.2025	.6485	.0876	1.498	2.311	1.5683	.45

336

1	2	3	4	5	6	7	8	9	10	11	12	13
Z	S	A	A/Z	A/ZS	A^2	Z^2	A^2/Z^2	$Z-A$	$A^2/(Z-A)$	$Z^2/(Z-A)$	e^Z	Z
.46	.6313	.3687	.8016	1.270	.1360	.2116	.6425	.0913	1.489	2.318	1.5841	.46
.47	.6250	.3750	.7979	1.277	.1406	.2209	.6366	.0950	1.480	2.325	1.6000	.47
.48	.6188	.3812	.7942	1.283	.1453	.2304	.6308	.0988	1.471	2.332	1.6161	.48
.49	.6126	.3874	.7906	1.291	.1501	.2401	.6250	.1026	1.463	2.340	1.6323	.49
.50	.6065	.3935	.7869	1.297	.1548	.2500	.6194	.1065	1.454	2.347	1.6487	.50
.51	.6005	.3995	.7833	1.304	.1596	.2601	.6136	.1105	1.444	2.354	1.6653	.51
.52	.5945	.4055	.7798	1.312	.1644	.2704	.6081	.1145	1.436	2.362	1.6820	.52
.53	.5886	.4114	.7762	1.319	.1692	.2809	.6025	.1186	1.427	2.368	1.6989	.53
.54	.5827	.4173	.7727	1.326	.1741	.2916	.5971	.1227	1.419	2.377	1.7160	.54
.55	.5769	.4230	.7691	1.333	.1789	.3025	.5915	.1270	1.409	2.382	1.7333	.55
.56	.5712	.4288	.7657	1.341	.1839	.3136	.5864	.1312	1.401	2.390	1.7507	.56
.57	.5655	.4345	.7622	1.348	.1888	.3249	.5811	.1355	1.393	2.398	1.7683	.57
.58	.5599	.4401	.7588	1.355	.1937	.3364	.5758	.1399	1.384	2.405	1.7860	.58
.59	.5543	.4457	.7554	1.363	.1986	.3481	.5707	.1443	1.377	2.412	1.8040	.59
.60	.5488	.4512	.7520	1.370	.2036	.3600	.5655	.1488	1.368	2.419	1.8221	.60
.61	.5434	.4566	.7486	1.378	.2085	.3721	.5603	.1534	1.359	2.426	1.8404	.61
.62	.5379	.4621	.7453	1.386	.2135	.3844	.5555	.1579	1.352	2.434	1.8589	.62
.63	.5326	.4674	.7419	1.393	.2185	.3969	.5504	.1626	1.344	2.441	1.8776	.63
.64	.5273	.4727	.7386	1.401	.2234	.4096	.5455	.1673	1.336	2.448	1.8965	.64
.65	.5220	.4780	.7353	1.409	.2285	.4225	.5408	.1720	1.328	2.456	1.9155	.65
.66	.5169	.4831	.7320	1.416	.2334	.4356	.5358	.1769	1.319	2.462	1.9348	.66
.67	.5117	.4883	.7288	1.424	.2384	.4489	.5312	.1817	1.312	2.471	1.9542	.67
.68	.5066	.4934	.7256	1.432	.2434	.4624	.5265	.1866	1.303	2.478	1.9739	.68
.69	.5016	.4984	.7224	1.440	.2484	.4761	.5217	.1916	1.296	2.485	1.9937	.69
.70	.4966	.5034	.7191	1.448	.2534	.4900	.5177	.1966	1.289	2.492	2.0138	.70
.71	.4916	.5084	.7160	1.456	.2585	.5041	.5127	.2016	1.282	2.500	2.0340	.71
.72	.4868	.5132	.7128	1.464	.2634	.5184	.5081	.2068	1.274	2.507	2.0544	.72
.73	.4819	.5181	.7097	1.473	.2684	.5329	.5037	.2119	1.267	2.515	2.0751	.73
.74	.4771	.5229	.7066	1.481	.2734	.5476	.4993	.2171	1.259	2.522	2.0959	.74
.75	.4724	.5276	.7035	1.489	.2784	.5625	.4949	.2224	1.252	2.530	2.1170	.75
.76	.4677	.5323	.7004	1.498	.2833	.5776	.4905	.2277	1.244	2.537	2.1383	.76
.77	.4630	.5370	.6974	1.506	.2884	.5929	.4864	.2330	1.238	2.545	2.1598	.77
.78	.4584	.5416	.6944	1.515	.2933	.6084	.4821	.2384	1.230	2.552	2.1815	.78
.79	.4538	.5462	.6913	1.523	.2983	.6241	.4780	.2438	1.224	2.560	2.2034	.79
.80	.4493	.5507	.6883	1.532	.3033	.6400	.4739	.2493	1.216	2.567	2.2255	.80
.81	.4449	.5551	.6854	1.541	.3081	.6561	.4698	.2548	1.209	2.575	2.2479	.81
.82	.4404	.5596	.6824	1.550	.3132	.6724	.4658	.2604	1.203	2.582	2.2705	.82
.83	.4360	.5640	.6795	1.558	.3181	.6889	.4618	.2660	1.196	2.590	2.2933	.83
.84	.4317	.5683	.6765	1.567	.3230	.7056	.4578	.2717	1.189	2.597	2.3164	.84
.85	.4274	.5726	.6736	1.576	.3279	.7225	.4538	.2774	1.182	2.605	2.3396	.85
.86	.4232	.5768	.6707	1.585	.3327	.7396	.4498	.2832	1.175	2.612	2.3632	.86
.87	.4190	.5810	.6679	1.594	.3376	.7569	.4460	.2890	1.168	2.619	2.3869	.87
.88	.4148	.5853	.6651	1.603	.3426	.7744	.4421	.2948	1.162	2.627	2.4109	.88
.89	.4107	.5893	.6622	1.612	.3472	.7921	.4383	.3007	1.155	2.634	2.4351	.89
.90	.4066	.5934	.6594	1.622	.3521	.8100	.4347	.3066	1.148	2.642	2.4596	.90

337

1	2	3	4	5	6	7	8	9	10	11	12	13
Z	S	A	A/Z	A/ZS	A^2	Z^2	A^2/Z^2	Z–A	$A^2/(Z–A)$	$Z^2/(Z–A)$	e^Z	Z
.91	.4025	.5975	.6566	1.631	.3570	.8281	.4311	.3125	1.142	2.650	2.4843	.91
.92	.3985	.6015	.6538	1.641	.3618	.8464	.4275	.3185	1.136	2.657	2.5093	.92
.93	.3946	.6054	.6510	1.650	.3665	.8649	.4239	.3246	1.129	2.665	2.5345	.93
.94	.3906	.6094	.6483	1.660	.3714	.8836	.4203	.3306	1.123	2.673	2.5600	.94
.95	.3867	.6133	.6455	1.669	.3761	.9025	.4167	.3367	1.117	2.680	2.5857	.95
.96	.3829	.6172	.6429	1.679	.3809	.9216	.4133	.3429	1.111	2.688	2.6117	.96
.97	.3791	.6209	.6401	1.688	.3855	.9409	.4097	.3491	1.104	2.695	2.6379	.97
.98	.3753	.6247	.6374	1.698	.3903	.9604	.4064	.3553	1.098	2.703	2.6645	.98
.99	.3716	.6284	.6348	1.708	.3949	.9801	.4029	.3616	1.092	2.710	2.6912	.99
1.00	.3679	.6321	.6321	1.718	.3996	1.0000	.3996	.3679	1.086	2.718	2.7183	1.00
1.01	.3642	.6358	.6295	1.728	.4042	1.0201	.3962	.3742	1.080	2.726	2.7456	1.01
1.02	.3606	.6394	.6269	1.738	.4088	1.0404	.3929	.3806	1.074	2.734	2.7732	1.02
1.03	.3570	.6430	.6243	1.749	.4134	1.0609	.3897	.3870	1.068	2.741	2.8011	1.03
1.04	.3535	.6465	.6217	1.759	.4180	1.0816	.3865	.3935	1.062	2.749	2.8292	1.04
1.05	.3499	.6501	.6191	1.769	.4226	1.1025	.3833	.3999	1.057	2.757	2.8577	1.05
1.06	.3465	.6535	.6166	1.780	.4271	1.1236	.3801	.4065	1.051	2.764	2.8864	1.06
1.07	.3430	.6570	.6140	1.790	.4316	1.1449	.3770	.4130	1.045	2.772	2.9154	1.07
1.08	.3396	.6604	.6115	1.801	.4361	1.1664	.3739	.4196	1.039	2.780	2.9447	1.08
1.09	.3362	.6638	.6090	1.811	.4406	1.1881	.3708	.4262	1.034	2.788	2.9743	1.09
1.10	.3329	.6671	.6065	1.822	.4450	1.2100	.3678	.4329	1.028	2.795	3.0042	1.10
1.11	.3296	.6704	.6040	1.833	.4494	1.2321	.3647	.4396	1.022	2.803	3.0344	1.11
1.12	.3263	.6737	.6015	1.843	.4539	1.2544	.3618	.4463	1.017	2.811	3.0649	1.12
1.13	.3230	.6770	.5991	1.855	.4583	1.2769	.3589	.4530	1.012	2.819	3.0957	1.13
1.14	.3198	.6802	.5966	1.866	.4627	1.2996	.3560	.4598	1.006	2.826	3.1268	1.14
1.15	.3166	.6834	.5942	1.877	.4670	1.3225	.3531	.4666	1.001	2.834	3.1582	1.15
1.16	.3135	.6865	.5918	1.888	.4713	1.3456	.3503	.4735	0.995	2.842	3.1899	1.16
1.17	.3104	.6896	.5894	1.899	.4755	1.3689	.3474	.4804	0.990	2.850	3.2220	1.17
1.18	.3073	.6927	.5870	1.910	.4798	1.3924	.3446	.4873	0.985	2.857	3.2544	1.18
1.19	.3042	.6958	.5847	1.922	.4841	1.4161	.3419	.4942	0.980	2.865	3.2871	1.19
1.20	.3012	.6988	.5823	1.933	.4883	1.4400	.3391	.5012	0.974	2.873	3.3201	1.20
1.21	.2982	.7018	.5800	1.945	.4925	1.4641	.3364	.5082	0.969	2.881	3.3535	1.21
1.22	.2952	.7048	.5777	1.957	.4967	1.4884	.3337	.5152	0.964	2.889	3.3872	1.22
1.23	.2923	.7077	.5754	1.969	.5008	1.5129	.3310	.5223	0.959	2.897	3.4212	1.23
1.24	.2894	.7106	.5731	1.980	.5050	1.5376	.3284	.5294	0.954	2.904	3.4556	1.24
1.25	.2865	.7135	.5708	1.992	.5091	1.5625	.3258	.5365	0.949	2.912	3.4903	1.25
1.26	.2837	.7163	.5685	2.005	.5131	1.5876	.3232	.5436	0.944	2.921	3.5254	1.26
1.27	.2808	.7192	.5663	2.017	.5172	1.6129	.3207	.5508	0.939	2.928	3.5609	1.27
1.28	.2780	.7220	.5640	2.029	.5213	1.6384	.3182	.5580	0.934	2.936	3.5966	1.28
1.29	.2753	.7247	.5618	2.041	.5252	1.6641	.3156	.5653	0.929	2.944	3.6328	1.29
1.30	.2725	.7275	.5596	2.054	.5293	1.6900	.3132	.5725	0.924	2.952	3.6693	1.30
1.31	.2698	.7302	.5574	2.066	.5332	1.7161	.3107	.5798	0.920	2.960	3.7062	1.31
1.32	.2671	.7329	.5552	2.079	.5371	1.7424	.3083	.5871	0.915	2.968	3.7434	1.32
1.33	.2645	.7355	.5530	2.091	.5410	1.7689	.3058	.5945	0.910	2.975	3.7810	1.33
1.34	.2618	.7382	.5509	2.104	.5449	1.7956	.3034	.6018	0.906	2.984	3.8190	1.34
1.35	.2592	.7408	.5487	2.117	.5488	1.8225	.3011	.6092	0.901	2.992	3.8574	1.35

338

1	2	3	4	5	6	7	8	9	10	11	12	13
Z	S	A	A/Z	A/ZS	A^2	Z^2	A^2/Z^2	Z–A	$A^2/(Z{-}A)$	$Z^2/(Z{-}A)$	e^Z	Z
1.36	.2567	.7433	.5466	2.129	.5525	1.8496	.2987	.6167	0.896	2.999	3.8962	1.36
1.37	.2541	.7459	.5444	2.142	.5564	1.8769	.2964	.6241	0.891	3.007	3.9354	1.37
1.38	.2516	.7484	.5423	2.155	.5601	1.9044	.2941	.6316	0.887	3.015	3.9749	1.38
1.39	.2491	.7509	.5402	2.169	.5639	1.9321	.2919	.6391	0.882	3.023	4.0149	1.39
1.40	.2466	.7534	.5381	2.182	.5676	1.9600	.2896	.6466	0.878	3.031	4.0552	1.40
1.41	.2441	.7558	.5361	2.196	.5712	1.9881	.2873	.6541	0.873	3.039	4.0960	1.41
1.42	.2417	.7582	.5340	2.209	.5749	2.0164	.2851	.6617	0.869	3.047	4.1371	1.42
1.43	.2393	.7607	.5319	2.223	.5787	2.0449	.2830	.6693	0.864	3.055	4.1787	1.43
1.44	.2369	.7631	.5299	2.237	.5823	2.0736	.2808	.6769	0.860	3.063	4.2207	1.44
1.45	.2346	.7654	.5279	2.250	.5858	2.1025	.2786	.6846	0.856	3.071	4.2631	1.45
1.46	.2322	.7678	.5259	2.265	.5895	2.1316	.2766	.6922	0.852	3.079	4.3060	1.46
1.47	.2299	.7701	.5239	2.279	.5931	2.1609	.2745	.6999	0.847	3.087	4.3492	1.47
1.48	.2276	.7724	.5219	2.293	.5966	2.1904	.2724	.7076	0.843	3.096	4.3929	1.48
1.49	.2254	.7746	.5199	2.306	.6000	2.2201	.2703	.7154	0.839	3.103	4.4371	1.49
1.50	.2231	.7769	.5179	2.321	.6036	2.2500	.2683	.7231	0.835	3.112	4.4817	1.50
1.51	.2209	.7791	.5160	2.336	.6070	2.2801	.2662	.7309	0.830	3.120	4.5267	1.51
1.52	.2187	.7813	.5140	2.350	.6104	2.3104	.2642	.7387	0.826	3.128	4.5722	1.52
1.53	.2165	.7835	.5121	2.365	.6139	2.3409	.2622	.7465	0.822	3.136	4.6182	1.53
1.54	.2144	.7856	.5101	2.379	.6172	2.3716	.2602	.7544	0.818	3.144	4.6646	1.54
1.55	.2122	.7878	.5082	2.395	.6206	2.4025	.2583	.7622	0.814	3.152	4.7115	1.55
1.56	.2101	.7899	.5063	2.410	.6239	2.4336	.2564	.7701	0.810	3.160	4.7588	1.56
1.57	.2080	.7920	.5044	2.425	.6273	2.4649	.2545	.7780	0.806	3.168	4.8066	1.57
1.58	.2060	.7941	.5026	2.440	.6306	2.4964	.2526	.7860	0.802	3.176	4.8550	1.58
1.59	.2039	.7961	.5007	2.456	.6338	2.5281	.2507	.7939	0.798	3.184	4.9037	1.59
1.60	.2019	.7981	.4988	2.470	.6370	2.5600	.2488	.8019	0.794	3.192	4.9530	1.60
1.61	.1999	.8001	.4970	2.486	.6402	2.5921	.2470	.8099	0.790	3.201	5.0028	1.61
1.62	.1979	.8021	.4951	2.502	.6434	2.6244	.2452	.8179	0.787	3.209	5.0531	1.62
1.63	.1959	.8041	.4933	2.518	.6466	2.6569	.2434	.8259	0.783	3.217	5.1039	1.63
1.64	.1940	.8060	.4915	2.534	.6496	2.6896	.2415	.8340	0.779	3.225	5.1552	1.64
1.65	.1920	.8080	.4897	2.549	.6529	2.7225	.2398	.8421	0.775	3.233	5.2070	1.65
1.66	.1901	.8099	.4879	2.566	.6559	2.7556	.2380	.8501	0.772	3.242	5.2593	1.66
1.67	.1882	.8118	.4861	2.583	.6590	2.7889	.2363	.8582	0.768	3.250	5.3122	1.67
1.68	.1864	.8136	.4843	2.598	.6619	2.8224	.2345	.8664	0.764	3.258	5.3656	1.68
1.69	.1845	.8155	.4825	2.615	.6650	2.8561	.2328	.8745	0.760	3.266	5.4195	1.69
1.70	.1827	.8173	.4808	2.632	.6680	2.8900	.2311	.8827	0.757	3.274	5.4739	1.70
1.71	.1809	.8191	.4790	2.648	.6709	2.9241	.2394	.8909	0.753	3.282	5.5290	1.71
1.72	.1791	.8209	.4773	2.665	.6739	2.9584	.2278	.8991	0.750	3.290	5.5845	1.72
1.73	.1773	.8227	.4756	2.682	.6768	2.9929	.2261	.9073	0.746	3.299	5.6407	1.73
1.74	.1755	.8245	.4738	2.700	.6798	3.0276	.2245	.9155	0.743	3.307	5.6973	1.74
1.75	.1738	.8262	.4721	2.716	.6826	3.0625	.2229	.9238	0.739	3.315	5.7546	1.75
1.76	.1720	.8280	.4704	2.735	.6856	3.0976	.2213	.9320	0.736	3.324	5.8124	1.76
1.77	.1703	.8297	.4687	2.752	.6884	3.1329	.2197	.9403	0.732	3.332	5.8709	1.77
1.78	.1686	.8314	.4670	2.770	.6912	3.1684	.2182	.9486	0.729	3.340	5.9299	1.78
1.79	.1670	.8330	.4654	2.787	.6939	3.2041	.2166	.9570	0.725	3.348	5.9895	1.79
1.80	.1653	.8347	.4637	2.805	.6967	3.2400	.2150	.9653	0.722	3.356	6.0496	1.80

1	2	3	4	5	6	7	8	9	10	11	12	13
Z	S	A	A/Z	A/ZS	A^2	Z^2	A^2/Z^2	$Z{-}A$	$A^2/(Z{-}A)$	$Z^2/(Z{-}A)$	eZ	Z
1.81	.1637	.8363	.4621	2.825	.6994	3.2761	.2135	.9736	0.718	3.364	6.1104	1.81
1.82	.1620	.8380	.4604	2.842	.7022	3.3124	.2120	.9820	0.715	3.373	6.1719	1.82
1.83	.1604	.8396	.4588	2.860	.7049	3.3489	.2105	.9904	0.712	3.381	6.2339	1.83
1.84	.1588	.8412	.4572	2.879	.7076	3.3856	.2090	.9988	0.708	3.390	6.2965	1.84
1.85	.1572	.8428	.4555	2.898	.7103	3.4225	.2075	1.0072	0.705	3.398	6.3598	1.85
1.86	.1557	.8443	.4539	2.915	.7128	3.4596	.2060	1.0157	0.702	3.406	6.4237	1.86
1.87	.1541	.8459	.4523	2.935	.7155	3.4969	.2046	1.0241	0.699	3.415	6.4883	1.87
1.88	.1526	.8474	.4508	2.954	.7181	3.5344	.2032	1.0326	0.695	3.423	6.5535	1.88
1.89	.1511	.8489	.4492	2.973	.7206	3.5721	.2017	1.0411	0.692	3.431	6.6194	1.89
1.90	.1496	.8504	.4476	2.992	.7232	3.6100	.2003	1.0496	0.689	3.439	6.6859	1.90
1.91	.1481	.8519	.4460	3.011	.7257	3.6481	.1989	1.0581	0.686	3.448	6.7531	1.91
1.92	.1466	.8534	.4445	3.032	.7283	3.6864	.1976	1.0666	0.683	3.456	6.8210	1.92
1.93	.1451	.8549	.4429	3.052	.7309	3.7249	.1962	1.0751	0.680	3.465	6.8895	1.93
1.94	.1437	.8563	.4414	3.072	.7332	3.7636	.1948	1.0837	0.677	3.473	6.9588	1.94
1.95	.1423	.8577	.4398	3.091	.7356	3.8025	.1935	1.0923	0.673	3.481	7.0287	1.95
1.96	.1409	.8591	.4383	3.111	.7381	3.8416	.1921	1.1009	0.670	3.490	7.0993	1.96
1.97	.1395	.8605	.4368	3.131	.7405	3.8809	.1908	1.1095	0.667	3.498	7.1707	1.97
1.98	.1381	.8619	.4353	3.152	.7429	3.9204	.1895	1.1181	0.664	3.506	7.2427	1.98
1.99	.1367	.8633	.4338	3.173	.7453	3.9601	.1882	1.1267	0.661	3.515	7.3155	1.99
2.00	.1353	.8647	.4323	3.195	.7477	4.0000	.1869	1.1353	0.659	3.523	7.3891	2.00
2.01	.1340	.8660	.4308	3.216	.7500	4.0401	.1856	1.1440	0.656	3.532	7.4633	2.01
2.02	.1327	.8673	.4294	3.237	.7523	4.0804	.1844	1.1527	0.653	3.540	7.5383	2.02
2.03	.1313	.8687	.4279	3.258	.7546	4.1209	.1831	1.1613	0.650	3.548	7.6141	2.03
2.04	.1300	.8700	.4265	3.280	.7568	4.1616	.1819	1.1700	0.647	3.557	7.6906	2.04
2.05	.1287	.8713	.4250	3.301	.7591	4.2025	.1806	1.1787	0.644	3.565	7.7679	2.05
2.06	.1275	.8725	.4236	3.323	.7613	4.2436	.1794	1.1874	0.641	3.573	7.8460	2.06
2.07	.1262	.8738	.4221	3.345	.7636	4.2849	.1782	1.1962	0.638	3.582	7.9248	2.07
2.08	.1249	.8751	.4207	3.368	.7657	4.3264	.1770	1.2049	0.635	3.591	8.0045	2.08
2.09	.1237	.8763	.4193	3.390	.7679	4.3681	.1758	1.2137	0.632	3.600	8.0849	2.09
2.10	.1225	.8775	.4179	3.412	.7701	4.4100	.1746	1.2225	0.630	3.607	8.1662	2.10
2.11	.1212	.8788	.4165	3.435	.7722	4.4521	.1734	1.2312	0.627	3.616	8.2482	2.11
2.12	.1200	.8800	.4151	3.458	.7743	4.4944	.1723	1.2400	0.624	3.624	8.3311	2.12
2.13	.1188	.8812	.4137	3.481	.7764	4.5369	.1711	1.2488	0.622	4.633	8.4149	2.13
2.14	.1177	.8823	.4123	3.504	.7785	4.5796	.1700	1.2577	0.619	3.641	8.4994	2.14
2.15	.1165	.8835	.4109	3.528	.7806	4.6225	.1689	1.2665	0.616	3.650	8.5849	2.15
2.16	.1153	.8847	.4096	3.551	.7826	4.6656	.1677	1.2753	0.614	3.658	8.6711	2.16
2.17	.1142	.8858	.4082	3.575	.7847	4.7089	.1666	1.2842	0.611	3.667	8.7583	2.17
2.18	.1130	.8870	.4069	3.599	.7867	4.7524	.1655	1.2930	0.608	3.675	8.8463	2.18
2.19	.1119	.8881	.4055	3.623	.7887	4.7961	.1644	1.3019	0.606	3.684	8.9352	2.19
2.20	.1108	.8892	.4042	3.648	.7907	4.8400	.1634	1.3108	0.603	3.692	9.0250	2.20
2.21	.1097	.8903	.4028	3.672	.7926	4.8841	.1623	1.3197	0.601	3.701	9.1157	2.21
2.22	.1086	.8914	.4015	3.697	.7946	4.9284	.1612	1.3286	0.598	3.709	9.2073	2.22
2.23	.1075	.8925	.4002	3.721	.7965	4.9729	.1602	1.3375	0.596	3.718	9.2999	2.23
2.24	.1065	.8935	.3989	3.747	.7984	5.0176	.1591	1.3465	0.593	3.726	9.3933	2.24
2.25	.1054	.8946	.3976	3.772	.8003	5.0625	.1581	1.3554	0.590	3.735	9.4877	2.25

1	2	3	4	5	6	7	8	9	10	11	12	13
Z	S	A	A/Z	A/ZS	A^2	Z^2	A^2/Z^2	Z–A	$A^2/(Z–A)$	$Z^2/(Z–A)$	e^Z	Z
2.26	.1044	.8956	.3963	3.798	.8022	5.1076	.1571	1.3644	0.588	3.744	9.583	2.26
2.27	.1033	.8967	.3950	3.823	.8040	5.1529	.1560	1.3733	0.585	3.752	9.679	2.27
2.28	.1023	.8977	.3937	3.849	.8059	5.1984	.1550	1.3823	0.583	3.761	9.777	2.28
2.29	.1013	.8987	.3925	3.876	.8077	5.2441	.1540	1.3913	0.581	3.770	9.875	2.29
2.30	.1003	.8997	.3912	3.902	.8095	5.2900	.1530	1.4002	0.578	3.778	9.974	2.30
2.31	.0993	.9007	.3899	3.928	.8113	5.3361	.1520	1.4093	0.576	3.786	10.074	2.31
2.32	.0983	.9017	.3887	3.955	.8131	5.3824	.1511	1.4183	0.573	3.795	10.176	2.32
2.33	.0973	.9027	.3874	3.982	.8149	5.4289	.1501	1.4273	0.571	3.804	10.278	2.33
2.34	.0963	.9037	.3862	4.009	.8166	5.4756	.1491	1.4363	0.569	3.812	10.381	2.34
2.35	.0954	.9046	.3849	4.036	.8184	5.5225	.1482	1.4454	0.566	3.821	10.486	2.35
2.36	.0944	.9056	.3837	4.064	.8201	5.5696	.1472	1.4544	0.564	3.829	10.591	2.36
2.37	.0935	.9065	.3825	4.092	.8218	5.6169	.1463	1.4635	0.562	3.838	10.697	2.37
2.38	.0926	.9074	.3813	4.120	.8235	5.6644	.1454	1.4726	0.559	3.847	10.805	2.38
2.39	.0916	.9084	.3801	4.148	.8251	5.7121	.1444	1.4816	0.557	3.855	10.913	2.39
2.40	.0907	.9093	.3789	4.176	.8268	5.7600	.1435	1.4907	0.555	3.864	11.023	2.40
2.41	.0898	.9102	.3777	4.205	.8284	5.8081	.1426	1.4998	0.552	3.872	11.134	2.41
2.42	.0889	.9111	.3765	4.234	.8301	5.8564	.1417	1.5089	0.550	3.881	11.246	2.42
2.43	.0880	.9120	.3753	4.263	.8317	5.9049	.1408	1.5180	0.548	3.890	11.359	2.43
2.44	.0872	.9128	.3741	4.292	.8333	5.9536	.1400	1.5272	0.546	3.898	11.473	2.44
2.45	.0863	.9137	.3729	4.322	.8349	6.0025	.1391	1.5363	0.543	3.907	11.588	2.45
2.46	.0854	.9146	.3718	4.352	.8364	6.0516	.1382	1.5454	0.541	3.916	11.705	2.46
2.47	.0846	.9154	.3706	4.382	.8380	6.1009	.1374	1.5546	0.539	3.924	11.822	2.47
2.48	.0837	.9163	.3695	4.412	.8395	6.1504	.1365	1.5637	0.537	3.933	11.941	2.48
2.49	.0829	.9171	.3683	4.442	.8410	6.2001	.1356	1.5729	0.535	3.942	12.061	2.49
2.50	.0821	.9179	.3672	4.473	.8426	6.2500	.1348	1.5821	0.533	3.950	12.182	2.50
2.51	.0813	.9187	.3660	4.504	.8441	6.3001	.1340	1.5913	0.530	3.959	12.305	2.51
2.52	.0805	.9195	.3649	4.535	.8456	6.3504	.1332	1.6005	0.528	3.968	12.429	2.52
2.53	.0797	.9203	.3638	4.567	.8470	6.4009	.1323	1.6097	0.526	3.977	12.554	2.53
2.54	.0789	.9211	.3626	4.598	.8485	6.4516	.1315	1.6189	0.524	3.985	12.680	2.54
2.55	.0781	.9219	.3615	4.630	.8499	6.5025	.1307	1.6281	0.522	3.994	12.807	2.55
2.56	.0773	.9227	.3604	4.662	.8514	6.5536	.1299	1.6373	0.520	4.003	12.936	2.56
2.57	.0765	.9235	.3593	4.695	.8528	6.6049	.1291	1.6465	0.518	4.011	13.066	2.57
2.58	.0758	.9242	.3582	4.728	.8542	6.6564	.1283	1.6558	0.516	4.020	13.197	2.58
2.59	.0750	.9250	.3571	4.760	.8556	6.7081	.1275	1.6650	0.514	4.029	13.330	2.59
2.60	.0743	.9257	.3560	4.794	.8570	6.7600	.1268	1.6743	0.512	4.038	13.464	2.60
2.61	.0735	.9265	.3550	4.827	.8583	6.8121	.1260	1.6835	.5098	4.046	13.599	2.61
2.62	.0728	.9272	.3539	4.861	.8597	6.8644	.1252	1.6928	.5078	4.055	13.736	2.62
2.63	.0721	.9279	.3528	4.895	.8610	6.9169	.1245	1.7021	.5059	4.064	13.874	2.63
2.64	.0714	.9286	.3518	4.929	.8624	6.9696	.1237	1.7114	.5039	4.073	14.013	2.64
2.65	.0707	.9293	.3507	4.964	.8637	7.0225	.1230	1.7206	.5020	4.081	14.154	2.65
2.66	.0699	.9300	.3496	4.999	.8650	7.0756	.1222	1.7299	.5000	4.090	14.296	2.66
2.67	.0693	.9307	.3486	5.034	.8663	7.1289	.1215	1.7392	.4981	4.099	14.440	2.67
2.68	.0686	.9314	.3476	5.069	.8676	7.1824	.1208	1.7486	.4962	4.108	14.585	2.68
2.69	.0679	.9321	.3465	5.105	.8688	7.2361	.1201	1.7579	.4943	4.116	14.732	2.69
2.70	.0672	.9328	.3455	5.141	.8701	7.2900	.1194	1.7672	.4924	4.125	14.880	2.70

341

1	2	3	4	5	6	7	8	9	10	11	12	13
Z	S	A	A/Z	A/ZS	A²	Z²	A²/Z²	Z–A	A²/(Z–A)	Z²/(Z–A)	e^Z	Z
2.71	.0665	.9335	.3444	5.177	.8714	7.3441	.1186	1.7765	.4905	4.134	15.029	2.71
2.72	.0659	.9341	.3434	5.213	.8726	7.3984	.1179	1.7859	.4886	4.143	15.180	2.72
2.73	.0652	.9348	.3424	5.250	.8738	7.4529	.1172	1.7952	.4867	4.152	15.333	2.73
2.74	.0646	.9354	.3414	5.287	.8750	7.5076	.1166	1.8046	.4849	4.160	15.487	2.74
2.75	.0639	.9361	.3404	5.325	.8762	7.5625	.1159	1.8139	.4831	4.169	15.643	2.75
2.76	.0633	.9367	.3393	5.362	.8774	7.6176	.1152	1.8233	.4812	4.178	15.800	2.76
2.77	.0627	.9373	.3384	5.400	.8786	7.6729	.1145	1.8327	.4794	4.187	15.959	2.77
2.78	.0620	.9380	.3374	5.438	.8798	7.7284	.1138	1.8420	.4776	4.196	16.119	2.78
2.79	.0614	.9386	.3364	5.477	.8809	7.7841	.1132	1.8514	.4758	4.204	16.281	2.79
2.80	.0608	.9392	.3354	5.516	.8821	7.8400	.1125	1.8608	.4740	4.213	16.445	2.80
2.81	.0602	.9398	.3344	5.555	.8832	7.8961	.1118	1.8702	.4723	4.222	16.610	2.81
2.82	.0596	.9404	.3335	5.595	.8843	7.9524	.1112	1.8796	.4705	4.231	16.777	2.82
2.83	.0590	.9410	.3325	5.634	.8855	8.0089	.1106	1.8890	.4687	4.240	16.945	2.83
2.84	.0584	.9416	.3315	5.674	.8866	8.0656	.1099	1.8984	.4670	4.249	17.116	2.84
2.85	.0578	.9422	.3306	5.715	.8877	8.1225	.1093	1.9078	.4653	4.257	17.288	2.85
2.86	.0573	.9427	.3296	5.756	.8887	8.1796	.1086	1.9173	.4635	4.266	17.462	2.86
2.87	.0567	.9433	.3287	5.797	.8898	8.2369	.1080	1.9267	.4618	4.275	17.637	2.87
2.88	.0561	.9439	.3277	5.838	.8909	8.2944	.1074	1.9361	.4601	4.284	17.814	2.88
2.89	.0556	.9444	.3268	5.880	.8919	8.3521	.1068	1.9456	.4584	4.293	17.993	2.89
2.90	.0550	.9450	.3258	5.922	.8930	8.4100	.1062	1.9550	.4568	4.302	18.174	2.90
2.91	.0545	.9455	.3249	5.964	.8940	8.4681	.1056	1.9645	.4551	4.311	18.357	2.91
2.92	.0539	.9461	.3240	6.007	.8950	8.5264	.1050	1.9739	.4534	4.320	18.541	2.92
2.93	.0534	.9466	.3231	6.050	.8961	8.5849	.1044	1.9834	.4518	3.328	18.728	2.93
2.94	.0529	.9471	.3222	6.094	.8971	8.6436	.1038	1.9929	.4501	4.337	18.916	2.94
2.95	.0523	.9477	.3212	6.138	.8981	8.7025	.1032	2.0023	.4485	4.346	19.106	2.95
2.96	.0518	.9482	.3203	6.182	.8990	8.7616	.1026	2.0118	.4469	4.355	19.298	2.96
2.97	.0513	.9487	.3194	6.226	.9000	8.8209	.1020	2.0213	.4453	4.364	19.492	2.97
2.98	.0508	.9492	.3185	6.271	.9010	8.8804	.1015	2.0308	.4437	4.373	19.688	2.98
2.99	.0503	.9497	.3176	6.316	.9020	8.9401	.1009	2.0403	.4421	4.382	19.886	2.99
3.00	.0498	.9502	.3167	6.362	.9029	9.0000	.1003	2.0498	.4405	4.391	20.086	3.00

APPENDIX II. CONFIDENCE LIMITS FOR VARIABLES (x) DISTRIBUTED IN A POISSON FREQUENCY DISTRIBUTION, FOR CONFIDENCE COEFFICIENTS ($=1 - P$) OF 0.95 and 0.99[1]

Confidence coefficient = 0.95						Confidence coefficient = 0.99				
x	Lower limit	Upper limit	x	Lower limit	Upper limit	x	Lower limit	Upper limit	x	Lower limit	Upper limit
0	0.0	3.7				0	0.0	5.3			
1	0.1	5.6	26	17.0	38.0	1	0.03	7.4	26	14.7	42.2
2	0.2	7.2	27	17.8	39.2	2	0.1	9.3	27	15.4	43.5
3	0.6	8.8	28	18.6	40.4	3	0.3	11.0	28	16.2	44.8
4	1.0	10.2	29	19.4	41.6	4	0.6	12.6	29	17.0	46.0
5	1.6	11.7	30	20.2	42.8	5	1.0	14.1	30	17.7	47.2
6	2.2	13.1	31	21.0	44.0	6	1.5	15.6	31	18.5	48.4
7	2.8	14.4	32	21.8	45.1	7	2.0	17.1	32	19.3	49.6
8	3.4	15.8	33	22.7	46.3	8	2.5	18.5	33	20.0	50.8
9	4.0	17.1	34	23.5	47.5	9	3.1	20.0	34	20.8	52.1
10	4.7	18.4	35	24.3	48.7	10	3.7	21.3	35	21.6	53.3
11	5.4	19.7	36	25.1	49.8	11	4.3	22.6	36	22.4	54.5
12	6.2	21.0	37	26.0	51.0	12	4.9	24.0	37	23.2	55.7
13	6.9	22.3	38	26.8	52.2	13	5.5	25.4	38	24.0	56.9
14	7.7	23.5	39	27.7	53.3	14	6.2	26.7	39	24.8	58.1
15	8.4	24.8	40	28.6	54.5	15	6.8	28.1	40	25.6	59.3
16	9.2	26.0	41	29.4	55.6	16	7.5	29.4	41	26.4	60.5
17	9.9	27.2	42	30.3	56.8	17	8.2	30.7	42	27.2	61.7
18	10.7	28.4	43	31.1	57.9	18	8.9	32.0	43	28.0	62.9
19	11.5	29.6	44	32.0	59.0	19	9.6	33.3	44	28.8	64.1
20	12.2	30.8	45	32.8	60.2	20	10.3	34.6	45	29.6	65.3
21	13.0	32.0	46	33.6	61.3	21	11.0	35.9	46	30.4	66.5
22	13.8	33.2	47	34.5	62.5	22	11.8	37.2	47	31.2	67.7
23	14.6	34.4	48	35.3	63.6	23	12.5	38.4	48	32.0	68.9
24	15.4	35.6	49	36.1	64.8	24	13.2	39.7	49	32.8	70.1
25	16.2	36.8	50	37.0	65.9	25	14.0	41.0	50	33.6	71.3

For larger values of x, E. S. Pearson's formulae are very close to correct:

For $1 - P = 0.95$: $x + 1.92 \pm 1.960\sqrt{x + 1.0}$

For $1 - P = 0.99$: $x + 3.32 \pm 2.576\sqrt{x + 1.7}$

More approximately, x can be considered a normally distributed variable with standard deviation $= \sqrt{x}$.

[1] From Ricker 1937, slightly modified.

APPENDIX III. CHARACTERISTICS AND RELATIONSHIPS FOR TWO TYPES OF RECRUITMENT CURVE, EACH EXPRESSED IN TWO DIFFERENT FORMS

Characteristics and relationships for two types of recruitment curve, each expressed in two different forms. For curves 1 and 3 the items marked by an (*) are valid regardless of the units in which R and P are expressed; for the remaining items it is necessary that R and P be in the same units and that $R = P$ at replacement abundance.

P   abundance of parents (stock)
R   abundance of progeny (recruits)
C   catch, in numbers
Y   mean number of fertilized eggs produced per mature fish (males and females together)
Z   total mortality rate (eggs to mature fish)
α, β parameters (different for the two types of curve)
a, A   parameters
r as subscript, indicates the condition of replacement abundance and recruitment
m as subscript, indicates the condition of maximum recruitment
E   as subscript, indicates steady-state conditions (mortality balanced by recruitment)
s as subscript, indicates the condition of maximum sustainable yield (MSY)
c as subscript, indicates the compensatory component of natural mortality
i as subscript, indicates the density-independent component of natural mortality

Most of the expressions are as given in Ricker (1973c), where their derivation is described; see also Paulik et al. (1967). Item 16 is a rearrangement of 13. Items 20 and 21 are from expressions A22 and A24 of Ricker (1958a), the former being a simpler form of No. 19 of the 1973 paper.

Appendix III (continued)

Curve No.	1	2	3	4
1.* Equation	$R = \alpha P e^{-\beta P}$	$R = P e^{a(1-P/P_r)}$	$R = \dfrac{P}{\alpha P + \beta}$	$R = \dfrac{P}{1 - A(1-P/P_r)}$
2.* Slope of the curve	$(1-\beta P)\alpha e^{-\beta P}$	$(1 - aP/P_r)e^{a(1-P/P_r)}$	$\dfrac{\beta}{(\alpha P + \beta)^2}$	$\dfrac{1-A}{(1 - A[1-P/P_r])^2}$
3.* Slope at the origin	α	e^a	$\dfrac{1}{\beta}$	$\dfrac{1}{1-A}$
4. Slope at maximum sustainable yield (MSY)	1	1	1	1
5.* Slope at maximum recruitment	0	0	0	0
6. Slope at replacement level of stock	$1 - \log_e \alpha$	$1 - a$	β	$1 - A$
7.* Slope at maximum level of stock	0	0	0	0
8. Replacement level of stock	$P_r = R_r = \dfrac{\log_e \alpha}{\beta}$	$\dfrac{P_r}{P_r} = \dfrac{R_r}{P_r} = 1$	$P_r = R_r = \dfrac{1-\beta}{\alpha}$	$\dfrac{P_r}{P_r} = \dfrac{R_r}{P_r} = 1$
9.* Spawners needed for maximum recruitment	$P_m = \dfrac{1}{\beta}$	$\dfrac{P_m}{P_r} = \dfrac{1}{a}$	$P_m = \infty$	$\dfrac{P_m}{P_r} = \infty$

346

10.* Maximum recruitment	$R_m = \dfrac{\alpha}{\beta e}$	$\dfrac{R_m}{P_r} = \dfrac{e^{\alpha-1}}{a}$	$R_m = \dfrac{1}{\alpha}$	$\dfrac{R_m}{P_r} = \dfrac{1}{A}$
11. Ratio of replacement spawners to maximum recruitment spawners	$P_r/P_m = \log_e \alpha$	$P_r/P_m = a$	not applicable	not applicable
12. Sustainable yield	$C_E = P_E(\alpha e^{-\beta P_E} - 1)$	$C_E = P_E(e^{\alpha(1-P_E/P_r)} - 1)$	$C_E = P_E\left(\dfrac{1}{\alpha P_E + \beta} - 1\right)$	$C_E = P\left(\dfrac{1}{1 - A(1-P/P_r)} - 1\right)$
13. Equilibrium rate of exploitation	$u_E = C_E/R_E = 1 - \dfrac{1}{\alpha e^{-\beta P}}$	$u_E = C_E/R_E = 1 - e^{-\alpha(1-P_E/P_r)}$	$u_E = C_E/R_E = 1 - \alpha P - \beta$	$u_E = C_E/R_E = A(1 - P/P_r)$
14. Limiting equilibrium rate of exploitation (when P → 0)	$1 - \dfrac{1}{\alpha}$	$1 - e^{-a}$	$1 - \beta$	A
15. Limiting equilibrium rate of fishing (when P → 0)	$\log_e \alpha$	a	$-\log_e \beta$	$-\log_e(1 - A)$
16. Spawners required to sustain an equilibrium rate of exploitation u_E	$P_E = \dfrac{\log_e(\alpha[1-u_E])}{\beta}$	$\dfrac{P_E}{P_r} = 1 + \dfrac{\log_e(1-u_E)}{a}$	$P_E = \dfrac{1 - u_E - \beta}{\alpha}$	$\dfrac{P_E}{P_r} = 1 - \dfrac{u_E}{A}$
17. Spawners needed for MSY	P_s is found by solving: $(1 - \beta P_s)\alpha e^{-\beta P_s} = 1$	P_s/P_r is found by solving: $(1 - aP_s/P_r)e^{a(1-P_s/P_r)} = 1$	$P_s = \dfrac{\sqrt{\beta} - \beta}{\alpha}$	$\dfrac{P_s}{P_r} = \dfrac{A - 1 + \sqrt{1-A}}{A}$
18. Recruitment at MSY	R_s is found by substituting P_s in (1)	R_s/P_r is found by substituting P_s/P_r in (1)	$R_s = \dfrac{1 - \sqrt{\beta}}{\alpha}$	$\dfrac{R_s}{P_r} = \dfrac{1 - \sqrt{1-A}}{A}$

347

Appendix III (concluded)

Curve No.	1	2	3	4
19. Maximum sustainable yield (MSY)	$C_s = P_s(\alpha e^{-\beta P_s} - 1)$	$C_s = P_s(e^{a(1 - P_s/P_r)} - 1)$	$C_s = \dfrac{(1 - \sqrt{\beta})^2}{\alpha}$	$\dfrac{C_s}{P_r} = \dfrac{(1 - \sqrt{1 - A})^2}{A}$
20. Rate of exploitation at MSY	$u_s = \beta P_s$	$u_s = a P_s/P_r$	$u_s = 1 - \sqrt{\beta}$	$u_s = 1 - \sqrt{1 - A}$
21. Curve parameters when u_s is known	$\log_e \alpha = u_s - \log_e(1 - u_s)$ $\beta = u_s/P_s$	$a = u_s - \log_e(1 - u_s)$	$\alpha = (1 - \beta)/P_r$ $\beta = (1 - u_s)^2$	$A = 1 - (1 - u_s)^2$
22. Rate of recruitment (a) actual	$R/P = \alpha e^{-\beta P}$	$R/P = e^{a(1 - P/P_r)}$	$\dfrac{R}{P} = \dfrac{1}{\alpha P + \beta}$	$\dfrac{R}{P} = \dfrac{1}{1 - A(1 - P/P_r)}$
(b) instantaneous	$\log_e \alpha - \beta P$	$a(1 - P/P_r)$	$-\log_e(\alpha P + \beta)$	$-\log_e(1 - A[1 - P/P_r])$
23. Rate of recruitment at minimal stock (a) actual	$R_0/P_0 = \alpha$	$R_0/P_0 = e^a$	$R_0/P_0 = \dfrac{1}{\beta}$	$R_0/P_0 = \dfrac{1}{1 - A}$
(b) instantaneous	$\log_e \alpha$	a	$-\log_e \beta$	$-\log_e(1 - A)$
24. Rate of recruitment at maximum recruitment (a) actual	$R_m/P_m = \alpha e^{-1}$	$R_m/P_m = e^{a-1}$	not applicable	not applicable
(b) instantaneous	$\log_e \alpha - 1$	$a - 1$		

25. Rate of recruitment at replacement				
(a) actual	$R_r/P_r = 1$	$R_r/P_r = 1$	$R_r/P_r = 1$	$R_r/P_r = 1$
(b) instantaneous	0	0	0	0
26. Instantaneous rate of compensatory mortality	$Z_c = \beta P$	$Z_c = aP/P_r$	$Z_c = \log_e(1 + \alpha P/\beta)$	$Z_c = \log_e\!\left(1 + \dfrac{AP}{P_r(1 - A)}\right)$
27. Instantaneous rate of density-independent mortality	$Z_i = \log_e Y - \log_e \alpha$	$Z_i = \log_e Y - a$	$Z_i = \log_e Y + \log_e \beta$	$Z_i = \log_e Y + \log_e(1 - A)$

APPENDIX IV. REGRESSION LINES

REGRESSIONS USED

The linear regression coefficients used in this Bulletin follow the rationale outlined by Ricker (1973a).

a) Ordinary "predictive" regressions are used when abscissal (X) observations are not subject to natural variability and are known accurately. They are also used in some instances where a functional regression is required but the measurement variability in X is probably small compared to that in Y, and the data are not extensive enough to obtain a useful estimate by one of the AM methods described below.

b) Functional regressions are used when X values are subject to natural variability and are a symmetrical (but not necessarily random) sample from a real or imaginary distribution of indefinite extent. Functional regressions are of either the GM or AM type described below.

c) With either (a) or (b) above the regression mentioned is used both for a description of the relationship and for prediction of Y from X.

FORMULAE

1. GM REGRESSIONS. Let Y and X represent the N pairs of variates, y and x the same variates measured from their means, and r the linear regression coefficient between Y and X.

Ordinary regression of Y on X:
$$b = \frac{\Sigma xy}{\Sigma x^2} \tag{a}$$

GM functional regression of Y on X:
$$v = \left(\frac{\Sigma y^2}{\Sigma x^2}\right)^{\frac{1}{2}} = \frac{b}{r} \tag{b}$$

Variance of b and of v:
$$\frac{\Sigma y^2 - (\Sigma xy)^2/\Sigma x^2}{(\Sigma x^2)(N-2)} \tag{c}$$

Confidence limits for v:
$$v(\sqrt{B+1} \pm \sqrt{B}) \tag{c'}$$

where $B = .F(1 - r^2)/(N - 2)$, F being the variance ratio for $n_1 = 1$ and $n_2 = N - 2$ degrees of freedom.

The GM regression is an unbiased estimate of the functional regression if either of the following conditions is met:

a) variability in Y and in X is almost wholly natural (little is due to error of measurement);

b) the ratio of the total variances of Y and X is approximately the same as the ratio of their measurement variances.

If neither of these conditions applies, one of the AM formulae below is usually used to estimate the functional regression.

2(a). WALD'S AM REGRESSION. Let the distribution be arranged in the numerical order of X, and be divided into 2 halves (the middle member of an odd series is omitted). Let ΣX_1, ΣY_1, and ΣX_2, ΣY_2 be the sums of the variates in the two halves. The series is now arranged in the order of Y and similar sums are computed; call them $\Sigma' X_1$, $\Sigma' Y_1$ and $\Sigma' X_2$ and $\Sigma' Y_2$. The symmetrical estimate of Wald's regression is:

Wald's functional AM
regression of Y on X:
$$\frac{(\Sigma Y_2 - \Sigma Y_1) + (\Sigma' Y_2 - \Sigma' Y_1)}{(\Sigma X_2 - \Sigma X_1) + (\Sigma' X_2 - \Sigma' X_1)} \tag{d}$$

351

2(b). NAIR–BARTLETT AM. REGRESSION. Let the distribution be arranged in the order of X and of Y, as above, and then be divided into 3 parts for each order, such that the two extremes are equal and the central part is as nearly equal to the others as possible.

Nair–Bartlett functional
AM regression of Y on X:

$$\frac{(\Sigma Y_3 - \Sigma Y_1) + (\Sigma'Y_3 - \Sigma'Y_1)}{(\Sigma X_3 - \Sigma X_1) + (\Sigma'X_3 - \Sigma'X_1)} \tag{e}$$

The Nair–Bartlett formula is slightly better than Wald's when observations are numerous and/or variability is small. To estimate confidence limits for these estimates, see Ricker (1973a) and the references therein.

3. INTERCEPTS. A regression line, of any of the above types, is made to pass through the point representing the means of all the observations of each sort $(\overline{X}, \overline{Y})$.

4. LINES THROUGH A FIXED POINT. The slope of the functional regression of a line that must pass through a fixed point (X_1, Y_1) is taken to be:

$$b = \frac{\Sigma(Y - Y_1)}{\Sigma(X - X_1)} \tag{f}$$

5. In several instances a parabola of the type:

$$Y = aX + bX^2 \tag{g}$$

has been fitted by the device of dividing through by X and computing the linear regression of Y/X on X, whereby the slope is an estimate of b and the Y-axis intercept is an estimate of a.

It has frequently been pointed out that a regression of the type Y/X against X is statistically suspect. With X appearing in the denominator of the first variable and in the numerator of the second, random variability will tend to generate a negative slope (curved, it is true) in the absence of any real relationship. However, when a relationship of some consequence does actually exist, the random component adds little to the slope of any *straight* line that is fitted. There is usually excellent (though not perfect) agreement between predictive regression lines fitted by this method and those fitted by direct minimization of the sum of squares of deviations from the parabola of expression (g).

AASS, P. 1972. Age determination and year-class fluctuation of cisco, *Coregonus albula* L., in the Mjøsa hydroelectric reservoir, Norway. Rep. Inst. Freshwater Res. Drottningholm 52: 5–22.

ABRAMSON, N. J. 1971. Computer programs for fish stock assessment. FAO (Food Agric. Organ. U.N.) Fish. Tech. Pap. 101: 1–154.

ADAMS, L. 1951. Confidence limits for the Petersen or Lincoln index used in animal population studies. J. Wildl. Manage. 15: 13–19.

ALLEN, K. R. 1950. The computation of production in fish populations. N.Z. Sci. Rev. 8: 89.

  1951. The Horokiwi Stream: a study of a trout population. N.Z. Mar. Dep. Fish. Res. Div. Bull. New Ser. 10: 231 p.

  1953. A method for computing the optimum size-limit for a fishery. Nature 172: 210.

  1954. Factors affecting the efficiency of restrictive regulations in fisheries management. I. Size limits. N.Z. J. Sci. Technol. Sect. B. 35: 498–529.

  1955. Factors affecting the efficiency of restrictive regulations in fisheries management. II. Bag limits. N.Z. J. Sci. Technol. Sect. B. 36: 305–334.

  1966a. Determination of age distribution from age–length keys and length distributions, IBM 7090, 7094, Fortran IV. Trans. Am. Fish. Soc. 95: 230–231.

  1966b. Some methods for the estimation of exploited populations. J. Fish. Res. Board Can. 23: 1553–1574.

  1967. Computer programs available at St. Andrews Biological Station. Fish. Res. Board Can. Tech. Rep. 20: 32 p. + Append.

  1968. Simplification of a method of computing recruitment rates. J. Fish. Res. Board Can. 25: 2701–2702.

  1969. An application of computers to the estimation of exploited populations. J. Fish. Res. Board Can. 26: 179–189.

  1973. Analysis of stock-recruitment relation in Antarctic fin whales (*Balaenoptera physalus*). Rapp. P.-V. Reun. Cons. Perm. Int. Explor. Mer 164: 132–137.

ANDREEV, N. N. 1955. Some problems in the theory of the capture of fish by gill nets. Tr. Vses. Nauchno-Issled. Inst. Morsk. Rybn. Khoz. Okeanogr. 30: 109–127. (In Russian)

ANON. 1962. Report on North American halibut stocks with reference to Articles III(1)(a) and IV of the International Convention for the High Seas Fisheries of the North Pacific Ocean. Int. North Pac. Fish. Comm. Bull. 7: 1–18.

BAILEY, N. J. J. 1951. On estimating the size of mobile populations from recapture data. Biometrika 38: 293–306.

  1952. Improvements in the interpretation of recapture data. J. Anim. Ecol. 21: 120–127.

BAJKOV, A. 1933. Fish population and productivity of lakes. Trans. Am. Fish. Soc. 62: 307–316.

BARANOV, F. I. 1914. The capture of fish by gillnets. Mater. Poznaniyu Russ. Rybolovstva 3(6): 56–99. (In Russian)

  1918. On the question of the biological basis of fisheries. Nauchn. Issled. Ikhtiologicheskii Inst. Izv. 1: 81–128. (In Russian)

  1926. On the question of the dynamics of the fishing industry. Byull. Rybn. Khoz. (1925) 8: 7–11. (In Russian)

  1927. More about the poor catch of vobla. Byull. Rybn. Khoz. 7(7). (In Russian)

  1948. Theory and assessment of fishing gear. 2nd ed. Pishchepromizdat, Moscow. 436 p. (In Russian)

BERST, A. H. 1961. Selectivity and efficiency of experimental gill nets in South Bay and Georgian Bay of Lake Huron. Trans. Am. Fish. Soc. 90: 413–418.

VON BERTALANFFY, L. 1934. Untersuchungen über die Gesetzlichkeit des Wachstums. I. Roux' Archiv 131: 613.

  1938. A quantitative theory of organic growth. Hum. Biol. 10: 181–213.

BEVERTON, R. J. H. 1953. Some observations on the principles of fishery regulation. J. Cons. Int. Explor. Mer 19: 56–68.

  1954. Notes on the use of theoretical models in the study of the dynamics of exploited fish populations. U.S. Fish. Lab., Beaufort, N.C., Misc. Contrib 2: 159 p.

---

[1]The works cited include those mentioned in the text and a number of others that are of general interest.

1962. Long-term dynamics of certain North Sea fish populations, p. 242–259. *In* E. D. LeCren and M. W. Holdgate [ed.] The exploitation of natural animal populations. Blackwell Scientific Publications, Oxford.

BEVERTON, R. J. H., AND S. J. HOLT. 1956. A review of methods for estimating mortality rates in fish populations, with special reference to sources of bias in catch sampling. Rapp. P.-V. Reun. Cons. Perm. Int. Explor. Mer 140: 67–83.

1957. On the dynamics of exploited fish populations. U.K. Min. Agric. Fish., Fish. Invest. (Ser. 2) 19: 533 p.

BEVERTON, R. J. H., AND B. B. PARRISH. 1956. Commercial statistics in fish population studies. Rapp. P.-V. Reun. Cons. Perm. Int. Explor. Mer 140: 58–66.

BISHOP, Y. M. M. 1959. Errors in estimates of mortality obtained from virtual populations. J. Fish. Res. Board Can. 16: 73–90.

BLACK, E. C. 1957. Alterations in the blood level of lactic acid in certain salmonoid fishes following muscular activity. I. Kamloops trout, *Salmo gairdneri*. J. Fish. Res. Board Can. 14: 117–134.

BOIKO, E. G. 1934. Estimation of the supply of Kuban zanders. Raboty Dono-Kubanskoi Nauchnoi Rybokhozyaistvennoi Stantsii 1: 1–43. (In Russian)

1964. Forecasting supplies and catches of Azov zanders. Trudy Vsesoyuznogo N.-I. Inst. Morsk. Rybn. Khoz. Okeanogr. 50: 45–88. (In Russian)

BRAATEN, D. O. 1969. Robustness of the DeLury population estimator. J. Fish. Res. Board Can. 26: 339–355.

BRETT, J. R. 1970. Fish — the energy cost of living, pages 37–52. *In* W. S. McNeil [ed.] Marine aquiculture. Oregon State University Press, Corvallis, Oreg.

BRODY, S. 1927. Growth rates. Univ. Missouri Agric. Exp. Sta. Bull. 97.
1945. Bioenergetics and growth. Reinhold Publishing Corp., New York, N.Y. 1023 p.

BURD, A. C., AND J. VALDIVIA. 1970. The use of virtual population analysis on the Peruvian anchoveta data. Instituto del Mar (Callao).

BURKENROAD, M. D. 1948. Fluctuation in abundance of Pacific halibut. Bull. Bingham Oceanogr. Collect. Yale Univ. 11(4): 81–129.

1951. Some principles of marine fishery biology. Publ. Inst. Mar. Sci. Univ. Tex. 2: 177–212.

CARLANDER, K. D. 1950. Handbook of freshwate fishery biology. Wm. C. Brown Co., Dubuqu Iowa. 281 p.

CARLANDER, K. D., AND L. E. HINER. 1943. Fisherie investigation and management report for Lak Vermillion, St. Louis County. Minn. Dep. Con serv., Div. Game Fish Sect. Res. Plann., Fish Invest. Rep. 54: 175 p.

CARLANDER, K. D., AND W. M. LEWIS. 1948. Som precautions in estimating fish populations. Pro Fish-Cult. 10: 134–137.

CARLANDER, K. D., AND R. R. WHITNEY. 1961. Ag and growth of walleyes in Clear Lake, Iow 1935–1957. Trans. Am. Fish. Soc. 90: 130–138.

CASSIE, R. M. 1954. Some uses of probability paper i the analysis of size frequency distribution Aust. J. Mar. Freshwater Res. 5: 513–522.

CHAPMAN, D. G. 1948. A mathematical study of co fidence limits of salmon populations calculate by sample tag ratios. Int. Pac. Salmon Fish Comm. Bull. 2: 67–85.

1951. Some properties of the hypergeometr distribution with applications to zoologic sample censuses. Univ. Calif. Publ. Stat. 1: 131 160.

1952. Inverse, multiple and sequential samp censuses. Biometrics 8: 286–306.

1954. The estimation of biological popula tions. Ann. Math. Stat. 25: 1–15.

1955. Population estimation based on chang of composition caused by a selective remova Biometrika 42: 279–290.

1961. Statistical problems in dynamics c exploited fisheries populations. Proc. 4th Be keley Symp. Math. Stat. and Probability. Cont Biol. and Probl. Med. 4: 153–168. Univ. Cali Press.

1973. Spawner-recruit models and estima tion of the level of maximum sustainable catc Rapp. P.-V. Reun. Cons. Perm. Int. Explor. Me 164: 325–332.

CHAPMAN, D. G., AND C. O. JUNGE. 1954. The estima tion of the size of a stratified animal populatio Ann. Math. Stat. 27: 375–389.

CHATWIN, B. M. 1953. Tagging of chum salmon i Johnstone Strait, 1945 and 1950. Bull. Fish. Re Board Can. 96: 33 p.

1958. Mortality rates and estimates o theoretical yield in relation to minimum size o lingcod (*Ophiodon elongatus*) from the Strait o Georgia, British Columbia. J. Fish. Res. Boar Can. 15: 831–849.

CHEVALIER, J. R. 1973. Cannibalism as a factor i first year survival of walleye in Oneida Lak Trans. Am. Fish. Soc. 102: 739–744.

354

CHRISTIE, W. J. 1963. Effects of artificial propagation and weather on recruitment in Lake Ontario whitefish fishery. J. Fish. Res. Board Can. 20: 597–646.

CHUGUNOV, N. L. 1935. An attempt at a biostatistical determination of the stocks of fishes in the North Caspian. Rybn. Khoz. 15(6): 24–29; 15(8): 17–21. (In Russian)

CHUGUNOVA, N. I. 1959. Handbook for the study of age and growth of fishes. Akademiya Nauk Press, Moscow. 164 p. (English version, 1963. Age and growth studies in fish. Office of Technical Services, Washington, D.C. 132 p.)

CLARK, F. N., AND J. F. JANSSEN. 1945a. Results of tagging experiments in California waters on the sardine (Sardinops caerulea.) Calif. Dep. Fish Game Fish Bull. 16: 1–42.

1945b. Measurement of the losses in the recovery of sardine tags. Calif. Dep. Fish Game Fish Bull. 61: 63–90.

CLARK, F. N., AND J. C. MARR. 1956. Population dynamics of the Pacific sardine. Calif. Coop. Oceanic Fish. Invest. Rep. July 1, 1953 – March 31, 1955. p. 11–48.

CLARKE, G. L., W. T. EDMONDSON, AND W. E. RICKER. 1946. Mathematical formulation of biological productivity. Ecol. Monogr. 16: 336–337.

CLAYDEN, A. D. 1972. Simulation of the changes in abundance of the cod (Gadus morhua L.) and the distribution of fishing in the North Atlantic. U.K. Min. Agr. Fish. Food, Fish. Invest. (Ser. 2) 27: 1–58.

CLOPPER, C. J., AND E. S. PEARSON. 1934. The use of confidence or fiducial limits applied to the case of the binomial. Biometrika 26: 404–413.

CLUTTER, R. I., AND L. E. WHITESEL. 1956. Collection and interpretation of sockeye salmon scales. Bull. Int. Pac. Salmon Fish. Comm. 9: 1–159.

COOPER, G. P., AND K. F. LAGLER. 1956. The measurement of fish population size. Trans. 21st N. Am. Wildl. Nat. Resour. Conf. p. 281–297.

CORMACK, R. M. 1969. The statistics of capture-recapture methods. Oceanogr. Mar. Biol. Ann. Rev. 6: 455–506.

COX, E. L. 1949. Mathematical bases for experimental sampling of size of certain biological populations. Va. Agric. Exp. Stn. Stat. Lab. Rep. 42 p.

CRAIG, C. C. 1953. Use of marked specimens in estimating populations. Biometrika 40: 170–176.

CREASER, C. H. 1926. The structure and growth of the scales in fishes, etc. Univ. Mich. Mus. Zool. Misc. Pub. 17: 82 p.

CROSSMAN, E. J. 1956. Growth, mortality and movements of a sanctuary population of maskinonge (Esox masquinongy Mitchill). J. Fish. Res. Board Can. 13: 599–612.

CUCIN, D., AND H. A. REGIER. 1966. On the dynamics and exploitation of lake whitefish in southern Georgian Bay. J. Fish. Res. Board Can. 23: 221–418.

CUSHING, D. H. 1968. Fisheries biology. A study in population dynamics. Univ. Wisconsin Press, Madison, Wis. 200 p.

1973. The dependence of recruitment on parent stock. J. Fish. Res. Board Can. 30(12, 2): 1965–1976.

CUSHING, D. H., AND J. G. K. HARRIS. 1973. Stock and recruitment and the problem of density-dependence. Rapp. P.-V. Reun. Cons. Explor. Perm. Int. Mer 164: 142–155.

DAHL, K. 1919. Studies of trout and trout-waters in Norway. Salmon Trout Mag. 18: 16–33.

1943. Ørret og ørretvann. Studier og forsok. New Edition: J. W. Coppeleus, Oslo. 182 p.

DAMAS, D. 1909. Contribution à la biologie des gadides. Rapp. P.-V. Reun. Cons. Perm. Int. Explor. Mer 10, App. 2: 277 p.

DANNEVIG, G. 1953. Tagging experiments on cod, Lofoten 1947–1952: some preliminary results. J. Cons. Int. Explor. Mer 19: 195–203.

DARROCH, J. N. 1958. The multiple-recapture census. I. Estimation of closed population. Biometrika 45: 343–359.

1959. The multiple-recapture census. II. Estimation when there is immigration or death. Biometrika 46: 336–351.

1961. The two-sample capture-recapture census when tagging and sampling are stratified. Biometrika 48: 241–260.

DAVIDSON, F. A. 1940. The homing instinct and age at maturity of pink salmon (Oncorhynchus gorbuscha). Bull. U.S. Bur. Fish. 48(15): 27–39.

DAVIDSON, F. A., E. VAUGHAN, S. J. HUTCHINSON, AND A. L. PRITCHARD. 1943. Factors affecting the upstream migration of the pink salmon (Oncorhynchus gorbuscha). Ecology 24: 149–168.

DAVIS, W. S. 1964. Graphic representation of confidence intervals for Petersen population estimates. Trans. Am. Fish. Soc. 93: 227–232.

DEASON, H. J., AND R. HILE. 1947. Age and growth of the kiyi, *Leucichthys kiyi* Koelz, in Lake Michigan. Trans. Am. Fish. Soc. 74: 88–142.

DeLURY, D. B. 1947. On the estimation of biological populations. Biometrics 3: 145–167.

1951. On the planning of experiments for the estimation of fish populations. J. Fish. Res. Board Can. 8: 281–307.

1958. The estimation of population size by a marking and recapture procedure. J. Fish. Res. Board Can. 15: 19–25.

DEMING, W. E. 1943. Statistical adjustment of data. 2nd ed. John Wiley and Sons, New York, N.Y. 261 p.

DERZHAVIN, A. N. 1922. The stellate sturgeon (*Acipenser stellatus* Pallas), a biological sketch. Byull. Bakinskoi Ikhtiologicheskoi Stantsii 1: 1–393. (In Russian)

DICKIE, L. M. 1955. Fluctuations in abundance of the giant scallop, *Placopecten magellanicus* (Gmelin), in the Digby area of the Bay of Fundy. J. Fish. Res. Board Can. 12: 797–857.

1968. Mathematical models of growth, p. 120–123. *In* Methods for assessment of fish production in fresh waters, W. E. Ricker [ed.]. International Biological Programme Handbook No. 3. Blackwell Scientific Publ., Oxford and Edinburgh. 313 p.

1974. Interaction between fishery management and environmental protection. J. Fish. Res. Board Can. 30: 2496–2506.

DICKIE, L. M., AND F. D. McCRACKEN. 1955. Isopleth diagrams to predict yields of a small flounder fishery. J. Fish. Res. Board Can. 12: 187–209.

DOI, T. 1955a. On the fisheries of "iwasi" (sardine, anchovy and round herring) in the Inland Sea. II. Relationship of yield to environments and prediction of fluctuations of fisheries. Bull. Jap. Soc. Sci. Fish. 21: 82–87.

1955b. Dynamical analysis on porgy (*Pagrosomus major* T. and S.) fishery in the Inland Sea of Japan. Bull. Jap. Soc. Sci. Fish. 21: 320–334.

1962. The predator–prey and competitive relationships among fishes caught in waters adjacent to Japan. Bull. Tokai Reg. Fish. Res. Lab. 32: 49–118. (Transl. from Japanese by Fish. Res. Board Can. Transl. Ser. No. 461, 1965).

DOMBROSKI, E. 1954. The sizes of Babine Lake sockeye salmon smolt emigrants, 1950–1953. Fish. Res. Board Can. Prog. Rep. (Pac.) 99: 30–34.

DOWDESWELL, W. H., R. A. FISHER, AND E. B. FORD. 1940. The quantitative study of population in the Lepidoptera. I. *Polyommatus icarus* Roth. Ann. Eugen. 10: 123–136.

DYMOND, J. R. 1948. European studies of the populations of marine fishes. Bull. Bingham Oceanogr. Collect. Yale Univ. 11: 55–80.

EDSER, T. 1908. Note on the number of plaice at each length, in certain samples from the southern part of the North Sea, 1906. J. R. Stat. Soc. 71: 686–690.

1926. Book review of: F. Baranov. On the question of the dynamics of the fishing industry. J. Cons. Int. Explor. Mer 1: 291–292.

ELSTER, H. J. 1944. Über das Verhältnis von Produktion, Bestand, Befischung and Ertrag sowie über die Möglichkeiten einer Steigerung der Erträge, untersucht am Beispiel der Blaufelchenfischerei des Bodensees. Z. Fisch. 42: 169–357.

FABENS, A. J. 1965. Properties and fitting of the von Bertalanffy growth curve. Growth 29: 265–289.

FISHER, R. A. 1937. The design of experiments. 2nd ed. Oliver and Boyd, London. 260 p.

1950. Statistical methods for research workers. 11th ed. Oliver and Boyd, London. 354 p.

FISHER, R. A., AND E. B. FORD. 1947. The spread of a gene in natural conditions in a colony of the moth *Panaxia dominula* L. Heredity 1: 143–174.

FLETCHER, I. R. 1973. A synthesis of deterministic growth laws. Univ. Rhode Island, School of Oceanography, Kingston, R.I. 54 p.

FOERSTER, R. E. 1935. The occurrence of unauthentic marked salmon. Biol. Board Can. Pac. Prog. Rep. 25: 18–20.

1936. The return from the sea of sockeye salmon (*Oncorhynchus nerka*), with special reference to percentage survival, sex proportions, and progress of migration. J. Biol. Board Can. 3: 26–42.

1954a. On the relation of adult sockeye salmon (*Oncorhynchus nerka*) returns to known smolt seaward migrations. J. Fish. Res. Board Can. 11: 339–350.

1954b. Sex ratios among sockeye salmon. J. Fish. Res. Board Can. 11: 988–997.

FORD, E. 1933. An account of the herring investigations conducted at Plymouth during the years from 1924–1933. J. Mar. Biol. Assoc. U.K. 19: 305–384.

FOX, W. W. 1970. An exponential yield model for optimizing exploited fish populations. Trans. Am. Fish. Soc. 99: 80–88.

FRASER, J. M. 1955. The smallmouth bass fishery of South Bay, Lake Huron. J. Fish. Res. Board Can. 12: 147–177.

FREDIN, R. A. 1948. Causes of fluctuations in abundance of Connecticut River shad. U.S. Fish Wildl. Serv. Fish. Bull. 54: 247–259.

1950. Fish population estimates in small ponds using the marking and recovery technique. Iowa State Coll. J. Sci. 24: 363–384.

FRIDRIKSSON, A. 1934. On the calculation of age distribution within a stock of cod by means of relatively few age-determinations as a key to measurements on a large scale. Rapp. P.-V. Cons. Reun. Perm. Int. Explor. Mer 86(6): 5 p.

FRY, F. E. J. 1949. Statistics of a lake trout fishery. Biometrics 5: 27–67.

FULTON, T. W. 1911. The sovereignty of the sea. Edinburgh and London.

1920. Report on the marking experiments of plaice, made by S/S Goldseeker, in the years 1910–1913. Sci. Invest. Fish. Scotl. (1919)1.

GARROD, D. J. 1967. Population dynamics of the arcto-Norwegian cod. J. Fish. Res. Board Can. 24: 145–190.

1968. "Schaefer-type" assessments of catch/effort relationships in the North Atlantic cod stocks. Int. Comm. Northwest Atl. Fish. Doc. 68/51: 17 p.

1969. Empirical assessments of catch/effort relationships in North Atlantic cod stocks. Int. Comm. Northwest Atl. Fish. Res. Bull. 6: 26–34.

GARWOOD, F. 1936. Fiducial limits for the Poisson distribution. Biometrika 28: 437–442.

GERKING, S. D. 1953a. Vital statistics of the fish population of Gordy Lake, Indiana. Trans. Am. Fish. Soc. 82: 48–67.

1953b. Evidence for the concepts of home range and territory in stream fishes. Ecology 34: 347–365.

1957. Evidence of ageing in natural populations of fishes. Gerontologia 1: 287–305.

1966. Annual growth cycle, growth potential, and growth compensation in the bluegill sunfish in northern Indiana lakes. J. Fish. Res. Board Can. 23: 1923–1956.

GILBERT, C. H. 1914. Contributions to the life history of the sockeye salmon, No. 1. Rep. B.C. Comm. Fish., 1913. p. 53–78.

GILBERT, R. O. 1973. Approximations of the bias in the Jolly–Seber capture-recapture model. Biometrics 29: 501–526.

GOODMAN, L. A. 1953. Sequential sampling tagging for population size problems. Ann. Math. Stat. 24: 56–69.

GRAHAM, M. 1929a. Studies of age-determination in fish. Part 1. A study of the growth rate of codling (Gadus callarias L.) on the inner herring trawling ground. U.K. Min. Agric. Fish., Fish. Invest. (Ser. 2) 11(2): 50 p.

1929b. Studies of age determination in fish. Part II. A survey of the literature. U.K. Min. Agric. Fish., Fish. Invest. (Ser. 2) 11(3): 50 p.

1935. Modern theory of exploiting a fishery. and application to North Sea trawling. J. Cons, Int. Explor. Mer 10: 264–274.

1938a. Rates of fishing and natural mortality from the data of marking experiments. J. Cons. Int. Explor. Mer 13: 76–90.

1938b. Growth of cod in the North Sea and use of the information. Rapp. P.-V. Reun. Cons. Perm. Int. Explor. Mer 108(10): 58–66.

1952. Overfishing and optimum fishing. Rapp. P.-V. Reun. Cons. Perm. Int. Explor. Mer 132: 72–78.

GRAHAM, M. [ed.] 1956. Sea fisheries: their investigation in the United Kingdom. Edward Arnold, London. 487 p.

GULLAND, J. A. 1953. Vital statistics of fish populations. World Fish. 2: 316–319.

1955a. Estimation of growth and mortality in commercial fish populations. U.K. Min. Agric. Fish., Fish. Invest. (Ser. 2) 18(9): 46 p.

1955b. On the estimation of population parameters from marked members. Biometrika 42: 269–270.

1956a. Notations in fish population studies. Contrib. Biarritz Meeting, the Int. Comm. Northwest Atl. Fish. 5 p. (mimeo).

1956b. The study of fish populations by the analysis of commercial catches. Rapp. P.-V. Reun. Cons. Perm. Int. Explor. Mer 140: 21–27.

1957. Review of: A. L. Tester. Estimation of recruitment and natural mortality rate from age composition and catch data. J. Cons. Int. Explor. Mer 22: 221–222.

1961. Fishing and the stocks of fish at Iceland. U.K. Min. Agric. Fish. Food, Fish. Invest. (Ser. 2) 23(4): 52 p.

1964a. Manual of methods for fish population analysis. FAO (Food Agric. Organ. U.N.) Fish. Tech. Pap. 40: 1–60.

1964b. The abundance of fish stocks in the Barents Sea. Rapp. P.-V. Reun. Cons. Perm. Int. Explor. Mer 155: 126–137.

1965. Estimation of mortality rates. Annex to Rep. Arctic Fish. Working Group, Int. Counc. Explor. Sea C.M. 1965(3): 9 p.

357

1966. Manual of sampling and statistical methods for fisheries biology. Part I. Sampling methods. (FAO) Food Agric. Organ., United Nations.

1968. Recent changes in the North Sea plaice fishery. J. Cons. Int. Explor. Mer 31: 305–322.

1969. Manual of methods for fish stock assessment. Part I. Fish population analysis. FAO (Food Agric. Organ. U.N.) Man. Fish. Sci. 4: 1–154.

1971. The fish resources of the ocean. Fishing News (Books) Ltd., London. 255 p.

GULLAND, J. A., AND D. HARDING. 1961. The selection of *Clarias mossambicus* (Peters) by nylon gill nets. J. Cons. Int. Explor. Mer 26: 215–222.

HALLIDAY, R. G. 1972. A yield assessment of the eastern Scotian Shelf cod stock complex. Int. Comm. Northwest Atl. Fish. Res. Bull. 9: 117–124.

HAMERSLEY, J. M. 1953. Capture–recapture analysis. Biometrika 40: 265–278.

HAMLEY, J. M. 1972. Use of the DeLury method to estimate gillnet selectivity. J. Fish. Res. Board Can. 29: 1636–1638.

HAMLEY, J. M., AND H. A. REGIER. 1973. Direct estimates of gillnet selectivity to walleye (*Stizostedion vitreum*). J. Fish. Res. Board Can. 30: 817–830.

HANCOCK, D. A. 1965. Graphical estimation of growth parameters. J. Cons. Int. Explor. Mer 29: 340–351.

HARDING, J. P. 1949. The use of probability paper for graphical analysis of polymodal frequency distributions. J. Mar. Biol. Assoc. U.K. 28: 141–153.

HART, J. L. 1931. The growth of the whitefish, *Coregonus clupeaformis* (Mitchill). Contrib. Can. Biol. Fish., N.S. 6: 427–444.

1932. Statistics of the whitefish (*Coregonus clupeaformis*) population of Shakespeare Island Lake, Ontario. Univ. Toronto Studies, Biol. Ser. 36: 1–28.

HASSELBLAD, V. 1966. Estimation of parameters for a mixture of normal distributions. Technometrics 8: 431–444.

HAYES, F. R. 1949. The growth, general chemistry and temperature relations of salmonid eggs. Q. Rev. Biol. 24: 281–308.

HAYNE, D. W. 1949. Two methods for estimating populations from trapping records. J. Mammal. 30: 399–411.

HEDERSTRÖM, H. 1959. Observations on the age of fishes. Rep. Inst. Freshwater Res. Drottningholm 40: 161–164. (Original version published in 1759)

HEINCKE, F. 1905. Das Vorkommen und die Verbreitung der Eier, der Larven und der verschiedenen Altersstufen der Nutzfische in der Nordsee. Rapp. P.-V. Reun. Cons. Perm. Int. Explor. Mer 3, Append. E. 41 p.

1913a. Investigations on the plaice. General report. I. Plaice fishery and protective measures. Preliminary brief summary of the most important points of the report. Rapp. P.-V. Reun. Cons. Perm. Int. Explor. Mer 16. 67 p.

1913b. Investigations on the plaice. General report. 1. Plaice fishery and protective regulations. Part I. Rapp. P.-V. Reun. Cons. Perm. Int. Explor. Mer 17A: 1–153.

HELLAND, A. 1913–1914. Rovdyrene i Norge. Tidsskrift for Skogbruk 1913–14. (cited from Hjort et al. 1933.)

HELLAND-HANSEN, B. 1909. Statistical research into the biology of the haddock and cod in the North Sea. Rapp. P.-V. Reun. Cons. Perm. Int. Explor. Mer 10, Append. 1. 62 p.

HENRY, K. A. 1953. Analysis of the factors affecting the production of chum salmon (*Oncorhynchus keta*) in Tillamook Bay. Oreg. Fish Comm. Contrib. 18. 37 p.

HICKLING, C. F. 1938. The English plaice-marking experiments, 1929–32. U.K. Min. Agric. Fish., Fish. Invest. (Ser. 2)16: 84 p.

HILE, R. 1936. Age and growth of the cisco *Leucichthys artedii* (LeSueur) in the lakes of the northeastern highlands, Wisconsin. Bull. U.S. Bur. Fish. 48(19): 211–317.

1941. Age and growth of the rock bass, *Ambloplites rupestris* (Rafinesque) in Nebish Lake, Wisconsin. Trans. Wis. Acad. Sci. 33: 189–337.

HJORT, J. 1914. Fluctuations in the great fisheries of northern Europe, viewed in the light of biological research. Rapp. P.-V. Reun. Cons. Perm. Int. Explor. Mer 20: 1–228.

HJORT, J., G. JAHN, AND P. OTTESTAD. 1933. The optimum catch. Hvalradets Skr. 7: 92–127.

HODGSON, W. C. 1929. Investigations into the age, length and maturity of the herring of the southern North Sea. Part III. The composition of the catches from 1923 to 1928. U.K. Min. Agric. Fish., Fish. Invest. (Ser. 2) 11(7): 75 p.

1933. Further experiments on the selective action of commercial drift nets. J. Cons. Int. Explor. Mer 8: 344–354.

HOFFBAUER, C. 1898. Die Altersbestimmung des Karpfens an seiner Schuppe. Allgemeine Fischereizeitung 23(19).

HOLLAND, G. A. 1957. Migration and growth of the dogfish shark, *Squalus acanthias* (Linnaeus), of the eastern North Pacific. Wash. Dep. Fish., Fish. Res. Pap. 2: 43–59.

HOLT, S. J. 1956. Contribution to working party no. 3, Population Dynamics. Int. Comm. Northwest Atl. Fish., Committee on Research and Statistics, Biarritz, 1956. 18 p.

      1963. A method of determining gear selectivity and its application. Int. Comm. Northwest Atlantic Fish. Spec. Bull. 5: 106–115.

HOWARD, G. V. 1948. A study of the tagging method in the enumeration of sockeye salmon populations. Int. Pac. Salmon Fish. Comm. Bull. 2: 9–66.

HUBBS, C. L., AND G. P. COOPER. 1935. Age and growth of the long-eared and green sunfishes in Michigan. Pap. Mich. Acad. Sci. Arts Lett. 20: 669–696.

HULME, H. R., R. J. H. BEVERTON, AND S. J. HOLT. 1947. Population studies in fisheries biology. Nature 159: 714.

HUNTSMAN, A. G. 1918. Histories of new food fishes. 1. The Canadian plaice. Bull. Biol. Board Can. 1:1–32.

HYLEN, A., A. JONSGÅRD, G. C. PIKE, AND J. T. RUUD. 1955. A preliminary report on the age composition of Antarctic fin-whale catches, 1945/46 to 1952/53, and some reflections on total mortality rates of fin whales. Norw. Whaling Gaz. 1955: 577-589.

ISHIDA, T. 1962. On the gill-net mesh selectivity curve. Bull. Hokkaido Reg. Fish. Res. Lab. 25: 20–25.

      1964. Gillnet mesh selectivity curves for sardine and herring. Bull. Hokkaido Reg. Fish. Res. Lab. 28: 56–60.

JACKSON, C. H. N. 1933. On the true density of tsetse flies. J. Anim. Ecol. 2: 204–209.

      1936. Some new methods in the study of *Glossina moristans*. Proc. Zool. Soc. Lond. 4: 811–896.

      1939. The analysis of an animal population. J. Anim. Ecol. 8: 238–246.

      1940. The analysis of a tsetse-fly population. Ann. Eugen. 10: 332–369.

JANSSEN, J. F., AND J. A. APLIN. 1945. The effect of internal tags upon sardines. Calif. Div. Fish Game. Fish Bull. 61: 43–62.

JENSEN, A. J. C. 1939. On the laws of decrease in fish stocks. Rapp. P.-V. Reun. Cons. Perm. Int. Explor. Mer 110: 85–96.

JENSEN, P. T. AND J. HYDE. 1971. Sex ratios and survival estimates among salmon populations. Calif. Fish Game 57: 90–98.

JOLLY, G. M. 1963. Estimates of population parameters from multiple recapture data with both death and dilution — deterministic model. Biometrika 50: 113–128.

      1965. Explicit estimates from capture-recapture data with both death and immigration — stochastic model. Biometrika 52: 225–247.

JONES, B. W. 1966. The cod and the cod fishery at Faroe. Fish. Invest. Minist. Agric. Fish. Food (G.B.) Ser. II, 24(5): 1–32.

JONES, R. 1956. The analysis of trawl haul statistics with particular reference to the estimation of survival rates. Rapp. P.-V. Reun. Cons. Perm. Int. Explor. Mer 140: 30–39.

      1957. A much simplified version of the fish yield equation. Doc. No. P. 21, presented at the Lisbon joint meeting of Int. Comm. Northwest Atl. Fish., Int. Counc. Explor. Sea, and Food Agric. Organ., United Nations. 8 p.

      1958. Lee's phenomenon of "apparent change in growth rate", with particular reference to haddock and plaice. Int. Comm. Northwest Atl. Fish. Spec. Publ. 1: 229–242.

      1961. The assessment of long-term effects of changes in gear selectivity and fishing effort. Mar. Res. (Scotland) 1961(2): 1–19.

      1968. Appendix to the Report of the North-West Working Group, Int. Counc. Explor. Sea 2 p.

      1973. Stock and recruitment with special reference to cod and haddock. Rapp. P.-V. Reun. Cons. Perm. Int. Explor. Mer 164: 156–173.

JUNGE, C. O., AND W. H. BAYLIFF. 1955. Estimating the contribution of a salmon production area by marked fish experiments. Wash. Dep. Fish., Fish. Res. Pap. 1(3): 51–58.

KAWAKAMI, T. 1952. On prediction of fishery based on variation of stock. Mem. Coll. Agric. Kyoto Univ. 62: 27–35.

KAWASAKI, T., and M. HATANAKA. 1951. Studies on the populations of the flatfishes of Sendai Bay. 1. *Limanda angustirostris* Kitahara. Tohoku J. Agric. Res. 2: 83–104.

KENNEDY, W. A. 1951. The relationship of fishing effort by gill nets to the interval between lifts. J. Fish. Res. Board Can. 8: 264–274.

1953. Growth, maturity, fecundity and mortality in the relatively unexploited whitefish, *Coregonus clupeaformis*, of Great Slave Lake. J. Fish. Res. Board Can. 10: 413–441.

1954a. Tagging returns, age studies and fluctuations in abundance of Lake Winnipeg whitefish, 1931–1951. J. Fish. Res. Board Can. 11: 284–309.

1954b. Growth, maturity and mortality in the relatively unexploited lake trout, *Cristivomer namaycush*, of Great Slave Lake. J. Fish. Res. Board Can. 11: 827–852.

1956. The first ten years of commercial fishing on Great Slave Lake. Bull. Fish. Res. Board Can. 107: 58 p.

KESTEVEN, G. L. 1946. A procedure of investigation in fishery biology. Aust. Counc. Sci. Ind. Res. Bull. 194: 1–31.

KESTEVEN, G. L., AND S. J. HOLT. 1955. A note on the fisheries resources of the northwest Atlantic. FAO (Food Agric. Organ, U.N.) Fish. Tech. Pap. 7: 11 p.

KETCHEN, K. S. 1950. Stratified subsampling for determining age distributions. Trans. Am. Fish. Soc. 79: 205–212.

1953. The use of catch-effort and tagging data in estimating a flatfish population. J. Fish. Res. Board Can. 10: 459–485.

1956. Factors influencing the survival of the lemon sole (*Parophrys vetulus*) in Hecate Strait, British Columbia. J. Fish. Res. Board Can. 13: 647–694.

1964. Measures of abundance from fisheries for more than one species. Rapp. P.-V. Reun. Cons. Perm. Int. Explor. Mer 155: 113–116.

KETCHEN, K. S., AND C. R. FORRESTER. 1966. Population dynamics of the petrale sole. Bull. Fish. Res. Board Can. 153: 195 p.

KILLICK, S. R. 1955. The chronological order of Fraser River sockeye salmon during migration, spawning and death. Int. Pac. Salmon Fish. Comm. Bull. 7: 95 p.

KNIGHT, W. 1968. Asymptotic growth: an example of nonsense disguised as mathematics. J. Fish. Res. Board Can. 25: 1303–1307.

1969. A formulation of the von Bertalanffy growth curve when the growth rate is roughly constant. J. Fish. Res. Board Can. 26: 3069–3072.

KRUMHOLZ, L. A. 1944. A check on the fin-clipping method for estimating fish populations. Pap. Mich. Acad. Sci. 29: 281–291.

KUBO, I., AND T. YOSHIHARA. 1957. Study of fisheries resources. Kyoritsu Publishing Co., Tokyo. 345 p. (In Japanese)

KUTKUHN, J. H. 1963. Estimating absolute age composition of California salmon landings. Calif. Dep. Fish Game Fish. Bull. 120: 1–47.

LAGLER, K. F. 1956. Freshwater fishery biology. Wm. C. Brown Co., Dubuque, Iowa. 421 p.

LAGLER, K. F., AND W. E. RICKER. 1942. Biological fisheries investigations of Foots Ponds, Gibson County, Indiana. Invest. Indiana Lakes Streams 2(3): 47–72.

LARKIN, P. A., AND W. E. RICKER. 1964. Further information on sustained yields from fluctuating environments. J. Fish. Res. Board Can. 21: 1–7.

LASSEN, H. 1972. User manual. Estimation of F and M. Danmarks Fiskeri- og Havundersøgelser 6: 1–13.

LATTA, W. C. 1959. Significance of trap-net selectivity in estimating fish population statistics. Pap. Mich. Acad. Sci. Arts Lett. 44: 123–138.

LECREN, E. D. 1947. The determination of the age and growth of the perch (*Perca fluviatilis*) from the opercular bone. J. Anim. Ecol. 16: 188–204.

LEE, R. M. 1912. An investigation into the methods of growth determination in fishes. Cons. Explor. Mer, Publ. de Circonstance 63: 35 p.

LESLIE, P. H. 1952. The estimation of population parameters from data obtained by means of the capture–recapture method. II. The estimation of total numbers. Biometrika 39: 363–388.

LESLIE, P. H., AND D. CHITTY. 1951. The estimation of population parameters from data obtained by means of the capture–recapture method. 1. The maximum likelihood equations for estimating the death rate. Biometrika 38: 269–292.

LESLIE, P. H., AND D. H. S. DAVIS. 1939. An attempt to determine the absolute number of rats on a given area. J. Anim. Ecol. 8: 94–113.

LI, C. C. 1956. The concept of path coefficient and its impact on population genetics. Biometrics 12: 190–210.

LINCOLN, F. C. 1930. Calculating waterfowl abundance on the basis of banding returns. U.S. Dep. Agric. Circ. 118: 4 p.

LINDNER, M. J. 1953. Estimation of growth rate in animals by marking experiments. U.S. Fish Wildl. Serv. Fish. Bull. 54(78): 65–69.

MAIER, H. N. 1906. Beiträge zur Altersbestimmung der Fische. I. Allgemeines. Die Altersbestimmung nach den Otolithen bei Scholle und Kabeljau. Wiss. Meeresuntersuchungen (Helgoland), N.F. 8(5): 57–115.

MANZER, J. I., AND F. H. C. TAYLOR. 1947. The rate of growth in lemon sole in the Strait of Georgia. Fish. Res. Board Can. Pac. Prog. Rep. 72: 24–27.

MARR, J. C. 1951. On the use of the terms abundance, availability and apparent abundance in fishery biology. Copeia 1951: 163–169.

McCOMBIE, A. M., AND F. E. J. FRY. 1960. Selectivity of gill nets for lake whitefish, *Coregonus clupeaformis*. Trans. Am. Fish. Soc. 89: 176–184.

MEEHAN, O. L. 1940. Marking largemouth bass. Prog. Fish-cult. 51: p. 46.

MERRIMAN, D. 1941. Studies on the striped bass (*Roccus saxatilis*) of the Atlantic coast. U.S. Fish Wildl. Serv. Fish. Bull. 50(35): 1–77.

MILLER, R. B. 1952. The role of research in fisheries management in the Prairie Provinces. Canadian Fish Culturist 12: 13–19.

　　1953. The collapse and recovery of a small whitefish fishery. J. Fish. Res. Board Can. 13: 135–146.

MILNE, D. J. 1955. The Skeena River salmon fishery with special reference to sockeye salmon. J. Fish. Res. Board Can. 12: 451–485.

MOISEEV, P. A. 1953. Cod and flounders of far-eastern seas. Izv. Tikhookean. Nauchno-Issled. Inst. Rybn. Khoz. Okeanogr. 40: 1–287. (In Russian)

　　1969. Biological resources of the world ocean. Pishchevaya Promyshlennost Press, Moscow. 340 p. (English version: National Technical Information Service, Springfield, Va., 334 p. 1971)

MONASTYRSKY, G. N. 1935. A method of making long-term forecasts for the vobla fishery of the North Caspian. Rybn. Khoz. 15(5): 11–18; 15(6): 18–23. (In Russian)

　　1940. The stock of vobla in the north Caspian Sea, and methods of assessing it. Tr. Vses. Nauchno-Issled. Inst. Rybn. Khoz. Okeanogr. 11: 115–170. (In Russian)

　　1952. Dynamics of the abundance of commercial fishes. Tr. Vses. Nauchno-Issled. Inst. Rybn. Khoz. Okeanogr. 21: 1–162. (In Russian)

MORAN, P. A. P. 1950. Some remarks on animal population dynamics. Biometrics 6: 250–258.

　　1951. A mathematical theory of animal trapping. Biometrika 38: 307–311.

　　1952. The estimation of death-rates from capture–mark–recapture sampling. Biometrika 39: 181–188.

MOTTLEY, C. M. 1949. The statistical analysis of creel-census data. Trans. Am. Fish Soc. 76: 290–300.

MURPHY, G. I. 1952. An analysis of silver salmon counts at Benbow Dam, South Fork of Eel River, California. Calif. Fish Game 38: 105–112.

　　1965. A solution of the catch equation. J. Fish. Res. Board Can. 22: 191–202.

NEAVE, F. 1954. Principles affecting the size of pink and chum salmon populations in British Columbia. J. Fish. Res. Board Can. 9: 450–491.

NEAVE, F., AND W. P. WICKETT. 1953. Factors affecting the freshwater development of Pacific salmon in British Columbia. Proc. 7th Pac. Sci. Congr. 4: 548–555.

NESBIT, R. A. 1943. Biological and economic problems of fishery management. U.S. Fish Wildl. Serv. Spec. Sci. Rep. Fish. 18: 23–53, 61–68.

NEYMAN, J. 1949. On the problem of estimating the number of schools of fish. Univ. Calif. Publ. Stat. 1: 21–36.

NICHOLSON, A. J. 1933. The balance of animal populations. J. Anim. Ecol. 2(suppl.): 132–178.

　　1954. An outline of the dynamics of animal populations. Aust. J. Zool. 2: 9–65.

NIKOLSKY, G. V. 1953. Concerning the biological basis of the rate of exploitation, and means of managing the abundance of fish stocks. Ocherki po Obshchim Voprosam Ikhtiologii, p. 306–318. Akademiya Nauk SSSR, Leningrad. (In Russian)

　　1965. Theory of fish population dynamics as the biological background for rational exploitation and management of fishery resources. Nauka Press, Moscow. 382 p. (English ed., Oliver and Boyd, Edinburgh, 1969, 323 p.)

OLSEN, S. 1959. Mesh selection in herring gill nets. J. Fish. Res. Board Can. 16: 339–349.

OMAND, D. N. 1951. A study of populations of fish based on catch-effort statistics. J. Wildl. Manage. 15: 88–98.

PALOHEIMO, J. E. 1958. A method of estimating natural and fishing mortalities. J. Fish. Res. Board Can. 15: 749–758.

　　1961. Studies on estimation of mortalities. I. Comparison of a method described by Beverton and Holt and a new linear formula. J. Fish. Res. Board Can. 18: 645–662.

PARKER, R. A. 1955. A method for removing the effect of recruitment on Petersen-type population estimates. J. Fish. Res. Board Can. 12: 447–450.

PARRISH, B. B., AND R. JONES. 1953. Haddock bionomics — I. The state of the haddock stocks in the North Sea, 1946–50, and at Faroe, 1914–50. Scott. Home Dep. Mar. Res. 4: 27 p.

PAULIK, G. J. 1973. Studies of the possible form of the stock-recruitment curve. Rapp. P.-V. Reun. Cons. Perm. Int. Explor. Mer 164: 302–315.

PAULIK, G. J., AND W. F. BAYLIFF. 1967. A generalized computer program for the Ricker model of equilibrium yield per recruitment. J. Fish. Res. Board Can. 24: 249–252.

PAULIK, G. J., AND L. E. GALES. 1964. Allometric growth and Beverton and Holt yield equation. Trans. Am. Fish. Soc. 93: 369–381.

PAULIK, G. J., A. S. HOURSTON, AND P. A. LARKIN. 1967. Exploitation of multiple stocks by a common fishery. J. Fish. Res. Board Can. 24: 2527–2537.

PEARSON, K. 1924. Tables for statisticians and biometricians. 2nd ed. Cambridge Univ. Press, Cambridge, Eng. 143 p.

PELLA, J. J., AND P. K. TOMLINSON. 1969. A generalized stock production model. Bull. Inter-Am. Trop. Tuna Comm. 13: 419–496.

PERLMUTTER, A. 1954. Age determination of fish. Trans. N.Y. Acad. Sci. 16: 305–311.

PETERSEN, C. G. J. 1892. Fiskensbiologiske forhold i Holboek Fjord, 1890–91. Beretning fra de Danske Biologiske Station for 1890 (91), 1: 121–183.

1896. The yearly immigration of young plaice into the Limfjord from the German Sea, etc. Rep. Dan. Biol. Sta. 6: 1–48.

1922. On the stock of plaice and the plaice fisheries in different waters. Rep. Dan. Biol. Sta. 29: 1–43.

PETERSON, A. E. 1954. The selective action of gill nets on Fraser River sockeye salmon. Int. Pac. Salmon Fish. Comm. Bull. 5: 1–101.

PIENAAR, L. V., AND W. E. RICKER. 1968. Estimating mean weight from length statistics. J. Fish. Res. Board Can. 25: 2743–2747.

PIENAAR, L. V., AND J. A. THOMSON. 1973. Three programs used in population dynamics: WVONB — ALOMA — BHYLD (Fortan 1130). Fish. Res. Board Can. Tech. Rep. 367: 33 p.

PONOMARENKO, V. P. 1968. The influence of the fishery upon the stock and recruitment of Barents Sea cod. Tr. Polyarn. Nauchno-Issled. Proektivnogo Inst. Morsk. Rybn. Khoz. Okeanogr. 23: 310–362. (In Russian)

POPE, J. A. 1956. An outline of sampling techniques. Rapp. P.-V. Reun. Cons. Perm. Int. Explor. Mer 140: 11–20.

POPE, J. G. 1972. An investigation of the accuracy of virtual population analysis using cohort analysis. Int. Comm. Northwest Atl. Fish. Res. Bull. 9: 65–74.

PRITCHARD, A. L., AND F. NEAVE. 1942. What did the tagging of coho salmon at Skeetz Falls, Cowichan River reveal? Fish. Res. Board Can. Pac. Prog. Rep. 51: 8–11.

REGIER, H. A. 1974. Sequence of exploitation of stocks in multispecies fisheries in the Laurentian Great Lakes. J. Fish. Res. Board Can. 30: 1992–1999.

REGIER, H. A., AND D. S. ROBSON. 1966. Selectivity of gill nets, especially to lake whitefish. J. Fish Res. Board Can. 23: 423–454.

REIBISCH, J. 1899. Über die Eizahl bei *Pleuronectes platessa* und die Altersbestimmung dieser Form aus den Otolithen. Wiss. Meeresuntersuchungen (Kiel) N.F. 4: 233–248.

RICHARDS, F. J. 1959. A flexible growth function for empirical use. J. Exp. Bot. 10: 290–300.

RICKER, W. E. 1937. The concept of confidence or fiducial limits applied to the Poisson frequency distribution. J. Am. Stat. Assoc. 32: 349–356.

1940. Relation of "catch per unit effort" to abundance and rate of exploitation. J. Fish Res. Board Can. 5: 43–70.

1942a. The rate of growth of bluegill sunfish in lakes of northern Indiana. Invest. Indiana Lakes Streams 2: 161–214.

1942b. Creel census, population estimates, and rate of exploitation of game fish in Shoe Lake, Indiana. Invest. Indiana Lakes Streams 2: 215–253.

1942c. The effect of reduction of predaceous fish on survival of young sockeye salmon at Cultus Lake. J. Fish. Res. Board Can. 5: 315–336.

1944. Further notes on fishing mortality and effort. Copeia 1944: 23–44.

1945a. Abundance, exploitation, and mortality of the fishes of two lakes. Invest. Indiana Lakes Streams 2: 345–448.

1945b. Natural mortality among Indiana bluegill sunfish. Ecology 26: 111–121.

1945c. A method of estimating minimum size limits for obtaining maximum yield. Copeia 1945(2): 84–94.

1946. Production and utilization of fish populations. Ecol. Monogr. 16: 373–391.

1948. Methods of estimating vital statistics of fish populations. Indiana Univ. Publ. Sci. Ser. 15: 101 p.

362

1949a. Mortality rates in some little-exploited populations of freshwater fishes. Trans. Am. Fish. Soc. 77: 114–128.

1949b. Effects of removal of fins on the growth and survival of spiny-rayed fishes. J. Wildl. Manage. 13: 29–40.

1954. Stock and recruitment. J. Fish. Res. Board Can. 11: 559–623.

1955a. Fish and fishing in Spear Lake, Indiana. Invest. Indiana Lakes Streams 4: 117–161.

1955b. Book reviews of: D. Lack. The natural regulation of animal numbers; and H. G. Andrewartha and L. C. Birch, The distribution and abundance of animals. J. Wildl. Manage. 19: 487–489.

1958a. Handbook of computations for biological statistics of fish populations. Bull. Fish. Res. Board Can. 119: 300 p.

1958b. Maximum sustained yields from fluctuating environments and mixed stocks. J. Fish. Res. Board Can. 15: 991–1006.

1958c. Production, reproduction and yield. Verh. Int. Verein. Limnol. 13: 84–100.

1962. Regulation of the abundance of pink salmon populations, p. 155–201. In N. J. Wilimovsky [ed.] Symposium on pink salmon. H. R. McMillan Lectures in Fisheries, Univ. British Columbia, Vancouver, B.C.

1963. Big effects from small causes: two examples from fish population dynamics. J. Fish. Res. Board Can. 20: 257–264.

1968. Background information and theory related to management of the Skeena sockeye salmon. Fish. Res. Board Can. MS Rep. 961: 38 p.

1969a. Effects of size-selective mortality and sampling bias on estimates of growth, mortality, production, and yield. J. Fish. Res. Board Can. 26: 479–541.

1969b. Food from the sea, p. 87–108. In Resources and man, the Report of the Committee on Resources and Man to the U.S. National Academy of Sciences. W. H. Freeman and Co., San Francisco. 259 p.

1970. Biostaticheskii metod A. N. Derzhavina. [A. N. Derzhavin's biostatistical method.] Rybn. Khoz. 46(10): 6–9; 46(11): 5–7.

1971a. [ed.] Methods for assessment of fish production in freshwaters. IBP Handbook No. 3, 2nd ed. Blackwell Scientific Publications, Oxford and Edinburgh. 348 p.

1971b. Derzhavin's biostatistical method of population analysis. J. Fish. Res. Board Can. 28: 1666–1672.

1971c. Sopostavlenie dvukh krivykh vosproizvodstva. (Comparison of two reproduction curves.) Rybn. Khoz. 47(3): 16–21, 47(4): 10–13.

1972. Hereditary and environmental factors affecting certain salmonid populations, p. 27–160. In The stock concept in Pacific salmon. MacMillan Lectures in Fisheries, Univ. British Columbia, Vancouver, B.C. 231 p.

1973a. Linear regressions in fishery research. J. Fish. Res. Board Can. 30: 409–434.

1973b. Two mechanisms that make it impossible to maintain peak-period yields from stocks of Pacific salmon and other fishes. J. Fish. Res. Board Can. 30: 1275–1286.

1973c. Critical statistics from two reproduction curves. Rapp. P.-V. Reun. Cons. Perm. Int. Explor. Mer 174: 333–340.

RICKER, W. E., AND R. E. FOERSTER. 1948. Computation of fish production. Bull. Bingham. Oceanogr. Collect. Yale Univ. 11: 173–211.

RICKER, W. E., AND K. F. LAGLER. 1942. The growth of spiny-rayed fishes in Foots Pond. Invest. Indiana Lakes Streams 2: 85–97.

RICKER, W. E., AND H. D. SMITH. 1975. A revised interpretation of the history of the Skeena River sockeye. J. Fish. Res. Board Can. 32. (In press)

RICKLEFS, R. E. 1967. A graphical method of fitting equations to growth curves. Ecology 48: 978–983.

ROBSON, D. S. 1963. Maximum likelihood estimation of a sequence of annual survival rates from a capture–recapture series. Int. Comm. Northwest Atl. Fish. Spec. Publ. 4: 330–335.

ROBSON, D. S., AND D. G. CHAPMAN. 1961. Catch curves and mortality rates. Trans. Am. Fish. Soc. 90: 181–189.

ROBSON, D. S., AND H. A. REGIER. 1964. Sample size in Petersen mark-recapture experiments. Trans. Am. Fish. Soc. 93: 215–226.

ROLLEFSEN, G. 1935. The spawning zone in cod otoliths and prognosis of stock. Fiskeridir. Skr. 4(11): 3–10.

1953. The selectivity of different fishing gear used in Lofoten. J. Cons. Int. Explor. Mer 19: 191–194.

ROTHSCHILD, B. J. 1967. Competition for gear in a multiple-species fishery. J. Cons. Int. Explor. Mer 31: 102–110.

ROUNSEFELL, G. A. 1949. Methods of estimating the runs and escapements of salmon. Biometrics 5: 115–126.

1958. Factors causing decline in sockeye salmon of Karluk River, Alaska. U.S. Fish Wildl. Serv. Fish. Bull. 58(130): 83–169.

ROUNSEFELL, G. A., AND W. H. EVERHART. 1953. Fishery science: Its methods and applications. John Wiley and Sons, New York, N.Y. 444 p.

ROUSSOW, G. 1957. Some considerations on sturgeon spawning periodicity in northern Quebec. J. Fish. Res. Board Can. 14: 553–572.

ROYCE W. F. 1972. Introduction to the fishery sciences. Academic Press, New York, N.Y. 351 p.

RUSSELL, F. S. 1931. Some theoretical considerations on the "overfishing" problem. J. Cons. Explor. Mer 6: 3–27.

SATOW, T. 1955. A trial for the estimation of the fishing mortality of a migrating fish procession. J. Shimonoseki Coll. Fish. 4: 159–166.

SCHAEFER, M. B. 1951a. Estimation of the size of animal populations by marking experiments. U.S. Fish Wildl. Serv. Fish Bull. 52: 189–203.

1951b. A study of the spawning populations of sockeye salmon in the Harrison River system, with special reference to the problem of enumeration by means of marked members. Int. Pac. Salmon Fish. Comm. Bull. 4: 207 p.

1954. Some aspects of the dynamics of populations important to the management of the commercial marine fisheries. Bull. Inter-Am. Trop. Tuna Comm. 1(2): 27–56.

1955. The scientific basis for a conservation program. Int. Tech. Conf. Conservation of the Living Resources of the Sea. Rome, April–May, 1955. p. 14–55.

1957. A study of the dynamics of the fishery for yellowfin tuna in the eastern tropical Pacific Ocean. Inter-Am. Trop. Tuna Comm. Bull. 2: 247–268.

SCHNABEL, Z. E. 1938. The estimation of the total fish population of a lake. Am. Math. Mon. 45: 348–352.

SCHUCK, H. A. 1949. Relationship of catch to changes in population size of New England haddock. Biometrics 5: 213–231.

SCHUMACHER, A. 1970. Bestimmung der fischereilichen Sterblichkeit beim Kabeljaubestand vor Westgrönland. Ber. Dtsch. Komm. Meeresforsch. 21(1–4): 248–259. (Transl. from German by Fish. Res. Board Can. Transl. Ser. No. 1690, 1971)

SCHUMACHER, F. X., AND R. W. ESCHMEYER. 1943. The estimate of fish population in lakes or ponds. J. Tenn. Acad. Sci. 18: 228–249.

SCOTT, D. C. 1949. A study of a stream population of rock bass. Invest. Indiana Lakes Streams 3: 169–234.

SEBER, G. A. F. 1962. The multi-sample single recapture census. Biometrika 49: 339–350.

1965. A note on the multiple-recapture census. Biometrika 52: 249–259.

1972. Estimating time-specific survival and reporting rates for adult birds from band returns. Biometrika 57: 313–318.

1973. The estimation of animal abundance. Hafner Press, New York, N.Y. 506 p.

SHIBATA, T. 1941. Investigations on the present state of trawl fisheries of the South China Sea. Tokyo. 75 p., (In Japanese)

SILLIMAN, R. P. 1943. Studies on the Pacific pilchard or sardine (Sardinops caerulea). 5. A method of computing mortalities and replacements. U.S. Fish Wildl. Serv. Spec. Sci. Rep. 24: 10 p.

1945. Determination of mortality rates from length frequencies of the pilchard or sardine (Sardinops caerulea). Copeia 1945(4): 191–196.

1966. Estimates of yield for Pacific skipjack and bigeye tunas. Proc. Governor's Conf. Central Pacific Fishery Resources. Honolulu, Hawaii, p. 243–249.

1967. Analog computer models of fish populations. Fish. Bull. (Washington, D.C.) 66: 31–46.

1969. Analog computer simulation and catch forecasting in commercially fished populations. Trans. Am. Fish. Soc. 98: 560–569.

1971. Advantages and limitations of "simple" fishery models in light of laboratory experiments. J. Fish. Res. Board Can. 28: 1211–1214.

SMITH, G. G. M. 1940. Factors limiting distribution and size in the starfish. J. Fish. Res. Board Can. 5: 84–103.

SNEDECOR, G. W. 1946. Statistical methods. 4th ed. Iowa State College Press. Ames, Iowa. 485 p.

SOLOMON, M. E. 1957. Dynamics of insect populations. Annu. Rev. Entomol. 2: 121–142.

STEVENSON, J. A., AND L. M. DICKIE. 1954. Annual growth rings and rate of growth of the giant scallop, Placopecten magellanicus (Gmelin) in the Digby area of the Bay of Fundy. J. Fish. Res. Board Can. 11: 660–671.

SUND, O. 1911. Undersøkelser over brislingen i norske farvand vaesentlig paa grundlag av "Michael Sars's" togt 1908. Aarsberetning Vedkommende Norges Fiskeri (1910) 3: 357–410. [Key passages translated in Ricker 1969a, append.)

TANAKA, S. 1953. Precision of age-composition of fish estimated by double sampling method using the length for stratification. Bull. Jap. Soc. Sci. Fish. 19: 657–670.

1954. The effect of restriction of fishing effort on the yield. Bull. Jap. Soc. Sci. Fish. 20: 599–603.

1956. A method of analyzing the polymodal frequency distribution and its application to the length distribution of porgy, *Taius tumifrons* (T.&S.). Bull. Tokai Reg. Fish. Res. Lab. 14: 1–12. (In Japanese, English summary)

TAYLOR, C. C. 1962. Growth equations with metabolic parameters. J. Cons. Int. Explor. Mer 27: 270–286.

TEMPLEMAN, W., AND A. M. FLEMING. 1953. Long-term changes in hydrographic conditions and corresponding changes in the abundance of marine animals. Int. Comm. Northwest Atl. Fish. 3: 79–86.

TERESHCHENKO, K. K. 1917. The bream (*Abramis brama* L.) of the Caspian-Volga region, its fishery and biology. Tr. Astrakhanskoi Ikhtiologicheskoi Laboratorii, 4(2): 1–159. (In Russian)

TESCH, F. W. 1971. Age and growth, p. 98–130. *In* W. E. Ricker [ed.] Methods for assessment of fish production in fresh waters, Int. Biol. Program, Handbook 3. 2nd ed. Blackwell Scientific Publications, Oxford and Edinburgh.

TESTER, A. L. 1937. The length and age composition of the herring (*Clupea pallasi*) in the coastal waters of British Columbia. J. Biol. Board Can. 3: 145–168.

1953. Theoretical yields at various rates of natural and fishing mortality in stabilized fisheries. Trans. Am. Fish. Soc. 82: 115–122.

1955. Estimation of recruitment and natural mortality rate from age composition and catch data in British Columbia herring populations. J. Fish. Res. Board Can. 12: 649–681.

THOMPSON, W. F. 1950. The effects of fishing on stocks of halibut in the Pacific. Fisheries Research Institute, Univ. Wash., Seattle, Wash. 60 p.

THOMPSON, W. F., AND F. H. BELL. 1934. Biological statistics of the Pacific halibut fishery. 2. Effect of changes in intensity upon total yield and yield per unit of gear. Rep. Int. Fish. (Pacific Halibut) Comm. 8: 49 p.

THOMPSON, W. F., H. A. DUNLOP, AND F. H. BELL. 1931. Biological statistics of the Pacific halibut fishery. 1. Changes in yield of a standardized unit of gear. Rep. Int. Fish. (Pacific Halibut) Comm. 6: 108 p.

THOMPSON, W. F., AND W. C. HERRINGTON. 1930. Life history of the Pacific halibut. 1. Marking experiments. Rep. Int. Fish. (Pacific Halibut) Comm. 2: 137 p.

TOMLINSON, P. K. 1970. A generalization of the Murphy catch equation. J. Fish. Res. Board Can. 27: 821–825.

TUTTLE, L., AND J. SATTERLY. 1925. The theory of measurements. Longmans Green, London. 255 p.

UNDERHILL, A. H. 1940. Estimation of a breeding population of chub suckers. Trans. N. Am. Wildl. Conf. 5: 251–255.

VAN OOSTEN, J. 1929. Life history of the lake herring (*Leucichthys artedii* LeSueur) of Lake Huron, as revealed by its scales, with a critique of the scale method. Bull. U.S. Bur. Fish. 44: 265–448.

1936. Logically justified deductions concerning the Great Lakes fisheries exploded by scientific research. Trans. Am. Fish. Soc. 65: 71–75.

VASNETSOV, V. V. 1953. Developmental stages of bony fishes, p. 207–217. *In* Ocherki po Obshchim Voprosam Ikhtiologii. Akademiya Nauk Press, Moscow. (In Russian)

VOEVODIN, I. N. 1938. On the problem of estimating the supply of zanders in the North Caspian. Tr. Pervoi Vsekaspiiskoi Nauchnoi Rybokhozyaistvennoi Konferentsii 2: 3–12. (In Russian)

WALFORD, L. A. 1946. A new graphic method of describing the growth of animals. Biol. Bull. 90(2): 141–147.

1958. Living resources of the sea: opportunities for research and expansion. Ronald Press, New York, N.Y. 321 p.

WALLACE, W. 1915. Report on the age, growth and sexual maturity of the plaice in certain parts of the North Sea. U.K. Board Agric. Fish., Fish. Invest. Ser. 2, 2(2): 79 pp.

WALTERS, C. J. 1969. A generalized computer simulation model for fish population studies. Trans. Am. Fish. Soc. 98: 505–512.

WATKIN, E. E. 1927. Investigations of Cardigan Bay herring. Univ. Coll. Wales, (Aberystwyth), Rep. Mar. Freshwater Invest. 2.

WATT, K. E. F. 1956. The choice and solution of mathematical models for predicting and maximizing the yield of a fishery. J. Fish. Res. Board Can. 13: 613–645.

WEATHERLEY, A. H. 1972. Growth and ecology of fish populations. Academic Press, London and New York. 293 p.

WIDRIG, T. M. 1954a. Definitions and derivations of various common measures of mortality rates relevant to population dynamics of fishes. Copeia 1954: 29–32.

1954b. Method of estimating fish populations with application to Pacific sardine. U.S. Fish Wildl. Serv. Fish. Bull. 56: 141–166.

WILIMOVSKY, N. J., AND E. C. WICKLUND. 1963. Tables of the incomplete beta function for the calculation of fish population yield. Univ. British Columbia, Vancouver, B.C. 291 p.

WINBERG, G. G. 1956. Rate of metabolism and food requirements of fishes. Nauchnye Trudy Beloruss-kogo Gosudarstvennogo Universiteta, Minsk. 253 p. (Transl. from Russian by Fish. Res. Board Can. Transl. Ser. No. 194, 1960)

WOHLSCHLAG, D. E. 1954. Mortality rates of whitefish in an arctic lake. Ecology 35: 388–396.

WOHLSCHLAG, D. E., AND C. A. WOODHULL. 1953. The fish population of Salt Springs Valley Reservoir, Calaveras County, California. Calif. Fish Game 39: 5–44.

YAMANAKA, I. 1954. The effect of size restriction on its yield (I). Annu. Rep. Jap. Sea Reg. F,sh. Res. Lab. 1954 (1): 119–126. (In Japanese)

YOSHIHARA, T. 1951. On the fitting, the summation and an application of the logistic curve. J. Tokyo Univ. Fish. 38: 181–195.

——— 1952. On Baranov's paper. Bull. Jap. Soc. Sci. Fish. 17: 363–366.

ZASOSOV, A. 1971. Notes to the third volume of F. I. Baranov's Collected Works, p. 268–302. Pishchevaya Promyshlennost Press, Moscow. (In Russian)

ZIPPIN, C. 1956. An evaluation of the removal method of estimating animal populations. Biometrics 12: 163–189.

INDEX

370

Fishing gear
  competition between 22
  efficiency 19
  kinds 19
  selectivity 70
Fishing grounds
  age of arrival on 43
  presence of younger fish on 147
Fishing intensity 3
Fishing mortality (*see* Rate of fishing, Rate of exploitation)
Fishing power 3
Fishing success (*see* Catch per unit effort)
Fishing success methods for estimating population 149(chapter 6)
Fishing-up effect 262, 300, 307
Flatfishes
  Lee's phenomenon 216
Fletcher, I. R.
  growth curves 232
Flow (discharge)
  related to salmon production 279
Foerster, R. E.
  absence of fins 90
  computation of production 240
  effect of removal of fins 86
  seasonal distribution of growth and mortality 18
Force of fishing and natural mortality 4
Ford, E.
  growth coefficient 6, 222
  growth compensation 232
  growth equation 222
  growth of herring 223
Ford, E. B.
  point censuses 76
  population statistics from marking 128
Ford–Walford relationship 223
Fork length 210
Forrester, C. R.
  Walford graph 229
Fox, W. W.
  Gulland–Fox exponential model 330
Fraser, J. M.
  vital statistics from tag recoveries and utilized stocks 189
  vulnerability and size 73
Fraser method 189
Fredin, R. A.
  randomness in sample censusing 91
  survival of shad 65
Fridriksson, A.
  stratified sampling for age 66

Fry, F. E. J.
  age frequency of lake trout 58
  catch and effort for lake trout 58
  selectivity of gillnets 71
  utilized stock method 181, 189
  vulnerability and size 73
Fulton, T. W.
  condition factor 209
Functional regressions 351

G

Gales, L. E.
  complex growth curve 225
  complex recruitment relationships 296
  computation of yield 241, 255
  effects of b-coefficients on yield estimates 256
Garrod, D. J.
  biological statistics for cod 173
  recruitment curve for cod 286, 287
Garwood, F.
  confidence limits 26
Generalized production model 323
Gerking, S. D.
  censusing river fishes 91
  Lee's phenomenon 216
  multiple census of redear sunfish 97
  randomness in sample censusing 91
Gilbert, C. H.
  growth compensation 232
Gilbert, R. O.
  Seber–Jolly population estimates 92
  statistical bias 134
Gillnets 55, 57, 70, 87
Glossary 2
Gompertz growth curve 232, 307
Goodman, L. A.
  population estimates from marking 128
Graham, M.
  age determination 203
  computation of equilibrium yield 251
  computation of surplus production 310
  estimate of effectiveness of fishing 120
  length–weight relationship 211
  survival from C/f 39
  survival from tagging experiments 114
Graham surplus-production model 310, 312, 314, 319, 322, 331
Graunt, John
  sample censusing 75
Growth 16, 25, 203(chapter 9)
  compensation 232
  computation 217
  depensation 232

.

www.ingramcontent.com/pod-product-compliance
Lightning Source LLC
Chambersburg PA
CBHW080700220326
41598CB00033B/5268